CONTENTS

The five optional topics, which are available online free of charge, are:

TO THE STUDENT

The aim of this book is to help make your study of advanced physics interesting and successful. It includes examples of modern applications, of new developments, and of how our scientific understanding has evolved.

Physics is our attempt to understand how the Universe works. Fortunately, there are some deep, underlying laws that simplify this ambitious task, but the concepts involved are often abstract and will be unfamiliar at first. Getting to grips with these ideas and applying them to solving problems can be daunting. There is no need to worry if you do not 'get it' straight away. Discuss ideas with other students, and of course check with your teacher or tutor. Most important of all, keep asking questions.

Chapters 1 to 10 of this book, together with Student Book 1, cover the Core content of the AQA A-level Physics specification. The five optional topics, of which you only need to study one, can be found at www.collins.co.uk/physics/sb2modules.

There are a number of features in the book to help you learn:

- Each Core chapter and each Option unit starts with a short outline of what you should have learned previously and what you will learn through the chapter. This is followed by a brief example of how the physics you will learn has been applied somewhere in the world.
- Important words and phrases are given in bold when used for the first time, with their meaning explained. There is also a glossary at the back of the book. If you are still uncertain, ask your teacher or tutor because it is important that you understand these words before proceeding.
- Throughout each chapter there are many questions, which enable you to quickly check your understanding. The answers are at the back of the book. If you get really stuck with a question, check the answer before you carry on.
- Similarly, throughout each chapter there are checklists of Key Ideas that summarise the main points you need to learn from what you have just read.
- Where appropriate, Worked examples are included to show how important calculations are done.
- There are many Assignments throughout the book. These tasks are designed to consolidate or extend your understanding of a topic. They give you a chance to apply the physics you have learned to new situations and to solve problems that require a mathematical approach. Some refer to practical work and will encourage you to think about scientific methods. The relevant Maths Skills (MS) and Practical Skills (PS) from the AQA A-level Physics specification are indicated.
- Some Core chapters have information about the 'Required practical' activities that you need to carry out during your course. These sections (printed on a beige background) provide the necessary information about the apparatus, equipment and techniques that you need to carry out the required practical work. There are questions about the use of equipment, techniques, improving accuracy in practical work, and data analysis.

At the end of each chapter there are Practice Questions, which are exam-style questions including some past paper questions. There are a number of sections, questions, Assignments and Practice Questions that have been labelled 'Stretch and challenge', which you should try to tackle. In places these go beyond what is required for your exams but they will help you to expand your knowledge and understanding of physics.

Good luck and enjoy your studies. We hope this book will encourage you to study physics further after you have completed your course.

PRACTICAL WORK IN PHYSICS

Practical work is a vital part of physics. Physicists apply their practical skills in a wide variety of contexts: from nuclear medicine in hospitals to satellite design; from testing new materials to making astronomical observations. In your A-level physics course you need to learn, practise and demonstrate that you have acquired these skills.

Physicists need to solve problems, such as design problems. This machine weaves superconducting wire into cable to produce powerful superconducting electromagnets for accelerators.

WRITTEN EXAMINATIONS

Your practical skills will be assessed in the written examinations at the end of the course. Questions on practical skills will account for about 15% of your marks. The practical skills that will be assessed in the written examinations are listed below. Throughout this book there are questions and longer assignments that will give you the opportunity to develop and practise these skills. The contexts of some of the exam questions will be based on the 'required practical activities' (see the final section of this chapter).

Practical skills assessed in written examinations:

Independent thinking

- Solve problems set in practical contexts
- Apply scientific knowledge to practical contexts.

Use and application of scientific methods and practices

- Comment on experimental design and evaluate scientific methods

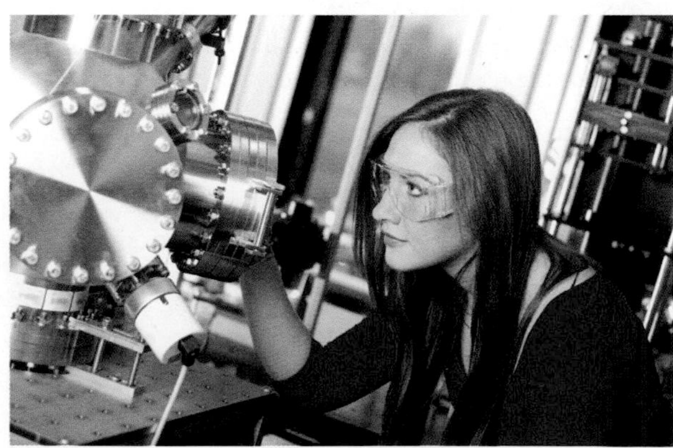

Physicists need to apply their knowledge when using practical equipment. This is a laser deposition chamber, in which a laser beam evaporates material in order to coat another surface.

› Present data in appropriate ways
› Evaluate results and draw conclusions with reference to measurement uncertainties and errors
› Identify variables, including those that must be controlled.

Numeracy and the application of mathematical concepts in a practical context

› Plot and interpret graphs
› Process and analyse data using appropriate mathematical skills
› Consider margins of error, accuracy and precision of data.

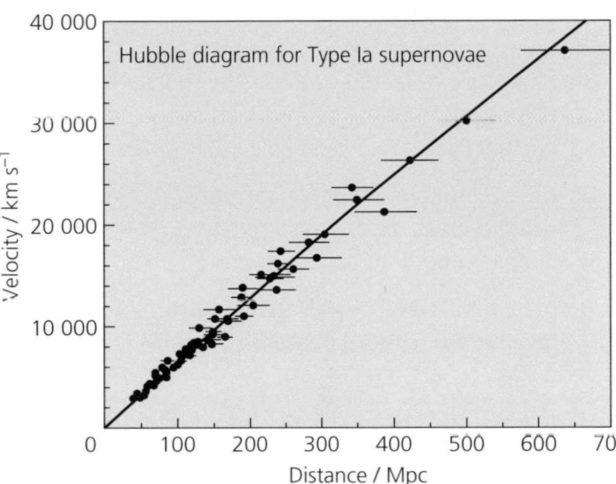

This graph of velocity against distance for supernova events, similar to that originally produced by Edwin Hubble, plots the distances with error bars because of the uncertainty in the values.

Instruments and equipment

› Know and understand how to use a wide range of experimental and practical instruments, equipment and techniques appropriate to the knowledge and understanding included in the specification

You will need to use a variety of equipment correctly and safely.

ASSESSMENT OF PRACTICAL SKILLS

Some practical skills, such as handling materials and equipment and making measurements, can only be practised when you are doing experiments. The following *practical competencies* will be assessed by your teacher when you carry out practical activities:

› Follow written procedures
› Apply investigative approaches and methods when using instruments and equipment
› Safely use a range of practical equipment and materials
› Make and record observations
› Research, reference and report findings.

You must show your teacher that you consistently and routinely demonstrate the competencies listed above during your course. The assessment will not contribute to your A-level grade, but will appear as a 'pass' alongside your grade on the A-level certificate.

These practical competencies must be demonstrated by using a specific range of *apparatus and techniques*. These are as follows:

› Use appropriate analogue apparatus to record a range of measurements (to include length/distance, temperature, pressure, force, angles and volume) and to interpolate between scale markings
› Use appropriate digital instruments, including electrical multimeters, to obtain a range of measurements (to include time, current, voltage, resistance and mass)
› Use methods to increase accuracy of measurements, such as timing over multiple oscillations, or use of a fiduciary marker, set square or plumb-line
› Use a stopwatch or light gates for timing
› Use calipers and micrometers for small distances, using digital or vernier scales
› Correctly construct circuits from circuit diagrams using dc power supplies, cells and a range of circuit components, including those where polarity is important
› Design, construct and check circuits using dc power supplies, cells and a range of circuit components
› Use signal generator and oscilloscope, including volts/division and time-base

- Generate and measure waves, using microphone and loudspeaker, or ripple tank, or vibration transducer, or microwave/radio wave source
- Use laser or light source to investigate characteristics of light, including interference and diffraction
- Use ICT such as computer modelling or data logger with a variety of sensors to collect data, or use software to process data
- Use ionising radiation, including detectors.

REQUIRED PRACTICAL ACTIVITIES

During the A-level course you will need to carry out 12 *required practical activities*. These are the main sources of evidence that your teacher will use to award you a 'pass' for your competency skills.

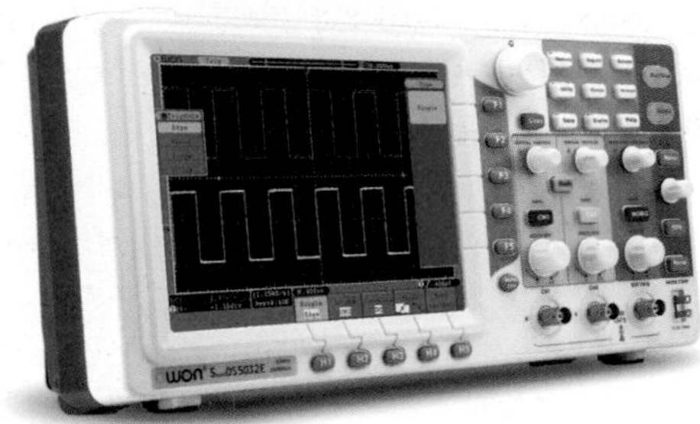

An oscilloscope

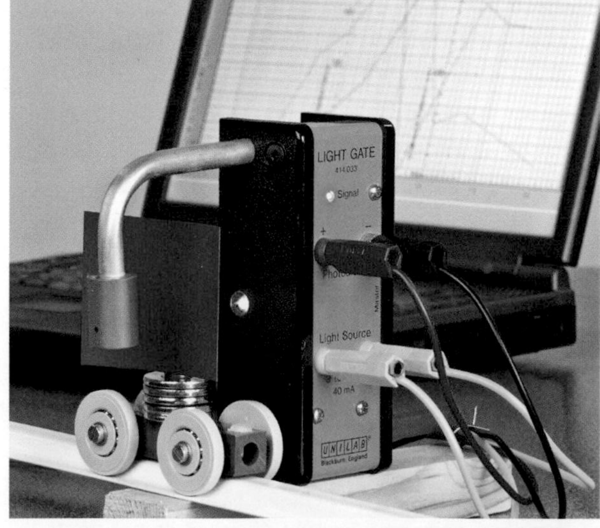

A motion experiment using a light gate

1 Investigation into the variation of the frequency of stationary waves on a string with length, tension and mass per unit length of the string

2 Investigation of interference effects to include the Young's slit experiment and interference by a diffraction grating

3 Determination of g by a free-fall method

4 Determination of the Young modulus by a simple method

5 Determination of resistivity of a wire using a micrometer, ammeter and voltmeter

6 Investigation of the emf and internal resistance of electric cells and batteries by measuring the variation of the terminal pd of the cell with current in it

7 Investigation into simple harmonic motion using a mass–spring system and a simple pendulum

8 Investigation of Boyle's (constant-temperature) law and Charles's (constant-pressure) law for a gas

9 Investigation of the charge and discharge of capacitors; analysis techniques should include log-linear plotting, leading to a determination of the time constant RC

10 Investigate how the force on a wire varies with flux density, current and length of wire using a top pan balance

11 Investigate, using a search coil and oscilloscope, the effect on magnetic flux linkage of varying the angle between a search coil and magnetic field direction

12 Investigation of the inverse-square law for gamma radiation.

Information about the apparatus, techniques and analysis of required practicals 1 to 6 are found in Student Book 1.

You will be asked some questions in your written examinations about these required practicals.

Practical skills are really important. Take time and care to learn, practise and use them.

1 CIRCULAR MOTION

PRIOR KNOWLEDGE

You will have learned how to combine several forces into a resultant and how to calculate the components of a single force (*see Chapter 10 of Year 1 Student Book*) and will have applied Newton's laws of motion to an object moving in a straight line (*see Chapter 11 of Year 1 Student Book*). You will be familiar with the equation for the circumference of a circle $(2\pi r)$ and may need to refresh your knowledge of the radian as a unit of angle measurement (*see Chapter 5 of Year 1 Student Book*).

LEARNING OBJECTIVES

In this chapter you will extend your knowledge of motion to consider objects moving in a circular path. You will learn about the velocity, acceleration and forces acting when cars, aircraft and rollercoasters, for example, are in circular motion. In particular, you will learn that a centripetal force is required to maintain circular motion.

(Specification 3.6.1.1)

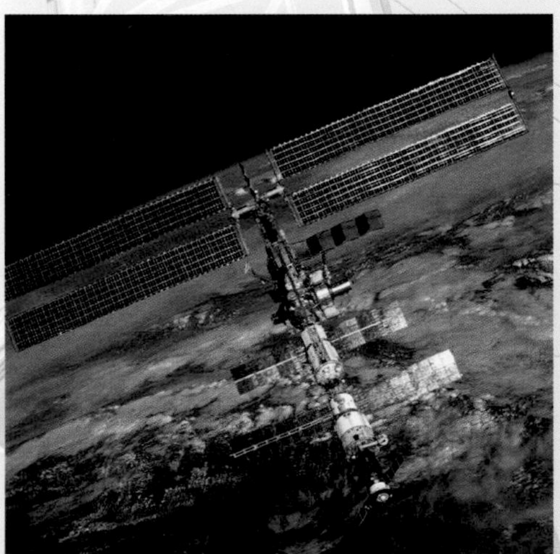

Figure 1 *The International Space Station in a circular orbit around the Earth.*

The *International Space Station* (Figure 1), the largest object ever flown in space, orbits the Earth in an approximately circular path with a speed of about 27 500 km h^{-1} at an altitude of 370–460 km. Covering an area the size of a football pitch, the *International Space Station* would have been far too large to be launched by one rocket, and so was assembled in space in stages, requiring over 40 missions organised by a partnership of space agencies from Europe, the USA, Russia, Japan and Canada.

A few more down-to-Earth examples of circular motion include driving a car around a bend in the road, drying the washing in the drum of a tumble dryer, rotating the crankshaft of a bicycle by pedalling, and riding on a big wheel (Figure 2) or on the many bends and loops of a rollercoaster. The mechanics of circular motion can be applied to these examples and many others.

Figure 2 (background) *The London Eye.*

1.1 MOTION IN A CIRCULAR PATH

Angular measurements in the study of circular motion are usually made using the SI unit, the **radian**. The radian is defined as the angle subtended at the centre of a circle by an arc that is equal in length to the radius (Figure 3). The radian is usually abbreviated to rad.

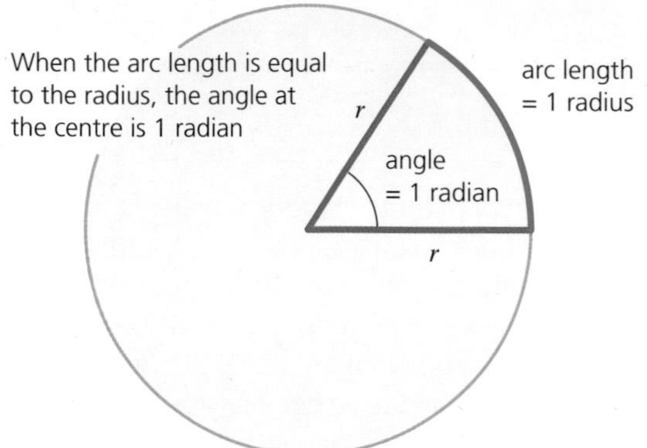

Figure 3 *Defining the radian*

Therefore, the arc of a circle that is equal in length to twice the radius subtends an angle of 2 rad. Similarly, an arc that is three times the length of the radius subtends an angle of 3 rad, and so on. We can therefore write an equation (illustrated in Figure 4) relating the angle subtended, θ, to the arc length s and the circle radius r:

$$\theta = \frac{s}{r}$$

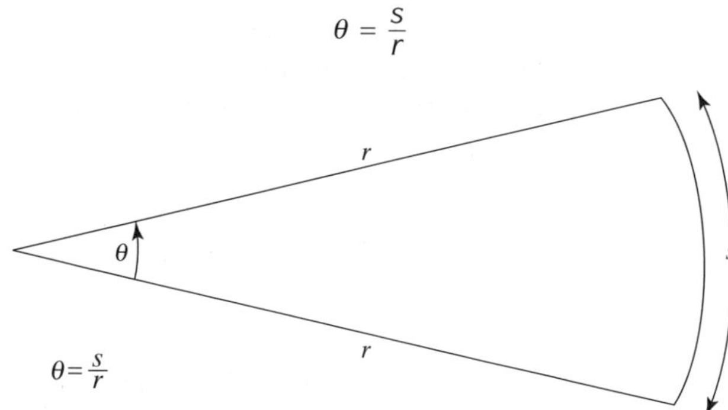

Figure 4 *The angle θ subtended at the centre of a circle by arc length s*

If an arc length is equal to the circumference of a circle, then substituting in the equation $\theta = \frac{s}{r}$ gives the angle $\theta = \frac{2\pi r}{r} = 2\pi$ rad. It is useful to be able recall that 2π rad is equivalent to $360°$ and is the angle subtended at the centre of a complete circle.

The angle described by a rotating or orbiting object is called the **angular displacement**, θ, and the time taken for an object to turn through a complete circle is called the **time period**, T. The number of revolutions per unit time is known as the **frequency**, f, the SI unit of which is the hertz (Hz), where 1 Hz is one revolution per second. The frequency and the time period are related by the equation:

$$f = \frac{1}{T}$$

QUESTIONS

1. The London Eye (Figure 2) completes one revolution in 30 minutes. Determine the angular displacement of one capsule in 10 minutes, in radians.

The **angular velocity**, or **angular speed**, ω, is the angular displacement per unit time and is measured in rad s^{-1}. Since there are 2π rad in one rotation, the angular speed is related to the time period by the equation:

$$\omega = \frac{2\pi}{T}$$

and, since $f = \frac{1}{T}$, the angular speed can also be written as $\omega = 2\pi f$.

The **linear speed**, v, or **orbital speed**, of an object moving in a circle is defined as the distance the object covers per unit time. This has a direction which is tangential to the circular path. The linear speed can be calculated by dividing the circumference of the circle by the time period:

$$v = \frac{2\pi r}{T}$$

and, since $\omega = \frac{2\pi}{T}$, this can be written as

$$v = r\omega$$

which can be rearranged to give the equation:

$$\omega = \frac{v}{r}$$

The Earth rotates about its axis once every 24 hours. Every point on the Earth's surface has the same value of angular speed caused by this rotation (Figure 5). This value of angular speed can be calculated from

$$\omega = \frac{2\pi}{T} = \frac{2\pi}{24 \times 3600} = 7.27 \times 10^{-5}\,\text{rad}\,\text{s}^{-1}$$

However, the value for linear speed at any particular point on the Earth's surface depends on the radius of the circle being described. Consequently, points at different latitudes (different displacements from the equator) have different values, with the speed at the equator being the greatest.

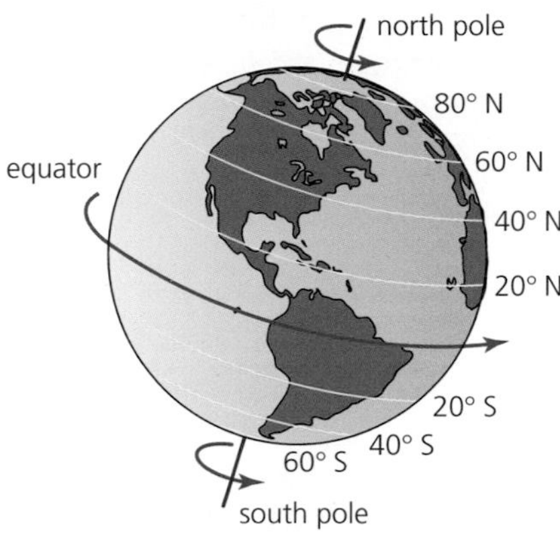

Figure 5 *The Earth rotating about its axis*

Worked example

The *International Space Station* (Figure 1) has an average linear speed of 27 500 km h^{-1} and orbits at an average height of 400 km above the Earth's surface. The radius of the Earth is 6.37×10^6 m. Calculate

a. the time period

b. the frequency

c. the angular speed

d. the number of complete orbits in one day.

a. Time period

$$T = \frac{\text{circumference of orbit}}{\text{linear speed}} = \frac{2\pi r}{v}$$

$$= \frac{2\pi \times (6.37 \times 10^6 + 400 \times 10^3)}{27500 \times 10^3 \div 3600}$$

$$= 5.57 \times 10^3 \text{ s} = \text{approx } 93 \text{ min}$$

b. Frequency

$$f = \frac{1}{5.57 \times 10^3} = 1.8 \times 10^{-4} \text{ Hz}$$

c. Angular speed

$$\omega = \frac{2\pi}{T} = \frac{2\pi}{5.57 \times 10^3} = 1.13 \times 10^{-3} \text{ rad s}^{-1}$$

d. Number of orbits in 1 day

$$N = \frac{\text{number of seconds in 1 day}}{T}$$

$$= \frac{24 \times 3600}{5.57 \times 10^3} = 15.5$$

QUESTIONS

2. A GPS (Global Positioning System) satellite orbits at a height of 20 200 km and completes two orbits of the Earth each day. Calculate the satellite's angular speed and orbital speed. [The radius of the Earth is 6.37×10^6 m]

3. The length of the minute hand of Big Ben is 4.3 m. Which of the options (A, B, C or D) in Table 1 correctly gives the average angular speed *and* the average linear speed for a point at the end of the minute hand?

	Angular speed / rad s^{-1}	Linear speed / mm s^{-1}
A	$\pi / 3600$	7.5
B	$2\pi / 3600$	7.5
C	$2\pi / 3600$	3.75
D	$2\pi / 1800$	3.75

Table 1

4. The European Space Agency launches its rockets in an eastward direction from the spaceport in French Guiana close to the equator in South America, so that the spinning Earth can impart extra velocity to the rocket. Calculate the speed of a point on the equator to determine the extra speed given to the rocket by the Earth's spin as the rocket is launched. [The radius of the Earth is 6.37×10^6 m]

Stretch and challenge

5. Lerwick in Shetland has a latitude of 60 °N (Figure 6). Determine the linear speed of the Earth's surface at Lerwick, due to the Earth's spin.

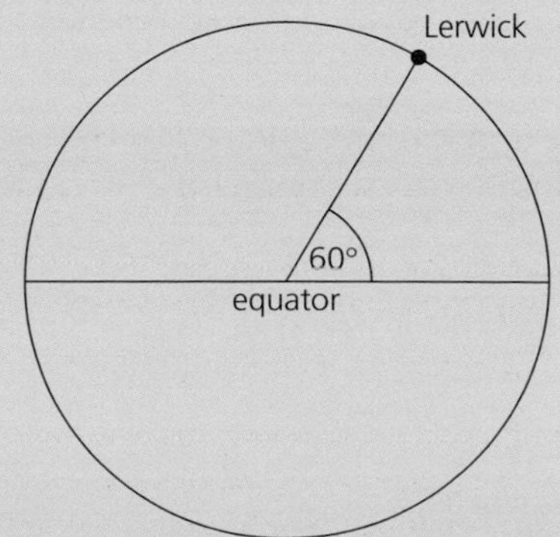

Figure 6

KEY IDEAS

- The SI unit of angular displacement is the radian, where 2π radians is equivalent to $360°$.
- The frequency f and the time period T of a circular motion are related by $f = \frac{1}{T}$.
- The angular displacement per unit time is the angular speed, $\omega = 2\pi f$, which is related to the tangential linear speed by the equation $\omega = \frac{v}{r}$.

1.2 CENTRIPETAL ACCELERATION AND FORCE

Newton's first law states that 'an object remains at rest, or continues at a constant velocity in a straight line, unless acted upon by a resultant force'. Consider a car being driven at a steady speed in a circular path (Figure 7) around a roundabout of radius r. The car's direction is continuously changing, and therefore its velocity must also be changing, since velocity is a vector quantity. It can be concluded that the car must have some acceleration, since acceleration is equal to rate of change of velocity. The acceleration a of an object moving at a steady speed in a circle is called the **centripetal acceleration**.

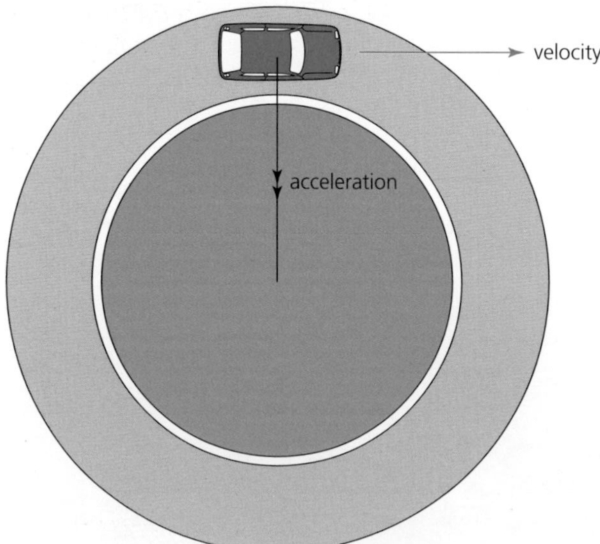

Figure 7 Acceleration and velocity of a car travelling at a steady speed around a roundabout

The direction of this centripetal acceleration is determined by the direction of the vector representing the velocity change (Figure 8). The velocity change from $\boldsymbol{v}_1$ to $\boldsymbol{v}_2$ is found by subtracting vector $\boldsymbol{v}_1$ from vector $\boldsymbol{v}_2$, which shows that the velocity change and therefore the centripetal acceleration is directed towards the centre of the circle at $90°$ to the object's linear velocity $\boldsymbol{v}$.

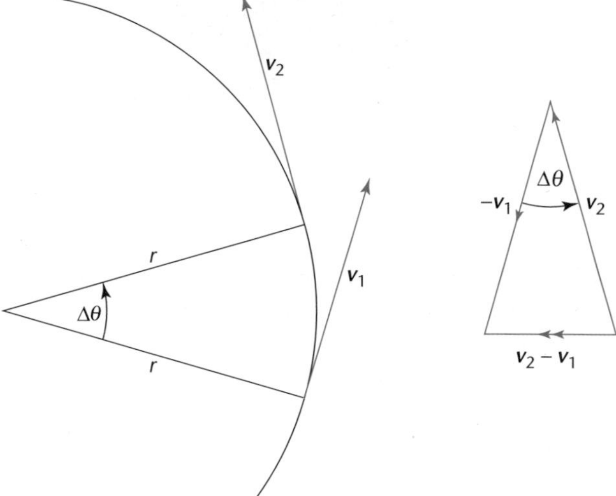

Figure 8 Velocity change of a point moving in a circle

Newton's first law tells us that there must be a force on the car causing it to move in a curve rather than a straight line. Newton's second law gives us the value of that force. It is summarised by the equation $F = ma$, where m is the mass being given an acceleration a by the resultant external force F. Since the car has an acceleration towards the centre of the roundabout, the force F must also be acting towards the centre. It is this **centripetal force** that prevents the car skidding off at a tangent to its path.

It can be shown that centripetal acceleration is given by

$$a = \frac{v^2}{r} = \omega^2 r$$

So, from $F = ma$, the centripetal force is

$$F = \frac{mv^2}{r} = m\omega^2 r$$

The term 'centripetal force' is simply the name given to the resultant force acting towards the centre of the circle. It is important to think about what real physical force or forces are present and contributing to that resultant force. This will be different in different physical situations. In the case of the car travelling around the roundabout, the centripetal force is created by lateral (sideways) friction between the car's tyres and the road (Figure 9).

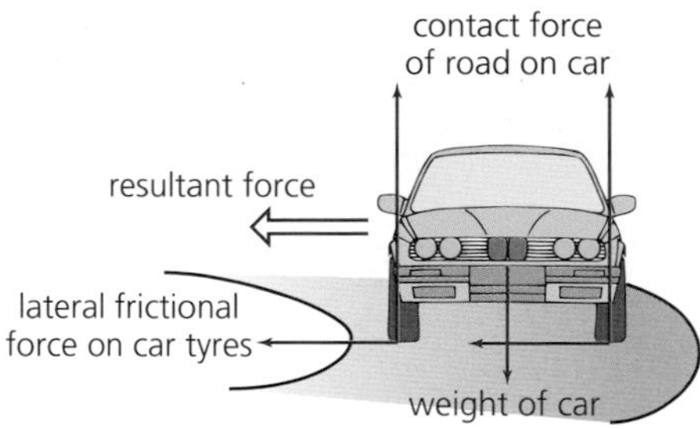

The resultant force is towards the centre of the circle.

Figure 9 *Forces on a car on a roundabout*

Often, curved paths are not part of circles. But the equations for circular motion can still be applied to a portion of the curve by considering a circle that matches the curvature at that point.

QUESTIONS

6. A typical family car of mass 900 kg is travelling at a speed of 30 km h^{-1} around a bend of radius 30 m.
   a. Calculate the size of the centripetal force that must be provided by friction at the tyre–road surface to prevent the car skidding.
   b. Determine the maximum speed at which the car could safely travel around the bend without skidding, given that the maximum lateral friction that can be achieved is about 5 kN.
7. a. What provides the centripetal force on the *International Space Station* in its orbit around the Earth?
   b. Determine the centripetal acceleration of and centripetal force on the *International Space Station.*

   [*International Space Station* mass 400 tonnes; average altitude 400 km; average linear speed 27 500 km h^{-1}; Earth's radius 6.37×10^6 m]
8. The gravitational field strength at a position above the surface of a planet (or other astronomical object), in N kg^{-1}, is numerically equal to the acceleration due to gravity at that point, in m s^{-2} (see Chapter 4). Mars orbits the Sun every 687 Earth days at a distance of 2.3×10^{11} m. Determine the centripetal acceleration of Mars and therefore the Sun's gravitational field strength in the region of Mars' orbit.

Stretch and challenge

9. The equation for centripetal acceleration a is $a = \frac{v^2}{r}$. It is derived by considering the velocity change of a point moving in a circle of radius r with speed $\mathbf{v}$ (see Figure 8). In time interval Δt, the object moves through an angle $\Delta\theta$ and the velocity changes from $\mathbf{v}_1$ to $\mathbf{v}_2$. The velocity change, $\Delta\mathbf{v}$, can be found with the use of the vector diagram, in which vector $\mathbf{v}_1$ is subtracted from vector $\mathbf{v}_2$. Given that acceleration is velocity change per unit time $\left(a = \frac{\Delta\mathbf{v}}{\Delta t}\right)$, and by treating Δt as a very short interval of time, show that the centripetal acceleration a is given by $a = \mathbf{v}\omega = \frac{\mathbf{v}^2}{r}$.

ASSIGNMENT 1: GOING ROUND THE BEND IN FORMULA ONE

(MS 0.1, MS 0.3, MS 1.1, MS 2.3)

The Formula One (F1) racing car (Figure A1), the ultimate racing machine, has been designed to take bends at speeds far in excess of the speeds that a typical family car could drive round a bend without skidding.

Figure A1 *Formula one cars rounding a bend*

The incredible speed of an F1 racing car is achieved by having a very powerful engine and a body that has expertly designed aerodynamic features. F1 aerodynamicists focus their efforts on minimising the drag forces that would reduce the speed that the racing car can achieve. They also focus on designing wings (also known as aerofoils) that create a downforce that pushes the car's tyres down onto the track to increase the amount of lateral friction that can be created to help with taking a bend at speed. The design of the aerofoils is essentially the same as for aircraft wings, but upside down to create downforce rather than lift (*see section 10.2 in Chapter 10 in Year 1 Student Book*).

Questions

A1 An F1 driver typically takes the Abbey bend at Silverstone (Figure A2) in second gear at a speed of $120\,\text{km}\,\text{h}^{-1}$. The mass of an F1 car is about 690 kg and the radius of the Abbey bend is about 70 m. Calculate the total lateral friction on the F1 car's tyres required to complete the bend at $120\,\text{km}\,\text{h}^{-1}$ without skidding.

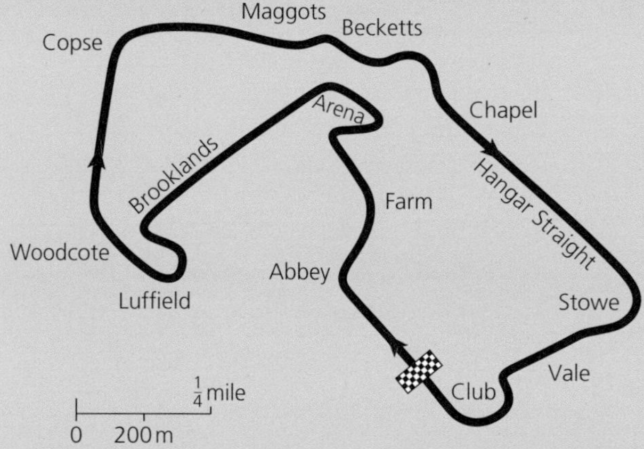

Figure A2 *Silverstone grand prix circuit*

A2 Research 'aerodynamic downforce' and describe briefly, with reference to air flow and air pressure, how it is achieved.

The considerable accelerations that the F1 drivers have to endure during a race are expressed as g-forces. A g-force is a measure of the effect of an acceleration on a body compared with the effect of gravity. The g-force on a person or any object when stationary on the Earth's surface is $1g$. The highest g-force that drivers are exposed to at Silverstone is $5g$, which means that the driver's body is experiencing an acceleration that makes him feel that his body weighs five times his weight when stationary. The value for g-force experienced when driving around a bend is the ratio of the centripetal acceleration to the acceleration due to gravity at the Earth's surface ($9.81\,\text{m}\,\text{s}^{-2}$).

Questions

A3 Determine the g-force typically experienced by a driver at the Abbey bend.

The motion of a conical pendulum

In some examples of circular motion, an object moves in a horizontal circle but is supported by a cable that is at an angle to the vertical, for example a conical pendulum (Figure 10). The pendulum bob of mass m is given an initial push so that it describes a horizontal circle of radius r. The centripetal force is provided by the horizontal component of the tension T in the string, and the weight of the pendulum bob is balanced by the vertical component of the tension.

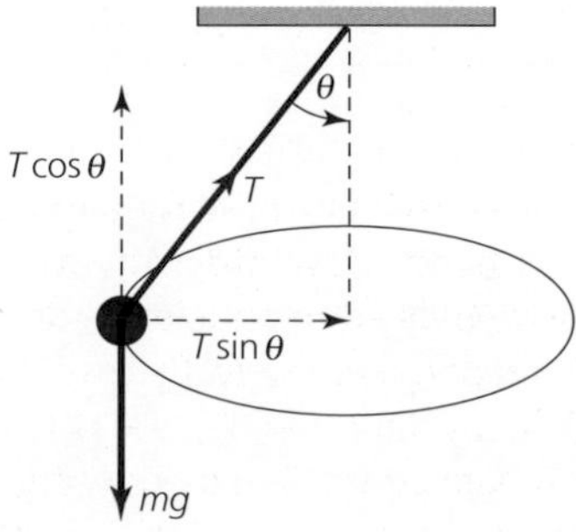

Figure 10 *A conical pendulum*

Therefore $T\sin\theta = \frac{mv^2}{r}$ and $T\cos\theta = mg$. Dividing the horizontal component of tension by the vertical component gives

$$\frac{T\sin\theta}{T\cos\theta} = \frac{mv^2/r}{mg} = \frac{mv^2}{r \times mg}$$

which rearranges to give

$$\tan\theta = \frac{v^2}{rg}$$

QUESTIONS

10. The fairground ride (Figure 11) rotates, making the chair with the rider move in a horizontal circle at a steady speed. The free-body force diagram shows the weight of the seat and rider and the tension T in the cable, along with the components of this tension. The cable attached to the chair makes an angle θ to the vertical. The mass of the chair and rider is 80 kg and the distance from the vertical axis of the ride to the centre of mass of the chair and rider is 3.3 m. Determine the linear speed of the rider when $\theta = 25°$.

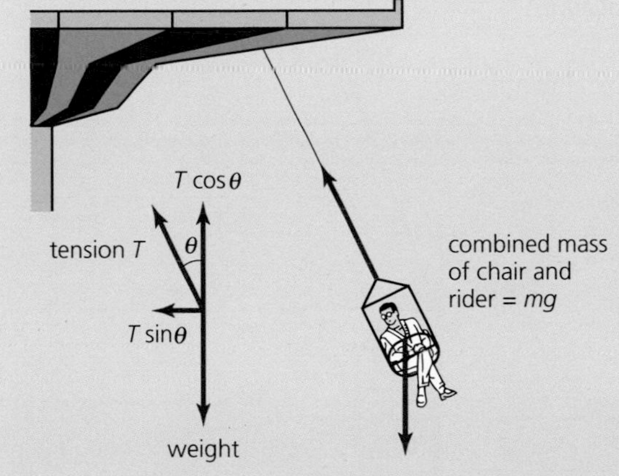

The resultant force, $T\sin\theta$, provides the centripetal force.

Figure 11 The horizontal component of tension provides the centripetal force required for circular motion in this fairground ride.

Banked tracks

The slip road used to enter or leave a motorway can be quite a tight bend, so, in order to avoid skidding, the road may be banked to enable the driver to complete the manoeuvre safely. Cyclists ride on a banked track in a velodrome (Figure 12), and athletes may race on a banked track in indoor athletics. A banked track makes curving easier and allows for a greater speed.

Figure 12 Banked track at a velodrome.

A simplified free-body force diagram (Figure 13) for a car or a cyclist on a banked track shows the normal contact force, N, and the lateral friction force, F, acting on the moving object. The components of both the normal contact force and the lateral friction force now contribute to create a resultant force towards the centre of the curve.

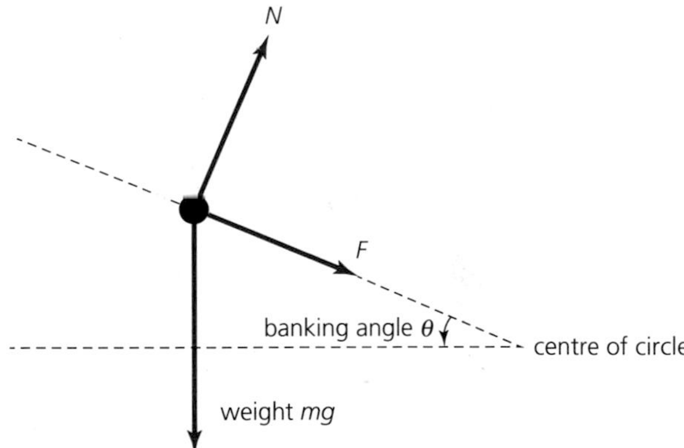

Figure 13 Free-body force diagram to illustrate the principle of banking

If the banking angle is represented by θ, then the component of the normal contact force acting towards the centre of the curve is equal to $N\sin\theta$. The component of the lateral friction force acting towards the centre of the curve is equal to $F\cos\theta$.

Therefore, the resultant force towards the centre is

$$\text{centripetal force} = \frac{mv^2}{r} = N\sin\theta + F\cos\theta$$

Although the contribution from lateral friction to the centripetal force is slightly reduced, this is more than compensated for by the extra contribution from the normal contact force. For example, a car going round a bend of radius 100 m could typically achieve a speed of 30 m s^{-1} without skidding if the bend is banked at an angle of 10° compared with 25 m s^{-1} for a flat road.

Worked example

A bobsleigh track has to be constructed with sharply banked curves (Figure 14), as the icy surface contributes little lateral friction towards the required centripetal force.

Figure 14 Sharply banked bobsleigh track

Determine the maximum speed, in km h^{-1}, that a bobsleigh could travel around a curved track of radius 20 m without exceeding a banking angle of 78°.

Assuming that the ice contributes no lateral friction, the expression for centripetal force

$$\frac{mv^2}{r} = N\sin\theta + F\cos\theta$$

becomes

$$N\sin\theta = \frac{mv^2}{r}$$

Assuming the bobsleigh is in vertical equilibrium, the weight, mg, is balanced by the vertical component of the normal contact force, $N\cos\theta$. Therefore

$$\frac{N\sin\theta}{N\cos\theta} = \frac{mv^2/r}{mg}$$

which gives

$$\tan\theta = \frac{v^2}{rg}$$

Hence the speed is

$$v = \sqrt{rg\tan\theta} = \sqrt{20 \times 9.81 \times \tan 78^\circ}$$

$$= 30.4\,\text{m s}^{-1} = 109\,\text{km h}^{-1}$$

Aircraft use banking in order to change direction (Figure 15).

Figure 15 An aircraft needs to bank in order to turn.

The horizontal component of the lift force (Figure 16) can now provide a resultant force towards the centre, X, of the curved path that the pilot wants to aircraft to follow.

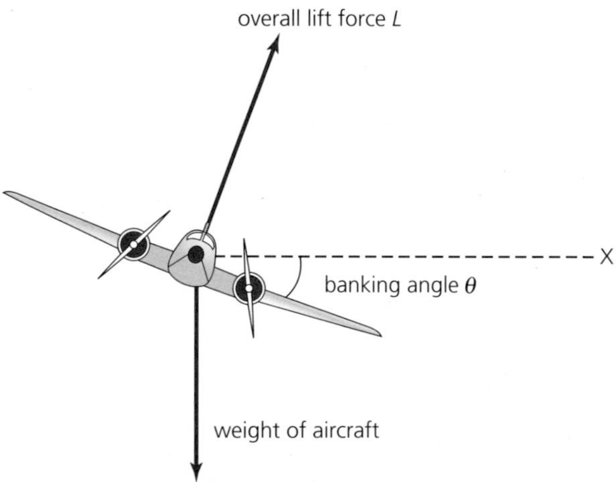

Figure 16 A banked aircraft.

The horizontal component of the lift force, $L\sin\theta$, where L is the overall lift force and θ is the banking angle, provides the centripetal force. The weight of the aircraft, mg, is balanced by the vertical component of the lift force equal to $L\cos\theta$. Therefore

$$\frac{L\sin\theta}{L\cos\theta} = \frac{mv^2/r}{mg}$$

which gives

$$\tan\theta = \frac{v^2}{rg}$$

QUESTIONS

11. Calculate the required banking angle for an aircraft to follow a horizontal curve of radius 5.0 km at a constant speed of 410 km h^{-1}.

KEY IDEAS

- An object moving at a constant speed in a circle experiences a continual change in direction and therefore a continual velocity change.
- The changing velocity of an object moving in a circle means it must have an acceleration. This 'centripetal acceleration' is directed towards the centre of the circle:

$$a = \frac{v^2}{r} = \omega^2 r$$

- A resultant force directed towards the centre of the circle, known as a 'centripetal force', is required if a mass is to follow a circular path and not continue along a tangent:

$$F = \frac{mv^2}{r} = m\omega^2 r$$

- The required centripetal force is provided by different means in different situations: it may, for example, be the component of tension in a cable or lateral friction at a tyre–road surface.
- Some curved tracks or roads are banked so that there is a horizontal component of the normal contact force, which provides extra centripetal force. This enables objects to complete the curve at a greater speed without skidding.

ASSIGNMENT 2: ESTIMATING CENTRIPETAL ACCELERATION AND FORCE

(MS 0.4, MS 1.1, MS 2.3)

Physicists need to be able to make reasonable estimates of physical quantities. This assignment involves estimating centripetal acceleration and force in some examples of circular motion. You will be given a minimal amount of data and will have to estimate other quantities that you may require in your calculations.

Questions

A1 Consider a fairground ride that moves the rider around in a horizontal circle – see Figure A1, which you can assume is approximately to scale. Given that one full circle is completed in about 10 s, estimate

a. the centripetal acceleration

b. the contact force exerted by the seat on the rider's back.

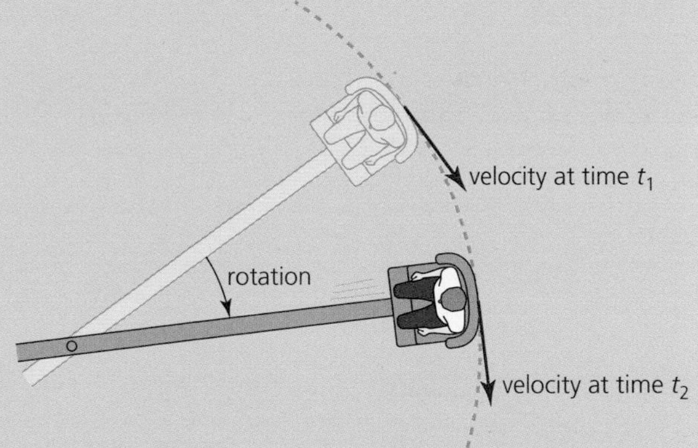

Figure A1 *A fairground ride moving in a horizontal circle*

A2 Some swimming pools provide spin dryers to extract water from swimming costumes. The cylindrical drum of the dryer rotates at about 3000 revolutions per minute in a horizontal circle about a vertical axis. The drum has small holes that allow water to travel out of the drum and be drained away. The drum diameter is approximately 30 cm.

a. Swimming costumes are flung to the sides of the drum. Explain why.

b. Estimate the centripetal acceleration of a point on the inside surface of the drum.

c. Estimate the size of the normal contact force exerted by the side of the drum on a swimming costume.

1.3 MOTION IN A VERTICAL CIRCLE

So far we have considered motion in horizontal circles. For an object moving in a vertical circle, the effect of the weight of the object on the resultant force towards the circle's centre has to be considered. When a rally car follows a track over the top of a small hill, it can lose contact with the surface of the track at the highest point of the hill (Figure 17). For the car to remain in contact with the track's surface, a resultant force towards the centre of curvature would be required (Figure 18).

Figure 17 *A rally car travelling too fast to stay in contact with the track*

contact forces at front and back wheels

direction of travel

mg

radius of curvature of the track

centre of curvature of the track

Figure 17 *The track over the hill follows an arc of a vertical circle.*

At the top of the hill, the centripetal force $\left(\frac{mv^2}{r}\right)$ is the resultant of the car's weight (mg) and the total normal contact force (N) of the track on the car's front and back wheels. So

$$\frac{mv^2}{r} = mg - N$$

This equation can be rearranged to give the normal contact force:

$$N = mg - \frac{mv^2}{r}$$

which shows that, if the car's speed, v, is too great, the normal contact force N will become zero when $\frac{mv^2}{r} = mg$. The car loses contact with the track.

QUESTIONS

12. A car is driving over a humped back bridge. The radius of curvature of the hump of the bridge is 30 m. Determine the maximum speed of the car if it is not to lose contact with the ground.

13. An object is whirled on the end of a string in a vertical circle. If the weight of the object is W and the tension in the string is T, which of the following expressions represents the centripetal force at the instant the object passes the lowest point of the circle? Explain your reasoning.

A $T + W$ **B** $T \times W$

C $T - W$ **D** $\frac{T}{W}$

KEY IDEAS

- For an object moving in a vertical circle, its weight must be considered when identifying the forces contributing to the required centripetal force.
- The centripetal force is provided by different means in different situations: it may, for example, be the resultant of the weight of the object and the tension in a cable, or the resultant of the weight of the object and the normal contact force at a surface. This will vary at different stages of the motion.

ASSIGNMENT 3: STAYING IN THE LOOP

(MS 0.1, MS 1.1, MS 2.2, MS 2.3)

The thrills of rollercoaster rides are created by speed, banked turns and accelerations causing the sensations of weightlessness and increased heaviness. Of course, the force of gravity acting on a person does not change, and their actual weight remains constant. So, what *does* change during the ride to cause the excitement and adrenalin rush that make rollercoaster rides so popular?

Consider first that part of the ride where the carriage is moving on the underside of the top of a vertical loop (Figure A1).

Figure A1 *Vertical loop section of a rollercoaster*

Questions

A1 Draw a free-body force diagram for a person in a carriage at the top of the loop. Note that, as the person is upside down and the seat is above them, the normal contact force acts downwards.

A2 a. Write an equation linking the normal contact force, N, the weight, mg, and the resultant $\frac{mv^2}{r}$ of these two forces required for circular motion at the top of the rollercoaster loop.

b. Using the data in Table A1, calculate the size of the normal contact force on the person in the carriage.

Mass of person in carriage	60 kg
Velocity of carriage at top of loop	8.8 m s^{-1}
Radius of curvature at the top of the loop	5.0 m

Table A1

A3 Now consider the forces in action as the carriage travels along the bottom section of a vertical loop. Using the data in Table A2, repeat questions **A1**, **A2 a** and **A2 b** for this situation.

Mass of person in carriage	60 kg
Velocity of carriage at bottom of a loop	15 m s^{-1}
Radius of curvature at the bottom of the loop	9.0 m

Table A2

A person cannot actually feel the force of gravity on their body. The force of gravity (weight) is experienced as a result of other contact forces acting on the human body as a consequence of the body's weight. The sensation of feeling lighter at the top of the loop and heavier at the bottom is caused by a change in the size of the normal contact force.

If a person was sitting in a rollercoaster carriage at rest, the normal contact force acting would have almost the same size as the person's weight. When experiencing accelerations on a rollercoaster, an increase in the normal contact force causes the sensation of heaviness and a decrease causes a feeling of lightness.

Questions

A4 Determine the speed of the carriage at the top of the loop that would create zero contact force and therefore give a feeling of weightlessness.

PRACTICE QUESTIONS

1. A mass of 50 g is whirled on the end of a piece of string in a vertical circle of radius 60 cm at a constant speed of $3.0\,\mathrm{m\,s^{-1}}$.
   - **a. i.** Draw a free-body force diagram for the mass corresponding to the instant that the mass is at the top of its path.
     - **ii.** Calculate the tension (T) in the string when the mass is at the top of its path.
   - **b. i.** Draw a free-body force diagram for the mass corresponding to the instant that the mass is at the bottom of its path.
     - **ii.** Calculate the tension in the string when the mass is at the bottom of its path.
   - **c.** Suggest at which position the string is most likely to snap. Explain your answer.
2. An aircraft is flying in a horizontal circle of radius 3.0 km at a constant speed of $280\,\mathrm{km\,h^{-1}}$.
   - **a.** Explain why the aircraft is accelerating even though it is travelling at a constant speed.
   - **b.** In order for the aircraft to follow a circular path, the pilot must bank the aircraft. Explain how this procedure creates a centripetal force.
   - **c.** Calculate:
     - **i.** the banking angle
     - **ii.** the centripetal acceleration of the aircraft.
3. In a neutral hydrogen atom, an electron orbits a proton at a distance of 0.053 nm. The centripetal force that maintains the electron's circular orbit is created by the electrostatic attraction between the positive proton and the negative electron, and is equal to 8.2×10^{-8} N. [The mass of the electron is 9.11×10^{-31} kg]

   Calculate:
   - **a.** the electron's orbital speed
   - **b.** the electron's angular speed
   - **c.** the time for one orbit.
4. A disc of diameter D is turning at a steady angular speed at frequency f about an axis through its centre (Figure Q1).

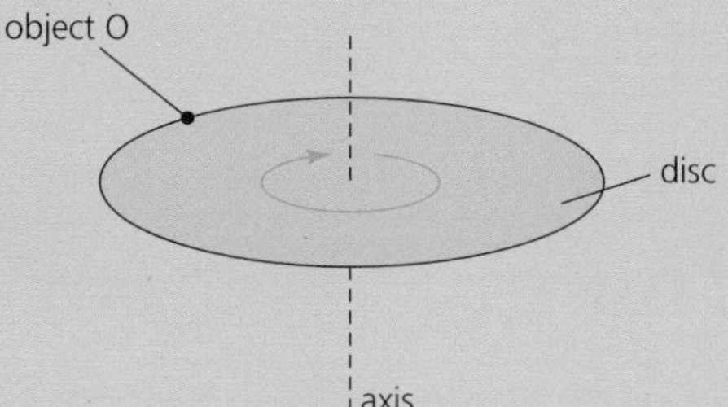

Figure Q1

What is the centripetal force on a small object O of mass m on the perimeter of the disc?

A $2\pi mfD$ **B** $2\pi mf^2D$

C $2\pi^2mf^2D$ **D** $2\pi mf^2D^2$

AQA Unit 4 January 2012 Q4

5. The Earth moves around the Sun in a circular orbit with a radius of 1.5×10^8 km. What is the Earth's approximate speed?

A $1.5 \times 10^3\,\text{m s}^{-1}$ B $5.0 \times 10^3\,\text{m s}^{-1}$

C $1.0 \times 10^4\,\text{m s}^{-1}$ D $3.0 \times 10^4\,\text{m s}^{-1}$

AQA Unit 4 June 2012 Q4

6. For a particle moving in a circle with uniform speed, which one of the following statements is correct?

A The kinetic energy of the particle is constant.

B The force on the particle is in the same direction as the direction of the motion of the particle.

C The momentum of the particle is constant.

D The displacement of the particle is in the direction of the force.

AQA Unit 4 June 2011 Q4

7. A revolving mountain top restaurant turns slowly, completing a full rotation in 50 minutes. A man is sitting in the restaurant 15 m from the axis of rotation. What is the speed of the man relative to a stationary point outside the restaurant?

A $\frac{\pi}{100}\,\text{m s}^{-1}$ B $\frac{3\pi}{5}\,\text{m s}^{-1}$

C $\frac{\pi}{200}\,\text{m s}^{-1}$ D $\frac{\pi}{1500}\,\text{m s}^{-1}$

AQA Unit 4 January 2013 Q5

8. A small body of mass m rests on a horizontal turntable at a distance r from the centre. If the maximum frictional force between the body and the turntable is $\frac{mg}{2}$, what is the angular speed at which the body starts to slip?

A $\sqrt{\frac{gr}{2}}$ B $\frac{g}{r}$

C $\sqrt{\frac{g}{2r}}$ D $\frac{1}{2}\sqrt{\frac{g}{r}}$

AQA Unit 4 June 2013 Q6

2 OSCILLATION

PRIOR KNOWLEDGE

You will be familiar with Hooke's law (*Chapter 12 of Year 1 Student Book*) and know what is meant by the spring constant. You will have experience of using equations for elastic potential energy, gravitational potential energy and kinetic energy, and understand the principle of conservation of energy (Chapter 11 of *Year 1 Student* Book). You may want to refresh your memory about stationary waves on strings and in pipes (see *section 5.6 of Chapter 5 of Year 1 Student Book*). You will need to know how to determine sine and cosine values for angles expressed in radians. You will have already studied circular motion (Chapter 1 of this book) as a type of periodic motion.

LEARNING OBJECTIVES

In this chapter you will extend your knowledge of periodic motion to include oscillations and resonance. You will learn that many natural systems oscillate with simple harmonic motion (SHM), and you will analyse such motion graphically. You will find out about forced vibration in mechanical systems, and the effect of damping. You will have the opportunity to learn how to analyse experimental data using logarithms.

(Specification 3.6.1.2 to 3.6.1.4)

Many people have heard of 'MRI' and know that it is a type of scan that a person may have in hospital to help in the diagnosis and treatment of an injury or illness (Figure 1). Some people will know that MRI stands for 'magnetic resonance imaging', but few will know that MRI is a medical application of an effect called 'nuclear magnetic resonance' (NMR). Resonance is all about efficiently transferring energy from one vibrating system to another. It occurs in many areas of physics and in everyday life, from lasers to musical instruments. In NMR, a nucleus, in the presence of a very strong magnetic field, is able to gain energy by absorbing radio-frequency electromagnetic radiation (typically 60 to 1000 MHz).

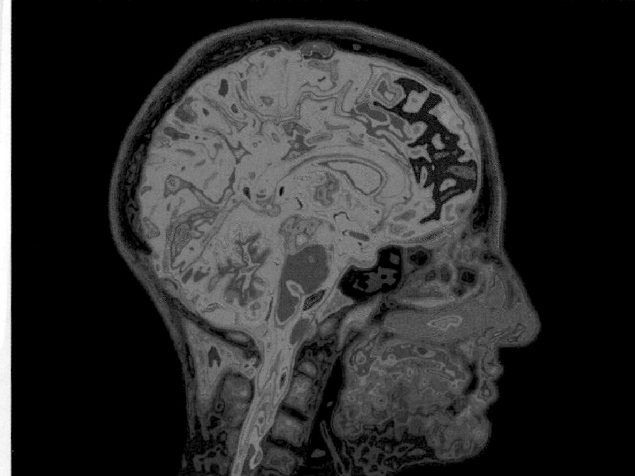

Figure 1 *MRI scan image of the human brain*

When a patient has an MRI scan, he or she is required to lie down inside a large tube-shaped chamber. A powerful magnetic field is activated in the chamber, enabling the transfer of energy from the radio-frequency radiation to hydrogen nuclei (protons) in specific body tissues. The controlled de-excitation of the protons enables the mapping of the tissue, in terms of the hydrogen atoms present.

2.1 SIMPLE HARMONIC MOTION

Oscillations and vibrations are a type of **periodic motion**. The motion repeats in a regular way as time passes. An object 'oscillates' when it repeatedly moves backwards and forwards about an equilibrium position. This **mechanical oscillation** (as opposed to the oscillation of fields, as in an electromagnetic wave) requires the action of a resultant force that is *always directed towards the equilibrium position*. The resultant force in this type of motion is often referred to as the **restoring force**, and the distance and direction of the oscillating object from its equilibrium position are its **displacement**, x.

Imagine holding a spring vertically, with a mass attached to its lower end. Initially, it is stationary – it is in equilibrium. If you pull the mass downwards, a restoring force acts upwards. When the mass moves up above its equilibrium position, the restoring force acts downwards. The result is an oscillatory motion in which the displacement x varies periodically.

A special kind of oscillation is **simple harmonic motion** (SHM):

> In simple harmonic motion (SHM), the restoring force is directly proportional to the displacement, and in the opposite direction.

The repeated up and down motion of a mass on a spring is an example of SHM and can be monitored using data logging equipment connected to a motion sensor (Figure 2). Computer analysis of the data shows that the displacement versus time graph is a cosine (or sine) curve.

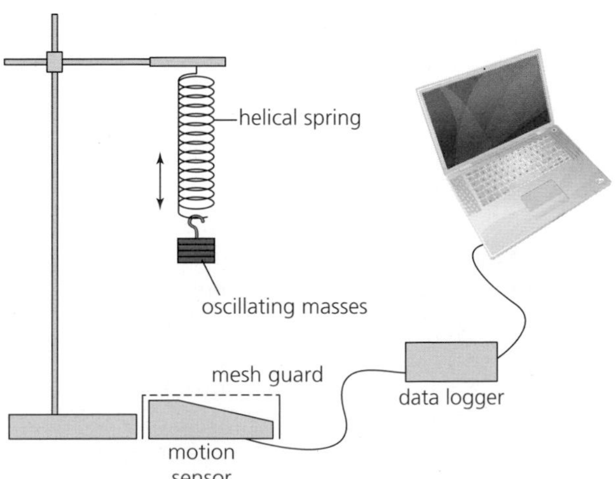

Figure 2 *A motion sensor with data logger and computer for analysis of the oscillatory motion of a mass on a spring*

The graph of displacement versus time for any SHM, in which the displacement is at its maximum value at $t = 0$, is shown in Figure 3. The equation of the graph is

$$x = A\cos(\omega t)$$

Here A is **amplitude** of the oscillation, which is the maximum displacement, and ω is a constant defined by $\omega = \frac{2\pi}{T}$, where T is the constant time period, which is the time for one **cycle** of oscillation. A cycle corresponds to the oscillating mass moving through any position then passing through that same position again in the same direction. For example, the mass at P in Figure 3 has displacement X. It passes through the equilibrium position, moves to one extreme, then to the other extreme, and then at Q has the same displacement X and moves in the same direction. Thus P to Q is one cycle. The number of cycles of oscillation per second is the frequency, f, measured in hertz (Hz = s^{-1}), which is related to the time period by $f = \frac{1}{T}$. Therefore ω can also be written as $\omega = 2\pi f$. It is called the **angular frequency** of the oscillation and is measured in rad s^{-1}.

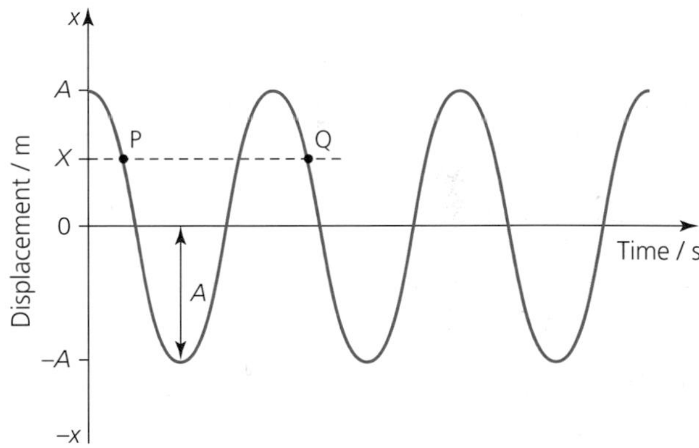

Figure 3 *Displacement versus time graph (cosine curve) for SHM with $x = A$ when $t = 0$*

The equation $x = A\cos(\omega t)$ representing the displacement of an object oscillating with SHM is correct provided the object is at an extreme position when the oscillation starts to be monitored – so, when $t = 0$, $x = A$. However, if the oscillating object is passing through its mean position when the oscillation starts to be monitored, then at $t = 0$, $x = 0$, and the graph of displacement versus time is a sine curve (Figure 4) $x = A\sin(\omega t)$, not a cosine curve.

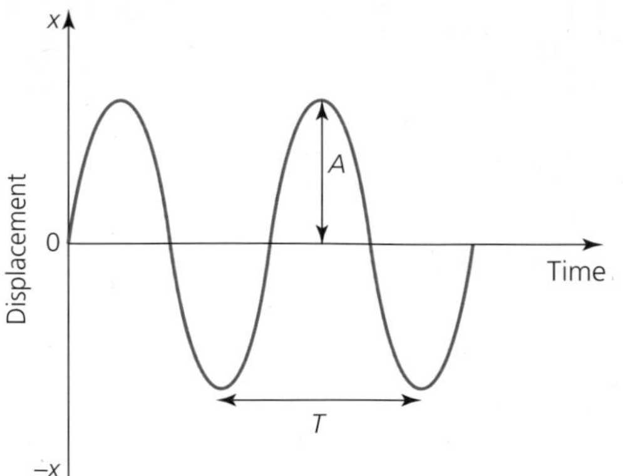

Figure 4 *Displacement versus time (sine curve) for SHM with x = 0 when t = 0*

QUESTIONS

1. Plot a displacement–time graph to show three cycles of an oscillation that has an amplitude of 6.0 mm and a frequency of 5000 Hz, assuming that displacement is a maximum when $t = 0$.
2. An object is oscillating with SHM. Assuming that its displacement $x = 0$ when $t = 0$, calculate the displacement after 0.24 ms if the amplitude is 6 mm and the frequency is 5000 Hz.
3. Sketch two cycles of a displacement versus time graph in which $x = 0$ when $t = 0$ for an oscillation of time period 0.5 s and amplitude 10 cm, showing appropriate scales on both axes.

Velocity and acceleration in SHM

Since velocity is defined as the rate of change of displacement, a graph of velocity versus time shows the variation with time of the gradient of the displacement–time graph (Figure 5). Similarly, an acceleration versus time graph can be obtained from the gradient of the velocity–time graph. A comparison of the graphs of displacement, velocity and acceleration versus time reveals the relationships between these three quantities during the oscillation (Figure 5).

- The velocity of the oscillating mass is zero when displacement is at a maximum, and at its greatest when displacement is zero.
- The acceleration is zero when the displacement is zero, and at its greatest when displacement is a maximum.

We can describe these differences in terms of **phase differences**. Phase difference can be expressed in terms of fractions of a cycle, or degrees or radians. To find the phase difference, for example between the displacement and the velocity, determine the time that elapses between each quantity being at a maximum. The phase difference in terms of a fraction of a cycle can then be found by dividing the time that elapses by the time period. Conversion to degrees or radians can be made by equating a full cycle to 360° or 2π radians. A comparison of the displacement and velocity graphs of Figure 5 shows a phase difference of a quarter of a cycle (90° or $\pi / 2$ rad) between the displacement and the velocity (*see section 5.2 in Chapter 5 of Year 1 Student Book*).

A comparison of the displacement and acceleration graphs reveals a phase difference of half of a cycle (180° or π rad) between these two quantities – they achieve their maximum values at the same instant but have opposite directions.

Simple harmonic motion is defined by the equation

$$a = -\omega^2 x$$

where a is the acceleration, x is the displacement of the oscillating object and ω is the angular frequency, which is a constant for the motion. The equation shows that the acceleration is directly proportional to displacement but in the opposite direction. The maximum value for the acceleration is given by

$$a_{max} = \omega^2 A$$

since the amplitude A is the maximum displacement.

The velocity of an object moving with SHM is given by

$$v = \pm 2\pi f \sqrt{A^2 - x^2}$$

with the maximum velocity $v_{max} = 2\pi f A = \omega A$, since the velocity is at its greatest at the equilibrium position which corresponds to $x = 0$.

A graph of acceleration a versus displacement x for an object oscillating with SHM (Figure 6) is a straight line through the origin with a gradient equal to $-\omega^2$. Since, for a constant mass, acceleration is directly proportional to the resultant force (from Newton's second law, $F = ma$), the restoring force is directly proportional to the displacement but in the opposite direction. Therefore a graph of the restoring force versus displacement would be a straight line through the origin with a gradient equal to $-m\omega^2$, where m is the mass of the oscillating object.

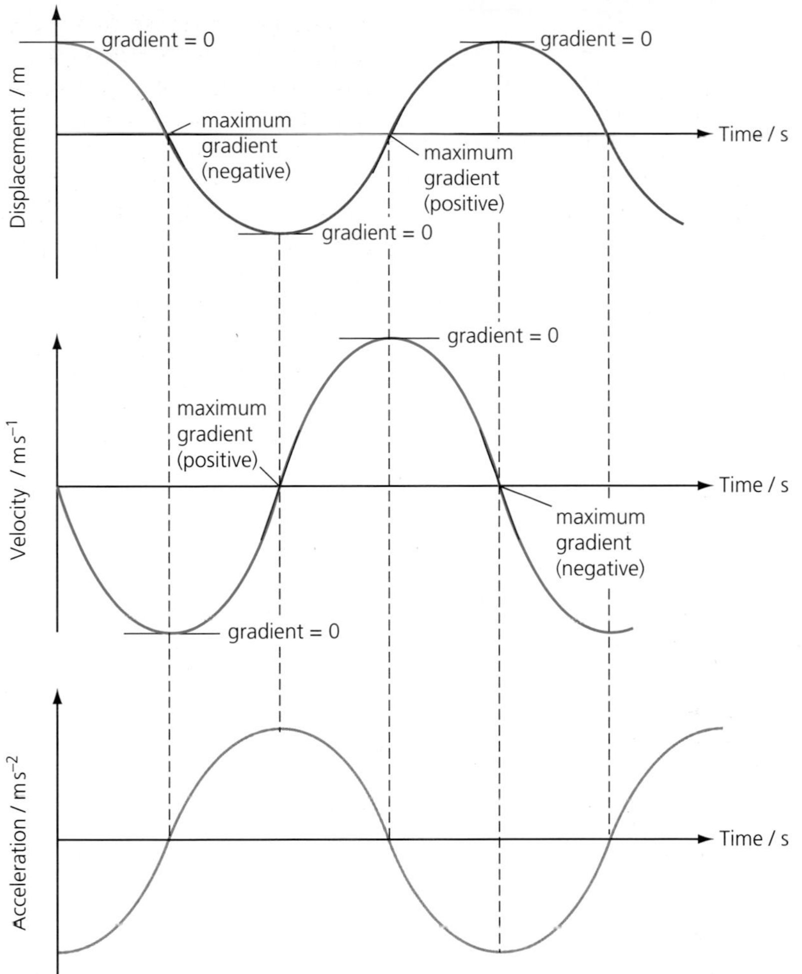

Figure 5 *Displacement, velocity and acceleration versus time graphs for a simple harmonic oscillator*

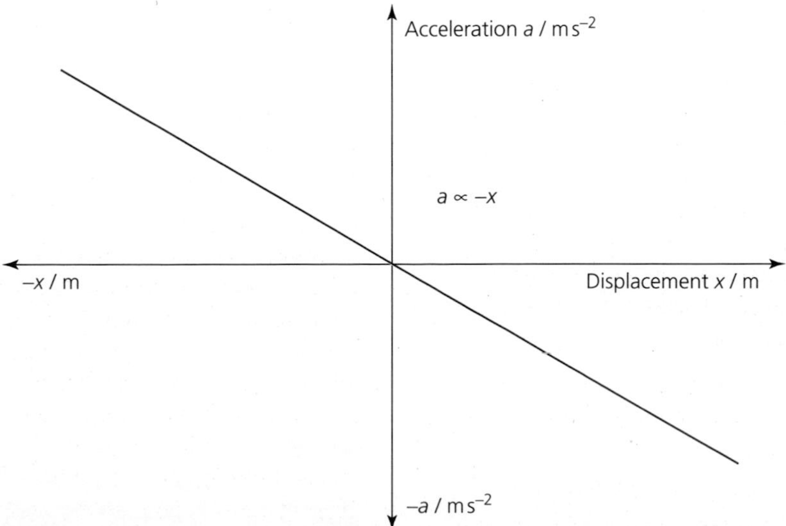

Figure 6 *Acceleration versus displacement for an object moving with SHM*

Worked example

A mass attached to a spring vibrates up and down, undergoing SHM with a time period of 1.6 s. The distance from the top to the bottom extreme positions is 8.0 cm. Determine the maximum acceleration and the speed of the mass as it passes through the equilibrium position.

Since, for SHM, acceleration $a = -\omega^2 x$, the maximum acceleration is

$$a_{max} = \omega^2 A = \left(\frac{2\pi}{T}\right)^2 A$$

The value of the amplitude A is half the distance from one extreme position to the other, so $A = 4.0$ cm. Therefore

$$a_{max} = \left(\frac{2\pi}{1.6}\right)^2 \times 0.040 = 0.62 \text{ms}^{-2}$$

Velocity $v = \pm 2\pi f \sqrt{A^2 - x^2}$, but at the equilibrium position, $x = 0$ and the velocity has its maximum value:

$$v_{max} = 2\pi f \sqrt{A^2} = \frac{2\pi}{T} A = \frac{2\pi}{1.6} \times 0.040 = 0.16 \text{ms}^{-1}$$

QUESTIONS

4. The piston of a car engine moves with a motion that is approximately SHM. One cycle of motion takes 0.017 s and the piston moves through a total distance of 100 mm. Calculate the maximum acceleration of the piston and its velocity at a distance of 20 mm from its equilibrium position.
5. Determine the acceleration of an object oscillating with SHM at a frequency of 2.0 Hz when its displacement is 0.20 m.
6. A graph of acceleration versus displacement for an object moving with SHM is shown in Figure 7.

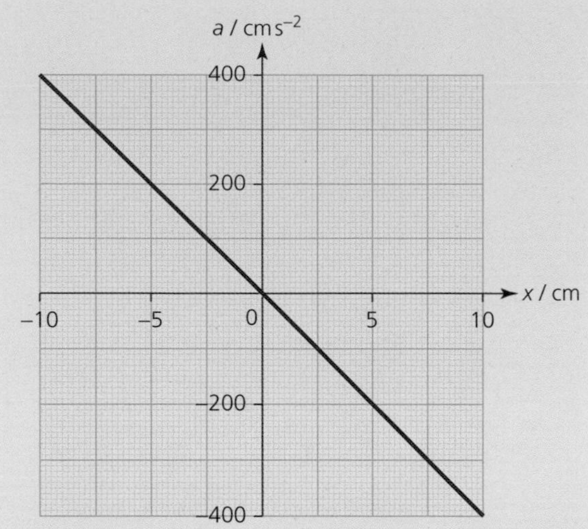

Figure 7

a. Determine the gradient of the line and write an equation showing the relationship between the object's acceleration and its displacement.

b. How would the magnitude of the gradient change if the oscillating system was changed so that the frequency was doubled?

KEY IDEAS

- SHM is an oscillation in which the acceleration (a) is directly proportional to the displacement (x) but in the opposite direction:

  $$a = -\omega^2 x$$

  where ω is the constant angular frequency of the motion.
- The displacement versus time graph of an oscillating object is a sine or cosine curve depending on whether the displacement is zero or at its maximum value at time $t = 0$.
- The velocity versus time graph can be determined from the gradient of the displacement versus time graph.
- The acceleration versus time graph can be determined from the gradient of the velocity versus time graph.

- The maximum velocity occurs as the object passes through the equilibrium position and is given by

$$v_{max} = \omega A$$

where A is the amplitude (maximum displacement) of the oscillation.

- The maximum acceleration occurs at the extremes of the oscillation and is given by

$$a_{max} = \omega^2 A$$

2.2 SIMPLE HARMONIC SYSTEMS

The mass–spring system

So why does a mass oscillating on the end of a spring move with SHM? First, here is a reminder of Hooke's law: 'The extension of a spring is directly proportional to the force applied provided the elastic limit has not been exceeded'. We can express this as

$$F = k\Delta l$$

where k is the spring constant and Δl is the extension (*see section 12.2 in Chapter 12 of Year 1 Student Book*).

Consider a spring that is initially suspended vertically but has no mass attached (Figure 8a). A mass m is then attached to the spring, extending it by an amount denoted by e (Figure 8b), with the mass then being at rest in its equilibrium position. Since the mass is at rest under the action of both the force of gravity, mg, and the tension in the spring ke, it follows that $ke = mg$.

Displacing the mass below its equilibrium position and then releasing it allows the mass to oscillate about its equilibrium position. The instant that the mass passes though the point that is a distance x below equilibrium is represented by Figure 8c. At this instant, the extension of the spring is given by $e + x$ and the tension in the spring is $k(e + x)$.

Ignoring the effects of air resistance on the oscillating mass, the resultant upward force F on the mass is equal to the tension minus the weight, which gives the equation

$$F = k(e + x) - mg$$

However, since $ke = mg$, the expression for the resultant force becomes $F = kx$.

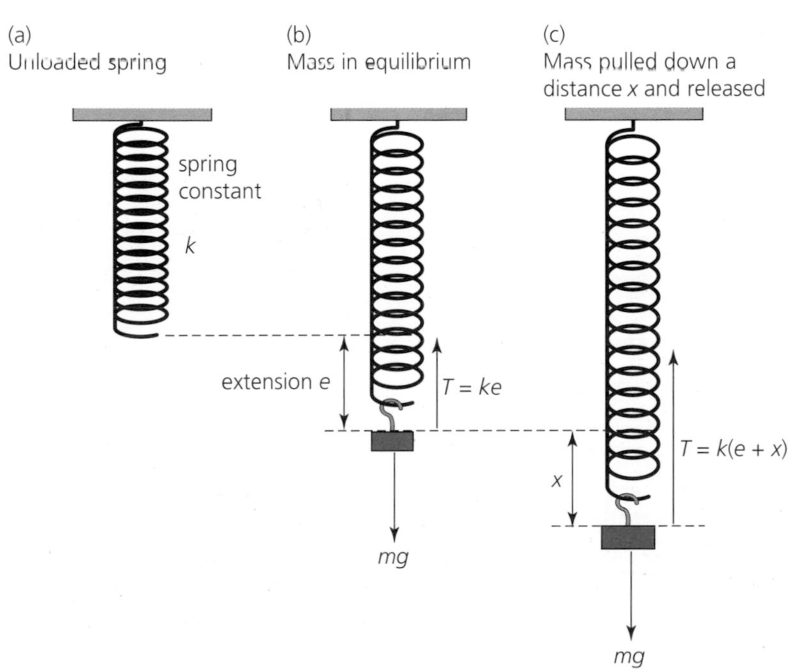

Figure 8 *The forces acting on a mass–spring system*

Since the resultant force is upwards when the displacement is downwards, the resultant force is written as

$$F = -kx$$

From Newton's second law ($F = ma$), the acceleration of the oscillating mass is therefore

$$a = -\frac{k}{m}x$$

Given that $\frac{k}{m}$ is constant, this shows that the acceleration of an oscillating mass is directly proportional to its displacement but in the opposite direction – which conforms to the definition of SHM.

Comparing the SHM defining equation $a = -\omega^2 x$ with $a = -\frac{k}{m}x$ shows that $\omega^2 = \frac{k}{m}$. Substituting $\omega = \frac{2\pi}{T}$ into the equation gives $\left(\frac{2\pi}{T}\right)^2 = \frac{k}{m}$, which rearranges to give the equation for the time period of a mass–spring system:

$$T = 2\pi\sqrt{\frac{m}{k}}$$

Worked example

A family car has a mass of 1000 kg when it is not loaded. This mass is supported equally by four identical springs. When the car is fully loaded, its mass goes up to 1250 kg and the springs compress by a further 2 cm. When the car goes over a bump in the road, it bounces on its springs. Find the time period of these oscillations.

The equation for the period of a mass–spring system is

$$T = 2\pi\sqrt{\frac{m}{k}}$$

We know the mass of the system, but we need to calculate the spring constant, k.

The extra weight of 250 kg × 9.81 N kg^{-1} = 2450 N will depress the four springs by 0.02 m.

Assuming each of the identical springs carries a quarter of the extra weight, 2450/4 = 613 N, we can use Hooke's law, $F = k\Delta l$, to find the spring constant k of one spring:

$$k = \frac{F}{\Delta l} = \frac{613}{0.02} = 3.06 \times 10^4\ \text{N m}^{-1}$$

Since the effective mass oscillating on each spring is 1250/4 = 313 kg, the time period T of the oscillation of the car is

$$T = 2\pi\sqrt{\frac{m}{k}} = 2\pi\sqrt{\frac{313}{3.06 \times 10^4}} = 0.64\ \text{s}$$

QUESTIONS

7. A mass of 250 g is attached to a spring with a spring constant of 30 N m^{-1}. Determine the time period if the mass was displaced so that the mass–spring system oscillated.
8. Four identical springs, each with a spring constant of 40 N m^{-1}, are arranged so that two are joined in series (Figure 9a) and the other two in parallel (Figure 9b). A mass of 500 g is attached to both arrangements.

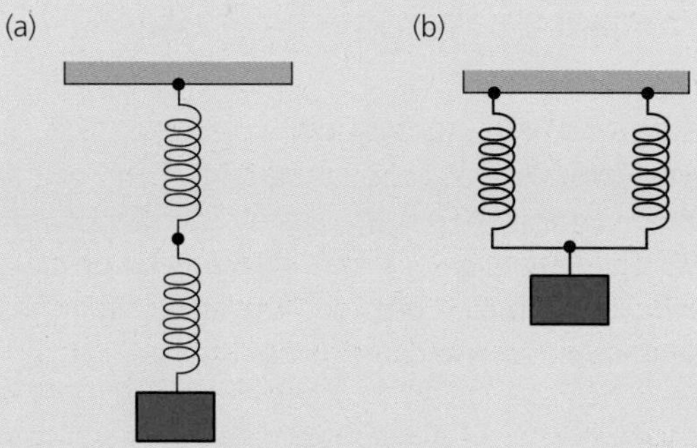

Figure 9 *Oscillating systems made up of springs (a) in series and (b) in parallel*

Determine the time period of oscillation of the arrangement of two springs

a. in series

b. in parallel.

[*Hint*: Combining two springs in series halves the spring constant. Combining two springs in parallel doubles the spring constant.]

9. Determine the phase difference in radians between two identical mass–spring systems, each with a time period of 1.2 s, if one system's oscillations are started 0.4 s before the other.
10. A mass of 200 g oscillates with a time period of 1.46 s when attached to a spring. Calculate the spring constant.
11. An oscillating mass–spring system has a time period T. If the mass is then doubled and the spring replaced so that the spring constant is halved, what is the new time period?

A $\sqrt{2}\,T$ **B** $2T$ **C** $\frac{T}{\sqrt{2}}$ **D** $\frac{T}{2}$

REQUIRED PRACTICAL: APPARATUS AND TECHNIQUES

Part 1: Investigation into simple harmonic motion using a mass–spring system

The aim of this practical is to test the equation $T = 2\pi\sqrt{\frac{m}{k}}$ and obtain a value for k, by carrying out measurements of time period for various masses attached to a helical spring. This practical gives you the opportunity to show that you can:

- use appropriate digital instruments to obtain a range of measurements (to include time)
- use methods to increase the accuracy of measurements, such as timing over multiple oscillations, or use of a fiduciary marker
- use a stopwatch for timing.

There are a number of ways of measuring the time period, including using digital or analogue stopwatches, light gates, or a motion sensor with a data logger. The method described here involves timing oscillations using a digital stopwatch and a fiducial marker.

Apparatus

A digital stopwatch with a precision of ±0.01 s is used, and a fiducial marker to help with counting oscillations.

A helical spring is clamped securely and supported by a stand (Figure P1). A fiducial marker in the form of an optical pin is inserted into a cork and supported in a clamp and stand. The purpose of this is to indicate the mean (equilibrium) position of the oscillation. A range of standard masses are used, accurate to within ±2 g and suitable for the choice of spring.

Techniques

Once a mass has been attached to the spring, the fiducial marker is aligned with the equilibrium position of the mass–spring system to enable accurate counting of cycles of oscillation. One cycle of oscillation corresponds to the mass passing the equilibrium position, moving to one extreme, passing the equilibrium position again, moving to the other extreme, then back to the equilibrium position. Although a full cycle also corresponds to the mass moving from one extreme to the other extreme and then back again, it is better to count oscillations with

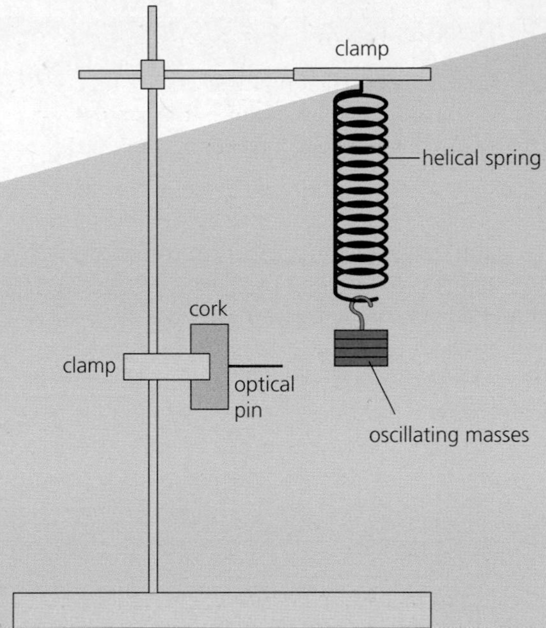

Figure P1 *Set-up for the oscillating mass–spring experiment*

respect to the equilibrium position because this does not change, whereas the extreme positions change as the oscillation loses energy.

The mass should be raised and then released so that it starts oscillating. Once the oscillation is established, the digital stopwatch should be started as the mass passes the fiducial marker. Twenty cycles of oscillation are then counted and the stopwatch should be stopped as the mass passes the fiducial marker completing its 20th cycle of oscillation. Two repeat measurements are taken and the average of the three measurements is calculated. The time period is determined by dividing the average time for 20 cycles of oscillations by 20. The process is repeated using a range of masses.

Analysis

The time period equation for the mass–spring system $T = 2\pi\sqrt{\frac{m}{k}}$ can rewritten as

$$T^2 = \frac{4\pi^2}{k}m$$

which can be compared with the equation of a straight line $y = mx + c$ (where m here is the constant gradient, not the variable mass). If a graph of T^2 as the y variable and mass m as the x variable is plotted with an origin (0, 0), then the theory predicts that the graph should be a straight line through the origin. The gradient of the line is equal to $\frac{4\pi^2}{k}$, which enables a value for the spring constant k to be determined.

The uncertainty in the time for 20 oscillations for a particular mass can be found from half of the range of the repeat measurements. The presence of reaction time error suggests that the uncertainty in the time for 20 oscillations cannot be less than typically ± 0.2 s. The uncertainty in time period T is found by dividing the uncertainty in 20 oscillations by 20. The percentage uncertainty in T can then be calculated. The percentage uncertainty in T^2 is twice the percentage uncertainty in T (*see section 1.4 in Chapter 1 of Year 1 Student Book*).

The uncertainty in the value of standard masses of 50 or 100 g is typically about ± 2 g. As more masses are added to the hanger, this uncertainty accumulates, but the percentage uncertainty for the total mass is unchanged. The uncertainties in T^2 and m enable error bars to added to the points on the T^2 versus m graph, so that the best line, and the steepest gradient and the shallowest gradient lines that fit within the error bars can be drawn. The percentage uncertainty in the value of k is equal to the percentage uncertainty in the gradient which can be calculated from

$$\text{percentage uncertainty} = \frac{\text{best gradient} - \text{worst gradient}}{\text{best gradient}} \times 100\%$$

where the worst gradient is whichever of the steepest or shallowest gradient values differs from the best gradient by the largest amount.

QUESTIONS

P1 Discuss whether reaction time error when measuring the time for 20 oscillations is random or systematic.

P2 A student measures time periods for a mass–spring system for various masses using the method described. One of her data sets is shown in Table P1. The repeat times for 20 cycles and the average time for 20 cycles have been recorded to the same precision as the digital stopwatch that was used to measure them.

a. i. Determine the uncertainty in the average time for 20 cycles from the range of the three $20T$ measurements.

ii. How does your value from part **i** compare with a typical reaction time error of ± 0.2 s?

b. Suggest an appropriate value for the uncertainty in time period T based on your answers to parts **a i** and **ii**, and record T to the appropriate number of significant figures.

c. i. The student decides to plot her full set of data in the form of a graph of T^2 versus mass and then obtain a value for the spring constant k. What would be the uncertainty in the plotted value of T^2 for the data set shown in Table P1?

ii. What source(s) of uncertainty would have the greatest effect on the maximum and minimum gradients, and hence the uncertainty in k?

iii. Would this still be the case if only 10 oscillations had been counted?

P3 Suggest whether or not a method using a motion sensor and data logger would give a more reliable test of the period–mass relationship and a more accurate value of k.

Mass attached, including hanger, m / kg	Time for 20 cycles, $20T_1$ / s	Repeat time for 20 cycles, $20T_2$ / s	Repeat time for 20 cycles, $20T_3$ / s	Average time for 20 cycles, $20T$ / s	Time period, T / s
0.150	9.86	9.64	9.72	9.74	

Table P1

The simple pendulum

The simple pendulum, consisting of a length of thread with a metal ball called a bob attached, usually made of lead or brass, has been an object of scientific study since the 17th century. Galileo is generally considered to be the first to observe that the time period of a swinging pendulum remained constant even as the oscillations died away (Figure 10). Another 17th century Italian astronomer, Giovanni Riccioli, devised a technique based on the oscillations of a pendulum to obtain the first accurate value for the acceleration due to gravity.

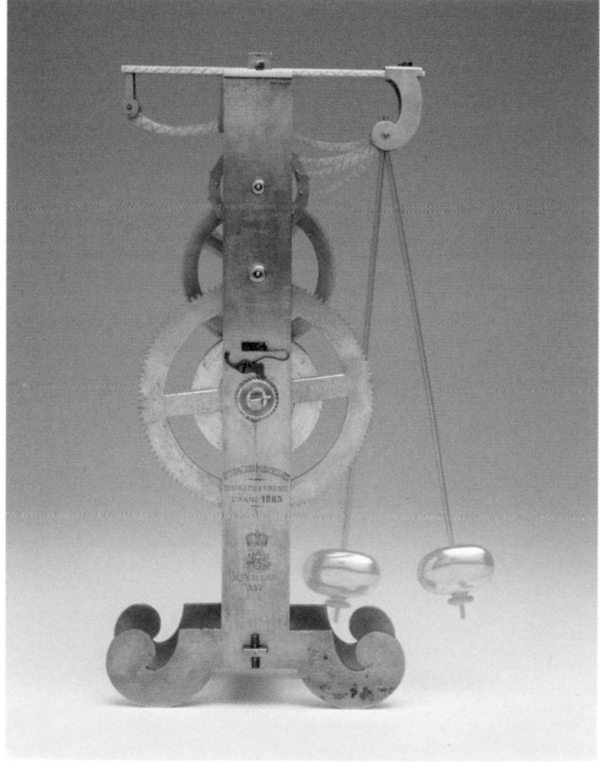

Figure 10 *A model of Galileo's proposed design for a pendulum clock. Galileo realised that a swinging pendulum could be used for time-keeping. This was a major step in the history of the development of clocks.*

A free-body force diagram of an oscillating simple pendulum (Figure 11) shows that the forces acting on the pendulum bob are the tension T in the string and the weight W of the bob, assuming that air resistance is negligible.

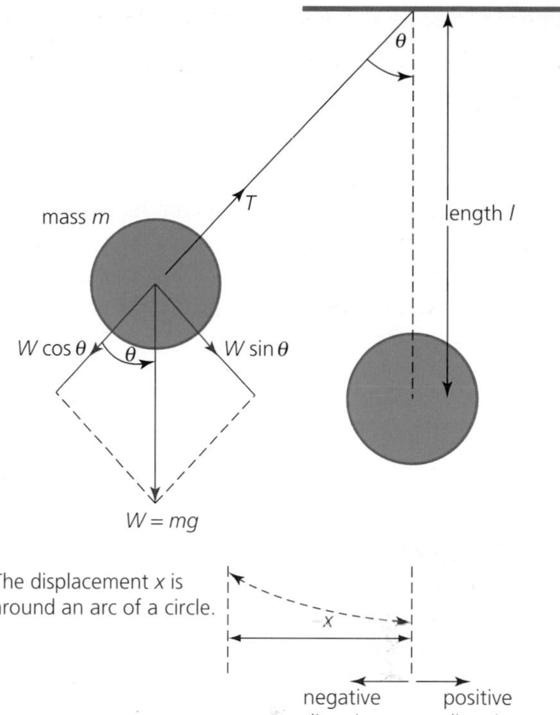

Figure 11 *The forces acting on the bob of a simple pendulum*

The weight of the bob is resolved into two components (*see section 10.3 in Chapter 10 of Year 1 Student Book*), one parallel to the tension and the other at 90° to the tension. Since the displacement x of the bob is actually along the arc of a circle with radius equal to the length l of the pendulum, the angle θ in radians is equal to $\frac{x}{l}$. The restoring force F is provided by the component of the bob's weight that acts at 90° to the tension, and therefore

$$F = W \sin\theta = mg \sin\theta$$

The small-angle approximation states that, for small angles ($<10°$), $\sin\theta \approx \theta$ in radians. If this is applied to the above equation, the expression for the restoring force becomes

$$F = mg\theta$$

Substituting $\theta = \frac{x}{l}$ gives $F = mg\frac{x}{l}$. But since the restoring force is always in the opposite direction to the displacement, the expression for the restoring force becomes

$$F = -mg\frac{x}{l}$$

From Newton's second law, the acceleration of the oscillating pendulum is therefore

$$a = -\frac{g}{l}x$$

This equation shows that the acceleration of an oscillating simple pendulum of constant length is directly proportional to and in the opposite direction to its displacement, *providing only small-amplitude oscillations are considered*. In other words, the simple pendulum oscillates with SHM.

Comparing $a = -\frac{g}{l}x$ with the defining equation for SHM, $a = -\omega^2 x$, shows that $\omega^2 = \frac{g}{l}$. Substituting $\omega = \frac{2\pi}{T}$ into the equation gives $\left(\frac{2\pi}{T}\right)^2 = \frac{g}{l}$, which rearranges to give the equation for the time period T of a simple pendulum:

$$T = 2\pi\sqrt{\frac{l}{g}}$$

The theory shows that the time period of a simple pendulum depends only on its length l provided angles of swing are less than about 10°. The mass of the pendulum bob has no effect on the time period provided the bob's mass is much greater than the mass of the string it is attached to. Although air resistance will remove energy from the oscillation, causing the amplitude to decrease, the time period of the oscillation is unaffected.

QUESTIONS

12. Calculate the length of a simple pendulum that would have a time period of 1.0 s.

13. A simple pendulum has a time period of T. What would be its new time period if its length is doubled?

A $2T$ **B** $\sqrt{2}\,T$ **C** $\frac{T}{\sqrt{2}}$ **D** $\frac{T}{2}$

KEY IDEAS

- The equation for time period T of a mass–spring system is

$$T = 2\pi\sqrt{\frac{m}{k}}$$

- The equation for the time period T of a simple pendulum for small-amplitude oscillations is

$$T = 2\pi\sqrt{\frac{l}{g}}$$

REQUIRED PRACTICAL: APPARATUS AND TECHNIQUES

Part 2: Investigation into simple harmonic motion using a simple pendulum

The two aims of this practical are to test the equation $T = 2\pi\sqrt{\frac{l}{g}}$ for small angle oscillations by obtaining measurements of the time period for various lengths l of a simple pendulum, and to use the data to obtain a value for the acceleration due to gravity g.

The practical gives you the opportunity to show that you can:

- use appropriate analogue apparatus to record a range of measurements (to include length)
- use appropriate digital instruments to obtain a range of measurements (to include time)
- use methods to increase the accuracy of measurements, such as timing over multiple oscillations, and the use of a fiduciary marker and set square
- use a stopwatch or light gates for timing.

Apparatus

The simple pendulum consists of a spherical brass bob attached to a length of string. The string of the simple pendulum is clamped securely between two wooden blocks (or the two halves of a split cork) and supported at a suitable height by a stand.

As in Part 1 of the Required Practical, there are a number of possible ways of measuring the time period, including using digital or analogue stopwatches, light gates or a data logger with a motion sensor. The method described here involves timing oscillations using a stopwatch and the use of a fiducial marker to help with counting oscillations. The fiducial marker is in the form of a vertical optical pin inserted into a wide block or cork that can stand on the bench to indicate the mean (equilibrium) position of the oscillation.

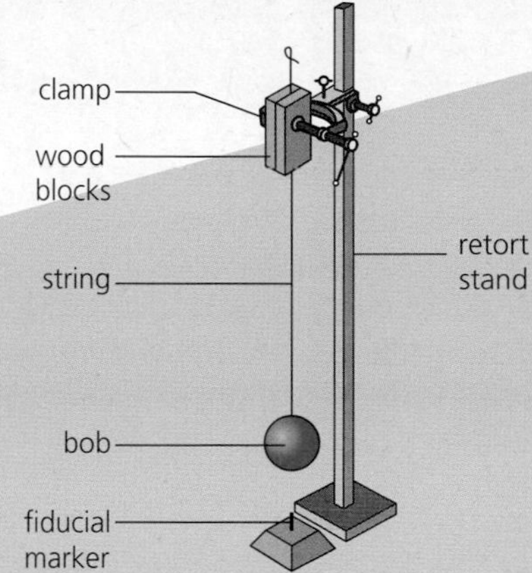

Figure P2 *Set-up for the oscillating pendulum experiment*

Techniques

The fiducial marker is aligned with the equilibrium position of the pendulum to enable accurate counting of cycles of oscillation (Figure P2).

The length l of the pendulum is measured from its point of support to the centre of the bob using a set square aligned to a clamped metre rule (Figure P3).

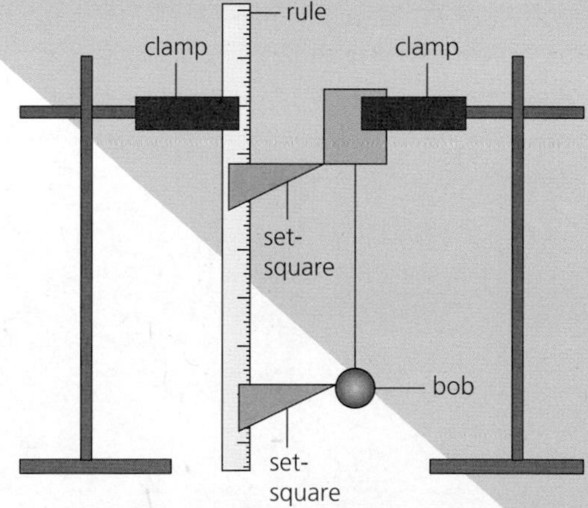

Figure P3 *Pendulum length measurement*

QUESTIONS

P4 Suggest the procedure for timing oscillations and analysing the data in order to meet the aims of the practical.

P5 How could an estimate be made of the uncertainty in the value of g obtained?

ASSIGNMENT 1: USING LOGARITHMS TO ANALYSE PENDULUM OSCILLATION DATA

(MS 0.5, MS 3.2, MS 3.3, MS 3.4, MS 3.10, MS 3.11)

In this assignment you will use logarithms to see if time period *T* and length *l* data for a simple pendulum support the theory that leads to the equation

$$T = 2\pi\sqrt{\frac{l}{g}}$$

Before you analyse the data, you will need to know (or recall) some things about logarithms, which we do now. A logarithm is basically a power. For example, since $1000 = 10^3$, we can write 'the logarithm of 1000 to base 10 is equal to 3', which is written mathematically as $\log_{10} 1000 = 3$. Similarly, $0.01 = 10^{-2}$ and can be written as $\log_{10} 0.01 = -2$. Logarithms do not have to be expressed to base 10, but this is the most common, and appears on your calculator as 'log'.

The three rules that apply to logarithms that you need to know are:

- Rule 1 is for any two numbers *A* and *B*: $\log(AB) = \log A + \log B$
- Rule 2 is for any two numbers *A* and *B*: $\log\left(\frac{A}{B}\right) = \log A - \log B$
- Rule 3 is number *A* raised to the power *n*: $\log A^n = n\log A$

For example, if two variables, *P* and *Q*, could be connected by an equation of the form

$$P = kQ^n$$

where *k* and *n* are both constants, then applying logarithms to both sides of the equation gives

$$\log P = \log(kQ^n)$$

Applying rule 1 to the right hand side of the equation gives

$$\log P = \log k + \log Q^n$$

Applying rule 3 to the second term on the right hand side gives

$$\log P = \log k + n\log Q$$

Rearranging the right hand side of the equation gives

$$\log P = n\log Q + \log k$$

This can be compared with the equation of a straight line, $y = mx + c$, to show that, if *P* and *Q* are related by $P = kQ^n$, and if log *P* data are plotted on the *y*-axis and log *Q* data are plotted on the *x*-axis, then the graph will be a straight line with gradient *n* and *y* intercept $\log k$.

Now let us return to our simple pendulum. Apply logarithms to the time period equation for the simple pendulum. First, rearrange the equation $T = 2\pi\sqrt{\frac{l}{g}}$ to give

$$T = \sqrt{l} \times \frac{2\pi}{\sqrt{g}} = l^{1/2} \times \frac{2\pi}{\sqrt{g}}$$

Taking logarithms of both sides of the equation and applying rule 1 gives

$$\log T = \log(l^{1/2}) + \log\left(\frac{2\pi}{\sqrt{g}}\right)$$

Applying rule 3 gives

$$\log T = \frac{1}{2}\log l + \log\left(\frac{2\pi}{\sqrt{g}}\right)$$

Comparing this with the equation of a straight line, $y = mx + c$, suggests that a graph of $\log T$ versus $\log l$ should have a gradient of $\frac{1}{2}$ (or 0.5) and a *y* intercept of $\log\left(\frac{2\pi}{\sqrt{g}}\right)$.

Questions

A1 Draw a two-column table with titles log(*l* / m) and log(*T* / s). Note that, although logarithms are essentially powers and do not have units, you are required to show the units of the quantity along with the symbol for the quantity in brackets, as shown with the titles just given.

A2 Length of pendulum and time period data are shown in Table A1.

Length of pendulum, l / m	Time for 20 cycles, $20T_1$ / s	Repeat time for 20 cycles, $20T_2$ / s	Repeat time for 20 cycles, $20T_3$ / s	Average time for 20 cycles, $20T$ / s	Time period, T / s
0.250	19.8	20.0	20.2	20.0	1.00
0.300	21.7	21.8	21.9	21.8	1.09
0.350	23.6	23.4	23.8	23.6	1.18
0.400	25.5	25.4	25.4	25.4	1.27
0.450	26.8	26.7	26.9	26.8	1.34
0.500	28.2	28.4	28.6	28.4	1.42
0.550	29.7	29.7	29.5	29.6	1.48
0.600	30.8	31.2	30.9	31.0	1.55

Table A1

a. Tabulate values of log(T / s) and log(l / m) using the data from Table A1 in the two-column table you drew for question **A1**. Plot a graph of log(l / m) versus log(T / s). Since you only require the gradient from this graph, you do not need to include an origin. Use sensible scales that extend over at least half of the graph paper.

b. Draw a line of best fit and determine the gradient of the graph using a gradient triangle that has sides of at least 8 cm. Compare your result with the theoretical prediction for the gradient of 0.5.

2.3 THE ENERGY OF AN OSCILLATING SYSTEM

Swinging pendulums, masses oscillating on springs and atoms vibrating in a solid are all examples of a periodic motion in which energy is continuously being changed from potential energy to kinetic energy and back again. For example, when a pendulum bob is moved sideways from the equilibrium position, it gains height and therefore also gains gravitational potential energy. As it is released, some of the gravitational potential energy is transferred to kinetic energy (Figure 12) and the bob accelerates, reaching its maximum speed as it passes through the equilibrium position. There is no resultant force acting on the bob at equilibrium, but its momentum means that is continues and starts to gain height, so transferring kinetic energy back to gravitational potential energy.

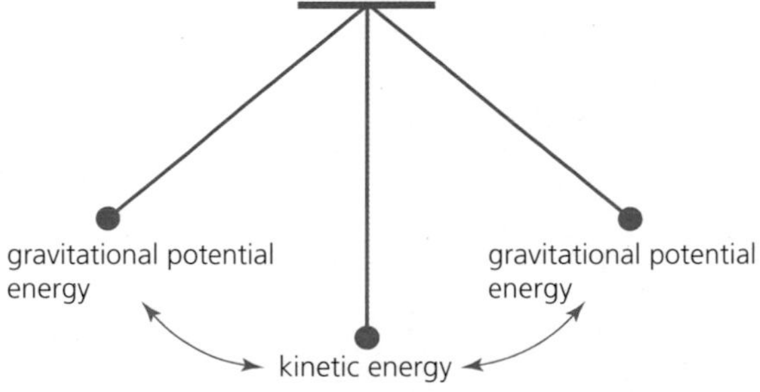

Figure 12 *Energy transfer during the oscillation of a simple pendulum*

A **free oscillation** is an idealised oscillation in which the amplitude of the oscillating object is constant, as no energy is input, and no energy is removed from the system by friction, for example.

The total energy of a free oscillation is the sum of the potential energy and kinetic energy and is constant throughout the oscillation.

The variation in potential energy E_p and kinetic energy E_k with displacement from equilibrium is shown in Figure 13.

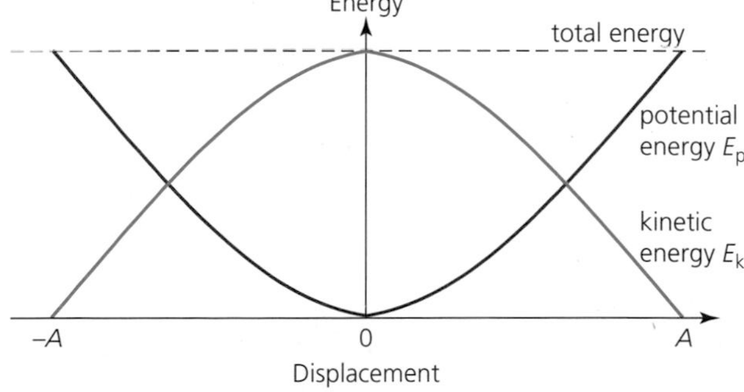

Figure 13 *Energy versus displacement for a free oscillation*

It can be shown that the kinetic energy of an object oscillating with SHM varies in proportion to $\sin^2(\omega t)$ and the potential energy to $\cos^2(\omega t)$, assuming that, at $t = 0$, the displacement (and hence potential energy) is a maximum (Figure 14).

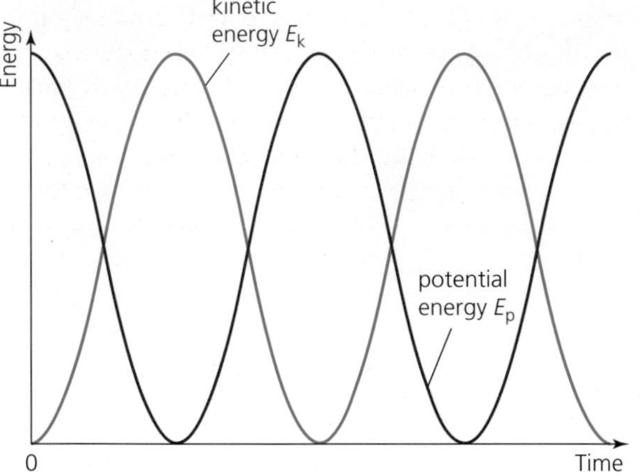

Figure 14 *The variation of kinetic energy and potential energy with time over one cycle of SHM, assuming that the displacement is a maximum at t = 0*

For a mass oscillating vertically on a spring, the potential energy consists of both elastic strain energy in the spring (maximum at the bottom extreme of the oscillation) and gravitational potential energy (maximum at the top extreme of the oscillation).

Worked example

A mass of 500 g is attached to a spring of spring constant 81.8 N m^{-1}, stretching it by 6.0 cm. The mass is pulled down a further 4.0 cm and released so that it moves upwards.

a. Describe the energy changes that take place as the mass moves from its release position to its upper extreme position.

b. Calculate the maximum kinetic energy of the mass (if necessary, *refer back to section 12.2 in Chapter 12 of Year 1 Sutdent Book*).

c. Sketch a graph showing how the kinetic energy varies with time for one cycle of oscillation, assuming that, at $t = 0$, the mass is at its extreme position.

a. Just before being released, there is elastic potential energy (elastic strain energy) stored in the spring. On being released, the mass rises and some of the elastic potential energy is converted to kinetic energy and gravitational potential energy. On passing through the equilibrium position, the kinetic energy reaches a maximum value. As the mass continues to its upper extreme position, the kinetic energy decreases to zero and the gravitational potential energy increases.

b. The elastic energy stored in the spring before the mass has been pulled down is

$$\frac{1}{2}F\Delta l = \frac{1}{2}k(\Delta l)^2 = \frac{1}{2}\times 81.8\times(0.06)^2 = 0.147\text{ J}$$

The mass is now pulled down a further 4.0 cm, increasing the elastic energy to

$$\frac{1}{2}k(\Delta l)^2 = \frac{1}{2}\times 81.8\times(0.1)^2 = 0.409\text{ J}$$

The mass is then released and, on passing through the equilibrium position, the elastic energy has decreased from 0.409 J to 0.147 J. The gravitational potential energy has increased by

$$mg\Delta h = 0.5\times 9.81\times 0.04 = 0.196\text{ J}$$

By conservation of energy, on passing through equilibrium:

kinetic energy = loss of elastic energy − gain in gravitational potential energy

Hence

kinetic energy = 0.409 − 0.147 − 0.196 = 0.066 J

c. Since the mass is at its extreme position at $t = 0$, its kinetic energy is zero at $t = 0$ and varies as a sine-squared graph (Figure 15).

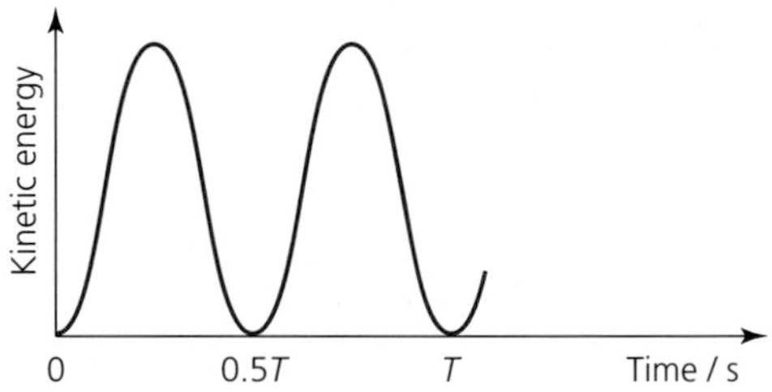

Figure 15

QUESTIONS

14. A pendulum is made from a 500 g mass hanging on a string. The mass is pulled back so that it is lifted through a height of 5.0 cm.

a. Describe the energy changes that occur from the mass being released and reaching its other extreme position.

b. Sketch a graph showing how the potential energy of the mass varies with time over one cycle of oscillation assuming that at $t = 0$ the mass is at its extreme position.

c. Calculate
   i. the additional gravitational potential energy that the mass gains as a result of being pulled back
   ii. the maximum kinetic energy
   iii. the maximum velocity reached by the mass.

15. A struck tuning fork vibrates with simple harmonic motion, causing the prongs to vibrate back and forth (Figure 16) at a constant frequency. The bottom end of the fork also vibrates up and down, so if held on a hard surface an audible sound is produced.

Figure 16 *Tuning fork vibration*

a. Describe the energy changes that occur during half a cycle of vibration.
b. If the fork frequency is 256 Hz, determine the maximum speed of the prongs when the amplitude is 1 mm.

Damping

In practice, most oscillations experience a resistive force that transfers energy away from the oscillating system. The resistive force, for example, friction or air resistance, always acts in the opposite direction to the motion of the oscillating object. The removal of energy from an oscillating system is known as **damping** and the extent of the damping determines how long it takes for an oscillation to die away. A displacement versus time graph for an oscillation that dies away after a few cycles is shown in Figure 17.

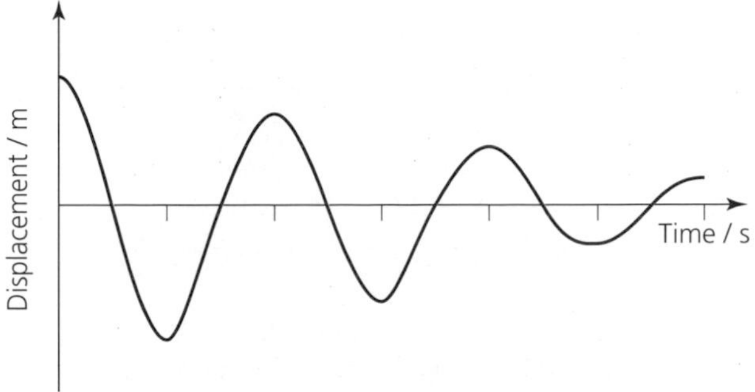

Figure 17 *Damped oscillations*

If damping is light, oscillations die away gradually. If resistive forces are very high, the system has heavy damping and the displaced object moves slowly back to its equilibrium position without oscillating. A system is said to be **critically damped** when the resistive forces are just enough to prevent oscillation and the object returns to equilibrium in the minimum possible time (Figure 18). Critical damping is the ideal state for many mechanical systems in order to prevent vibration damage (see section 2.4).

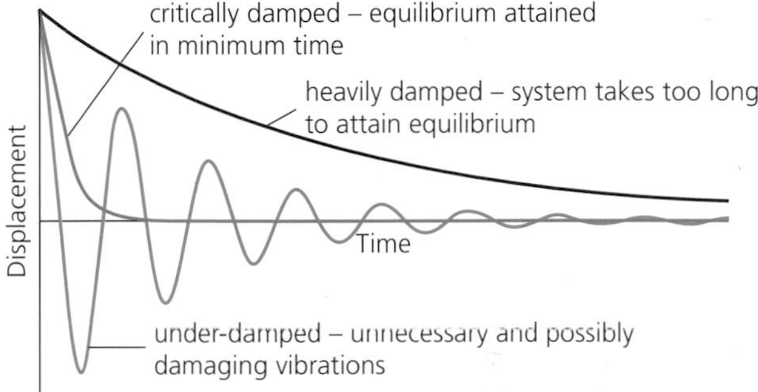

Figure 18 *The effects of damping on an oscillating system*

An example of a critically damped system is the front suspension on a mountain bike (Figure 19). The combination of the suspension spring and a shock absorber ensure a smoother ride.

Figure 19 *Suspension systems include a shock absorber to damp oscillations.*

KEY IDEAS

- In mechanical oscillations, there is a continuous interchange between potential energy and kinetic energy.
- Oscillating systems that lose energy to their surroundings are described as damped.

2.4 FORCED VIBRATIONS AND RESONANCE

When a child's swing is pulled back and released, it will oscillate at its **natural frequency** f_0, which, as for a simple pendulum (see section 2.2), depends only on the length of the swing. This is an example of a free oscillation in which, after the initial displacement, the oscillation is not subject to an external force that transfers energy to the system. However, a swing being repeatedly pushed is an example of a **forced oscillation** in which energy is repeatedly transferred to the oscillation. The frequency of the forced oscillation is equal to that of the pushes, or the **driving frequency**. A person's eardrum being forced to oscillate by sound waves and a loudspeaker being forced to oscillate by the electrical signal supplied to it are two examples of forced oscillations.

If a situation occurs such that the driving frequency has the same value as the natural frequency of the oscillator, the energy transfer to the oscillator occurs at its maximum efficiency, causing the amplitude of the forced oscillation to increase. The amplitude will continue to get bigger until the energy supplied by the driving system is equal to the energy lost by damping. This effect of producing a large-amplitude oscillation by matching the driving frequency to the natural frequency is called **resonance**.

Resonance can cause problems in mechanical systems, so engineers have to take this into account when designing bridges, tall buildings and machinery, as the effects of resonance can be catastrophic (Figure 20). However, resonance does have its advantages, as it is key to the operation of most musical instruments. Other useful applications of resonance include the microwave cooker and MRI scanning (see the introduction to this chapter).

Figure 20 The resonant oscillations of the Tacoma Narrows Bridge led to its collapse.

Resonance can easily be demonstrated with a mass attached to a spring. If a person moves the free end of the spring up and down at a low frequency (Figure 21) the mass–spring system starts to vibrate in phase with the person's hand. However, if the frequency of vibration is increased, there comes a point at which the amplitude becomes very large and the vibrations so wild that the mass will probably fall off the spring.

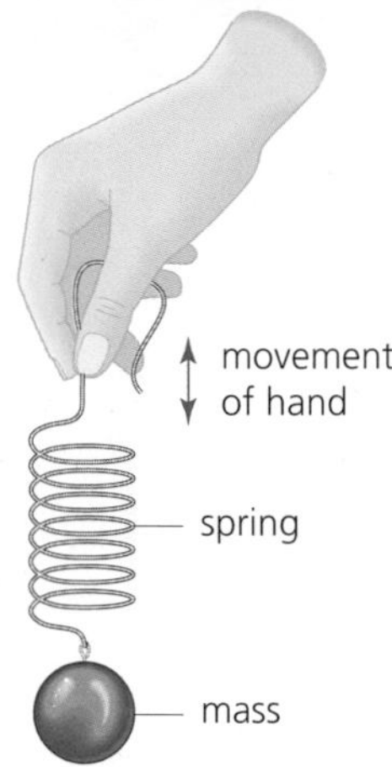

Figure 21 Forced oscillation of a mass–spring system

A clamped plastic rod, such as a ruler, can be made to under go resonance by applying a driving force from a vibration generator connected to a signal generator (Figure 22). The frequency at which resonance occurs, called the **resonant frequency**, can be determined by adjusting the signal generator until the largest amplitude of oscillation is found. Continuing to increase the driving frequency can show a second, higher, resonant frequency.

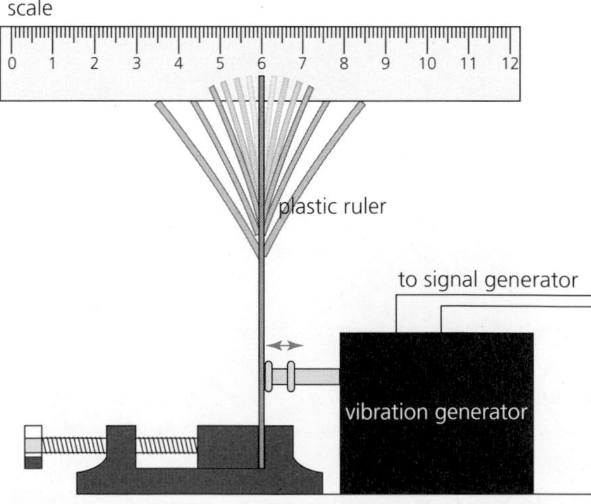

Figure 22 Forced vibrations of a clamped plastic ruler

Resonance can also be demonstrated using the apparatus known as Barton's pendulums, shown in Figure 23. A heavy pendulum, the driver, is attached to a string, which also supports a number of other, lighter, pendulums of different lengths. When the driver is set in motion, some of the energy is transferred to the other pendulums, making them oscillate at the same frequency as the driver. Energy is most efficiently transferred from the driver to the lighter pendulum that has the same length and therefore the same natural frequency as the driver. This lighter pendulum oscillates with a large amplitude as it undergoes resonance with a phase difference of 90°, or $\pi/2$ radians, behind the driving pendulum. The shorter pendulums are almost in phase with the driver; the longer ones are almost 180° or π radians out of phase with the driver.

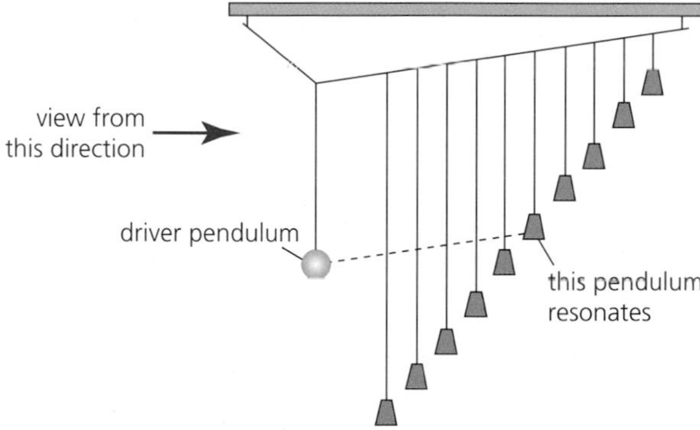

Figure 23 *Barton's pendulums*

Many musical instruments rely on resonance for their production of sound. **Stationary waves** are formed on the strings of an instrument, such as a guitar or a violin, when the string is made to vibrate, by being plucked or scraped with a bow. Superpositions of the reflections from either end produce a stationary wave on the string (*see section 5.6 in Chapter 5 of Year 1 Student Book*). The pitch of the note is determined by the first harmonic. But higher harmonics can occur simultaneously with the first harmonic, and the number and amplitude of these determine the quality of the sound (Figure 24).

The strings themselves make almost no sound, since the thin wires cut through the air easily and cause very little vibration of the air molecules. In an acoustic guitar, the vibration of a string is transferred to the body of the guitar via the bridge (Figure 25). The guitar body has been designed so that its own range of harmonic frequencies matches those of the strings, enabling the energy of vibration of the strings to be efficiently transferred to the guitar body. The guitar body **resonates**, and causes the air within it and surrounding it to vibrate, producing a travelling sound wave of sufficient amplitude to be heard.

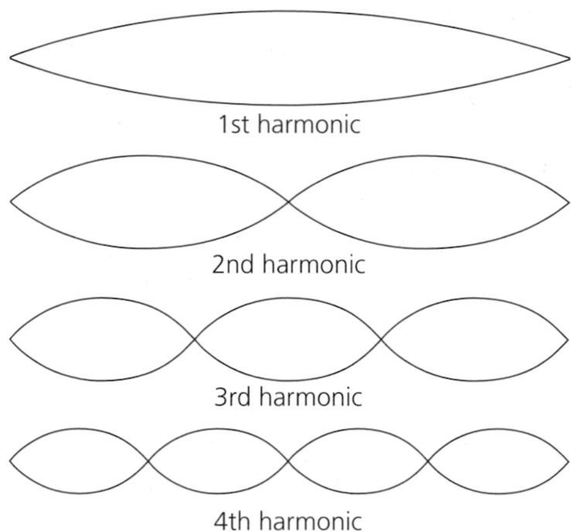

Figure 24 *Harmonics on a guitar string*

Figure 25 *The bridge of the guitar transmits the vibration of the strings to the body of the guitar.*

The effect of damping on resonance

The amplitude at resonance depends on the degree of damping that is present – the more energy lost to the surroundings, the smaller the maximum amplitude (Figure 26). There are three key features to note from the amplitude versus frequency graph:

- With no or very little damping, the largest value for the maximum amplitude occurs when the frequency of the driving oscillation equals the natural frequency f_0.
- The resonant frequency decreases as the degree of damping increases.
- The resonance curve becomes less sharp as the damping is increased.

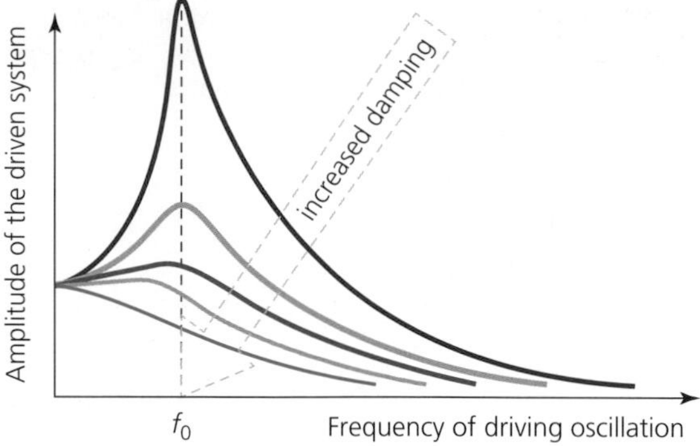

Figure 26 *The effects of damping on resonance*

Damping is crucial in structures or machinery that have the potential for resonance. For example, if a helicopter's damping systems are not properly maintained, when it lands it can undergo 'ground resonance' due to an imbalance in the rotation of the helicopter rotor blades. This may cause it to vibrate so violently that it breaks up.

Worked example

The casing of a washing machine vibrates as the motor driving the drum spins round. As the motor speeds up, the vibrations increase, until, at a certain motor speed, the casing vibrates violently. At higher motor speeds, the vibrations die away again.

a. Explain these observations.

b. Washing machines often have a large block of concrete bolted to the casing. Explain why.

c. How could the amplitude of the vibrations be decreased?

a. The casing is resonating. When the driving frequency of the oscillations caused by the motor matches the natural vibration frequency of the casing, there are large-amplitude vibrations. At lower and higher motor speeds, there is less energy transferred to the casing during each cycle and the vibrations get smaller.

b. The concrete increases the mass of the vibrating system. As with a mass on a spring, this lowers its natural frequency, so that resonance does not occur in the motor's normal range of running speed.

c. Adding damping to the system would decrease the amplitude. (A shock absorber can be fitted to the casing. The shock absorber is a piston moving in a cylinder filled with oil. As the casing vibrates, oil is forced through small holes in the piston, dissipating energy and reducing the vibrations.)

QUESTIONS

16. What is the phase difference, as a fraction of the vibration cycle, and in radians, between a forcing vibration and the displacement of the oscillating body that is undergoing resonance?

17. How is the resonant frequency of a mechanical system affected if the amount of damping is increased?

KEY IDEAS

- Resonance is a large-amplitude oscillation that occurs when a system is forced to oscillate at a resonant frequency.
- The resonant frequency of a freely oscillating system is equal to its natural frequency.
- With increasing damping, the resonant frequency decreases slightly.
- Damping also reduces the maximum amplitude at resonance and makes the resonance less sharp.

PRACTICE QUESTIONS

1. A student is required to determine the time period of a simple pendulum using a stopwatch to count 20 oscillations. A fiducial marker aligned with the equilibrium position of the oscillation is used to help with the counting of cycles of oscillation.
   a. Suggest an additional element of the procedure for finding the time period that the student could adopt to make her measurements as accurate as possible.
   b. Suggest a reason why it is better to locate the fiducial marker at the equilibrium position rather than one of the extreme positions of the oscillation.
2. a. Predict the value of the time period of oscillation of a mass–spring system if the mass attached to the spring is 500 ± 10 g and the spring constant is $50 \pm 2\,\text{N m}^{-1}$.
   b. Determine the uncertainty in the value predicted for the time period.
   c. A data logger and motion sensor are used to monitor the oscillation of the mass–spring system, and a sample rate of 20 Hz was selected.
      i. What would be the time interval between measurements?
      ii. How many measurements would be made in the time for eight cycles of oscillation?
3. A trolley of mass 0.80 kg rests on a horizontal surface attached to two identical stretched springs, as shown in Figure Q1. Each spring has a spring constant of $30\,\text{N m}^{-1}$, can be assumed to obey Hooke's law, and to remain in tension as the trolley moves.

Figure Q1

   a. i. The trolley is displaced to the left by 60 mm and then released. Show that the magnitude of the resultant force on it at the moment of release is 3.6 N.
      ii. Calculate the acceleration of the trolley at the moment of release and state its direction.
   b. i. The oscillating trolley performs simple harmonic motion. State two conditions which have to be satisfied to show that a body performs simple harmonic motion.
      ii. The frequency f of oscillation of the trolley is given by

$$f = \frac{1}{2\pi}\sqrt{\frac{2k}{m}}$$

      where m = mass of trolley and k = spring constant of one spring. Calculate the period of oscillation of the trolley, stating an appropriate unit.
   c. Copper ions in a crystal lattice vibrate in a similar way to the trolley, because the inter-atomic forces act in a similar way to the forces exerted by the springs. Figure Q2 shows how this model of a vibrating atom can be represented.

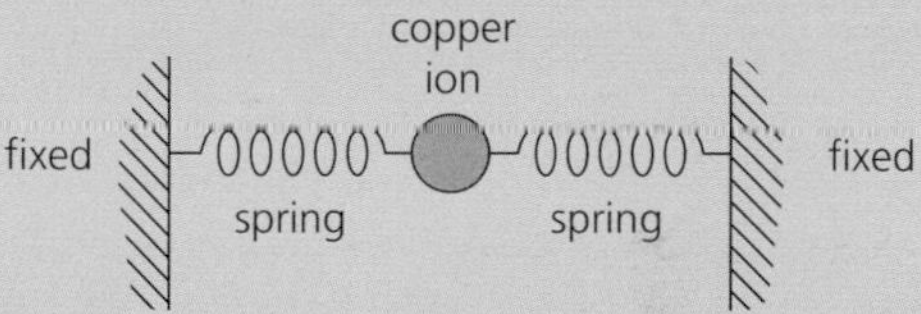

Figure Q2

      i. The spring constant of each inter-atomic 'spring' is about $200\,\text{N m}^{-1}$. The mass of the copper ion is 1.0×10^{-25} kg. Show that the frequency of vibration of the copper ion is about 10^{13} Hz.
      ii. If the amplitude of vibration of the copper ion is 10^{-11} m, estimate its maximum speed.
      iii. Estimate the maximum kinetic energy of the copper ion.

AQA Unit 4 Section B June 2011 Q2

4. A simple pendulum consists of a brass bob of mass 20.0 g on the end of a thin piece of string of length 600 mm. The bob is displaced to one side of its equilibrium position so that the string makes an angle of 10° with the vertical and then released so that it undergoes simple harmonic motion.

a. State the energy changes that take place as the bob swings from one extreme to the other.

b. Calculate the time period of the pendulum.

c. i. Determine the initial amplitude of the oscillation.

ii. Calculate the maximum speed of the pendulum bob during its oscillation.

iii. Determine the magnitude of the tension in the string as the bob passes through the equilibrium position.

5. Which one of the graphs in Figure Q3 shows how the acceleration, *a*, of a body moving with simple harmonic motion varies with its displacement, *x*?

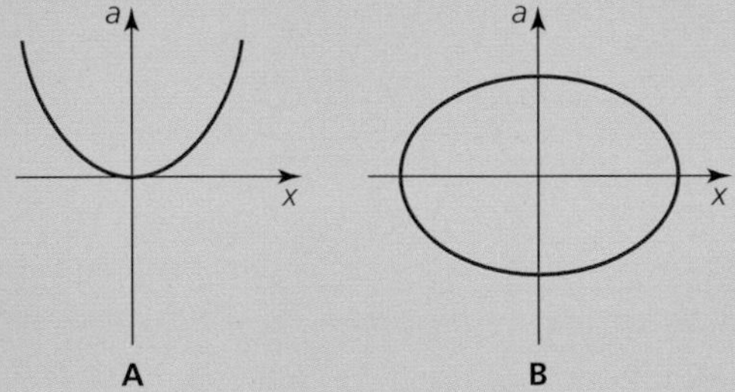

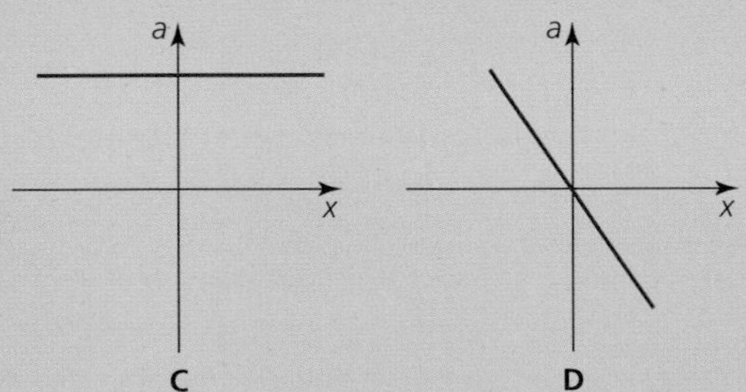

Figure Q3

AQA Unit 4 Section A January 2011 Q4

6. The tip of each prong of a tuning fork emitting a note of 320 Hz vibrates in simple harmonic motion with an amplitude of 0.50 mm. What is the speed of each tip when its displacement is zero?

A zero **B** $0.32\pi \, mm s^{-1}$

C $160\pi \, mm s^{-1}$ **D** $320\pi \, mm s^{-1}$

AQA Unit 4 Section A June 2014 Q8

7. Which one of the following gives the phase difference between the object's displacement and velocity in simple harmonic motion?

A $\frac{\pi}{4}$ **B** $\frac{\pi}{2}$

C $\frac{3\pi}{4}$ **D** π

8. A mass moving with simple harmonic motion has an amplitude of 0.2 m and a time period of 1 s. What is the speed of the mass, in $m s^{-1}$, when its displacement is 0.1 m?

A 1 **B** 2

C 3 **D** 4

3 THERMAL PHYSICS

PRIOR KNOWLEDGE

You will be familiar with states of matter, changes of state and the particle model of matter. You will have experience of calculating kinetic energy and potential energy, and will understand the principle of conservation of energy (*Chapter 11 of Year 1 Student Book*). You will have an understanding of momentum and of applications of Newton's laws of motion (*Chapter 11 of Year 1 Student Book*). You may recall that the absolute temperature scale is in degrees kelvin (K) and that 0 K (absolute zero) is equivalent to −273.15 °C.

LEARNING OBJECTIVES

In this chapter you will learn about the internal energy of an object and how this relates to its temperature and to the state of matter. You will learn the importance of the theoretical concept of an ideal gas, and how it leads to mathematical descriptions of the behaviour of a gas in terms of the motion of its constituent molecules. You will gain experience in solving problems related to heating, cooling and change of state, and see how experimental work and theoretical work go hand in hand to extend scientific knowledge.

(Specification 3.6.2.1 to 3.6.2.3)

The coldest place on Earth is the Antarctic. The lowest ground temperature recorded, measured by a NASA satellite, was −94.7 °C. Despite such low temperatures, there are several research stations in the Antarctic, including the Amundsen–Scott South Pole Station (Figure 1), which is a US scientific research station operating the South Pole Telescope. Whilst the South Pole is far from the ideal place for humans, it is a great place to view space, particularly at submillimetre (radio) wavelengths. This is partly because of the high altitude and the absence of water vapour in the atmosphere. But also, above the South Pole is a region in the sky called the Southern Hole, which is unusually free from interstellar radiation and dust, and enables excellent viewing into very deep space and therefore back to the early Universe.

Figure 1 *The Amundsen–Scott South Pole Station is just 250 m from the South Pole.*

Consistently one of the hottest places on Earth is Death Valley in California, USA (Figure 2), where a ground temperature of 93.9 °C has been recorded.

Currently the highest artificial temperature was created at CERN's Large Hadron Collider in 2012, where lead ions travelling at 99% of the speed of light smashed together in the ALICE detector to produce a quark–gluon plasma at a temperature of 5.5×10^{12} K in an attempt to recreate the conditions of the very early Universe. The lowest artificial temperature currently is 450×10^{-12} K achieved by a team of scientists at the Massachusetts Institute of Technology, USA.

Figure 2 (background) *Temperatures in Death Valley, California, reach the highest on Earth.*

3.1 INTERNAL ENERGY

Consider the energy in a cup of hot tea in terms of the atoms and molecules that are present (Figure 3). The tea itself is a seething mass of high-speed molecules travelling in random directions whilst vibrating, rotating and exerting mutually attractive forces on each other of an electrical nature. The kinetic energy of the molecules' translational, rotational and vibrational motion (see section 3.5), added to their mutual potential energy caused by the forces between them, make up the **internal energy** of the liquid.

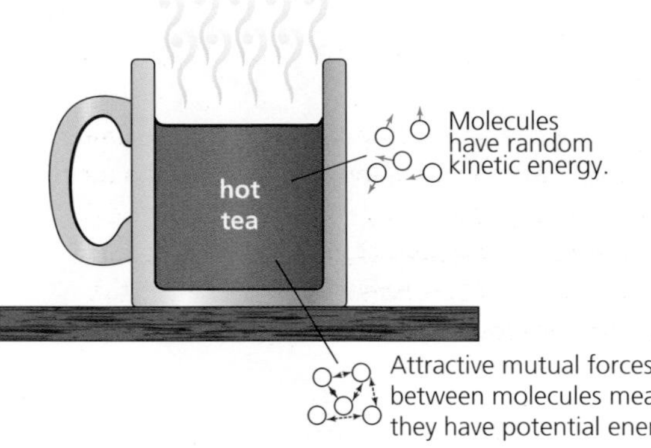
hot
tea
Molecules have random kinetic energy.
Attractive mutual forces between molecules means they have potential energy.

Figure 3 *Molecular energy in a hot cup of tea*

Internal energy is defined as the sum of the randomly distributed kinetic energy and potential energy of all the particles in a body.

There are two ways of increasing the internal energy of a system – by *heating* and by *doing work* on the system. The difference between heating and working is in the way that the energy is transferred.

- Heating involves an energy transfer caused by a temperature difference, called a **thermal energy transfer**.
- Doing work involves an energy transfer as a result of a force moving.

For example, the internal energy of a metal nail can be raised by putting it in contact with a body at a higher temperature such as the flame of a Bunsen burner. Alternatively, the nail can be repeatedly hit with a hammer, again resulting in an increase in its internal energy. Both methods result in a rise in temperature of the nail.

In some circumstances a system itself does work against an external force. For example, if carbon dioxide is contained at high pressure in a cylinder and then some of the gas is allowed to escape, the gas expands rapidly, doing work as it pushes back the atmosphere. The work done by the gas is at the expense of its own internal energy, resulting in the gas cooling – in this case so drastically that some of it will solidify to produce bits of solid carbon dioxide (also known as dry ice).

The equivalence of the effects of heat and work on the internal energy of a system was established by experiments devised during the 19th century by James Prescott Joule, scientist and brewer from Salford, near Manchester. With his famous paddle wheel experiment (Figure 4), Joule was able to demonstrate that doing work on a liquid by rotating paddles raised the liquid's temperature just as applying heat to a liquid would have done. His experimental measurements enabled him to determine the amounts of mechanical work and heat that would raise the temperature of a specific mass of water by the same amount. Joule's work establishing the equivalence of heat and work is the reason why these two quantities are measured in the same unit, the joule.

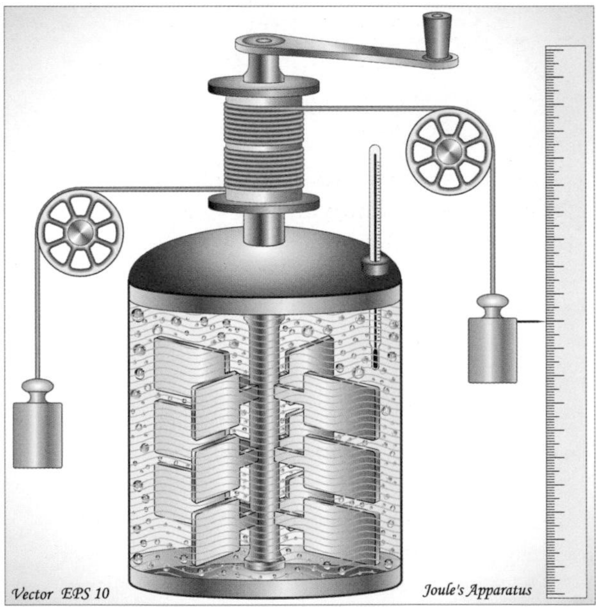
Vector EPS 10
Joule's Apparatus

Figure 4 *Joule's paddle wheel experiment*

This equivalence of heating and working and their effect on internal energy is formalised in the **first law of thermodynamics**. The law is expressed mathematically in different ways but can be summarised as follows:

- The increase in internal energy of a system is equal to the sum of the energy transferred to the system by heating and the energy transferred to the system by work done on it by an external force.

- The decrease in internal energy of a system is equal to the sum of the energy transferred away from the system by cooling and the energy transferred away from the system as a result of the system doing work against an external force.

We will see later in this chapter (section 3.6) that it is the kinetic component of the internal energy of a body that is related to the body's temperature. So if energy transferred to a substance by heat or work results in a temperature rise, this indicates that there has been an increase in the kinetic energy of the particles.

However, if any expansion or change of state (melting or vaporisation) occurs, this indicates an increase in the potential energy component of the substance's internal energy. We will consider this in section 3.3.

KEY IDEAS

- Internal energy is the sum of the randomly distributed kinetic energy and the potential energy of all the particles in a body.
- The internal energy of a system increases if the system is heated or work is done on the system by an external force.
- The internal energy of a system decreases if the system is cooled or the system itself does work against an external force.

3.2 SPECIFIC HEAT CAPACITY

Consider heating 1 kg of copper and 1 kg of water so that the temperature of each increases by 1 °C. Since the temperature increase is the same for each, the increase in the kinetic energy component of the internal energy increases by the same amount. However, to achieve the 1 °C rise in temperature, 1 kg of water requires over 10 times the amount of thermal energy input than is required by 1 kg of copper. This is because, when water is heated, most of the energy supplied increases the random potential energy of the water molecules, and this does not affect the temperature.

The property of a particular material to exhibit a change in temperature as a result of a thermal energy transfer is quantified by its **specific heat capacity**, which is given the symbol c. Values of specific heat capacity for some substances are shown in Table 1.

Specific heat capacity is defined as the energy needed to raise the temperature of 1 kg of the material by 1 °C without any change of state, and is measured in the unit $\text{J kg}^{-1}\,°\text{C}^{-1}$ (or the equivalent $\text{J kg}^{-1}\,\text{K}^{-1}$).

Material	Specific heat capacity, c / $\text{J kg}^{-1}\,°\text{C}^{-1}$
Air	993
Water	4 190
Copper	385
Concrete	3 350
Gold	135
Hydrogen	14 300

Table 1 *Some values of specific heat capacity*

The greater the specific heat capacity, the smaller the temperature change for a particular amount of thermal energy transferred. Suppose energy Q is transferred to a mass m of a particular material, resulting in a temperature rise of $\Delta\theta$. Then the specific heat capacity c can be determined from

$$c = \frac{Q}{m\Delta\theta}$$

which rearranges to give

$$Q = mc\,\Delta\theta$$

The specific heat capacity of water is $4190\,\text{J kg}^{-1}\,°\text{C}^{-1}$. This is a particularly large value compared with most other materials (see Table 1). It means that a relatively large amount of energy has to be transferred to (or from) water to significantly raise (or lower) its temperature. Consequently, when heated or cooled, the temperature of a large mass of water changes very slowly. At the beach on a hot day, dry sand, with its specific heat capacity of about $800\,\text{J kg}^{-1}\,°\text{C}^{-1}$, is hot under-foot, but if you paddle in the sea the water feels much cooler.

Determination of specific heat capacity by an electrical method

The specific heat capacity of a solid can be found by an electrical method. In this method, an immersion heater is placed in one cavity in a block of the material and a thermometer is placed in another cavity. The block is heated for a measured time and the temperature rise is recorded. The electrical power P of the immersion heater is given by $P = VI$, where V represents the voltmeter reading and I the current reading (*see section 13.3 in Chapter 13 of Year 1 Student Book*).

Worked example 1

The specific heat capacity of brass is to be found by an electrical method. An immersion heater is used to heat a 1.00 kg block of brass for 600 s. The voltmeter and ammeter readings are 12.5 V and 2.60 A, respectively. The thermometer records a temperature rise of 51.0 °C. Use the data to determine a value for the specific heat capacity of brass, stating any assumptions you make.

Assuming that all the energy supplied to the immersion heater is transferred to the brass block, the energy supplied to the block is $Q = VIt$, where t is the time for which the power supply is switched on.

Assuming that the brass block does not transfer any of its internal energy to its surroundings, the specific heat capacity of brass is

$$c = \frac{Q}{m\Delta\theta} = \frac{VIt}{m\Delta\theta} = \frac{12.5 \times 2.60 \times 600}{1.00 \times 51.0}$$
$$= 382\,\mathrm{J\,kg^{-1}\,{}^{\circ}C^{-1}}$$

The accepted value for the specific heat capacity of brass is $377\,\mathrm{J\,kg^{-1}\,{}^{\circ}C^{-1}}$, which is slightly lower than the value calculated in the worked example, so it is likely that some of the energy supplied to the block was transferred to the surroundings.

ASSIGNMENT 1: MEASURING THE SPECIFIC HEAT CAPACITY OF AN ALUMINIUM BLOCK

(PS 1.2, PS 2.1, PS 2.3, PS 3.2, PS 3.3, MS 1.1, MS 1.5, MS 2.3)

The aim of this assignment is to determine a value for the specific heat capacity of aluminium, along with its uncertainty. Energy is transferred to a metal block using an electrical immersion heater (Figure A1). When the immersion heater is switched on, the vibrational kinetic energy of the atoms of the immersion heater increases, resulting in an increase in temperature. The immersion heater is now at a higher temperature than the aluminium block, so energy is transferred from the immersion heater to the block by thermal conduction, making the block's atoms vibrate more, resulting in a rise in temperature of the block.

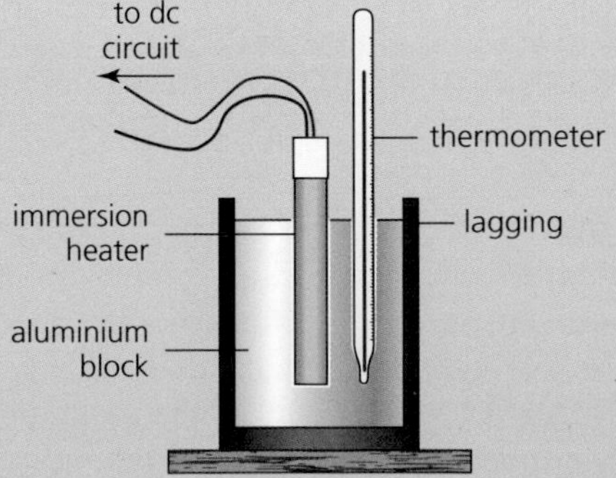

Figure A1 *Heating an aluminium block using an immersion heater*

The formula $c = \frac{Q}{m\Delta\theta}$ shows that to determine the specific heat capacity of aluminium requires the measurement of mass m of the block, its temperature rise $\Delta\theta$ and the energy Q supplied to the block by the immersion heater.

There are two methods for measuring Q. The voltage across and the current supplied to the immersion heater can be measured with a voltmeter and ammeter, respectively, and the heat energy supplied determined from VIt, where t is the heating time. Alternatively, a joulemeter can be used to give a direct reading of the energy consumption of the heater. The ammeter and voltmeter may be analogue or digital, whilst the joulemeter is most likely to be digital. Whatever the type, the resolution of the instrument gives a reasonable estimate of the absolute uncertainty in the corresponding measurement.

Whether a digital stopwatch (typical resolution ±0.01 s) or analogue stopwatch (typical resolution ±0.1 s) is used to measure the heating time t, the absolute uncertainty in t is determined by the reaction time of the experimenter, which typically could be ±0.2 s.

If the temperature is measured with a thermometer with graduations of 1 °C, then ±0.5 °C is a reasonable assessment of the absolute uncertainty. However, since the experiment requires measurement of temperature *rise* $\Delta\theta$, two temperature measurements must be subtracted, giving the absolute uncertainty in $\Delta\theta$ as ±1 °C.

The mass of the aluminium block could be measured with a balance, the resolution of which would give an appropriate value for the absolute uncertainty in the mass measurement.

If the ammeter–voltmeter method is used to determine the specific heat capacity c of aluminium, then the percentage uncertainty in the value obtained for $c = \frac{VIt}{m\Delta\theta}$ is equal to the sum of the percentage uncertainties of the mass, temperature rise, current, voltage and time measurements.

For this electrical method to generate an accurate value for the specific heat capacity of aluminium, the block must be very effectively insulated to minimise energy transfer to the surroundings. Only then can it be assumed that all the energy supplied by the immersion heater is transferred to the internal energy of the block.

Questions

A1 A student undertakes a similar experiment to determine the specific heat capacity of copper, and obtains the measurements and absolute uncertainty values shown in Table A1.

Determine

a. the percentage uncertainty in each quantity

b. the value for the specific heat capacity of copper

c. the percentage uncertainty in the specific heat capacity of copper

d. the absolute uncertainty in the specific heat capacity of copper.

Quantity	Value of quantity with uncertainty
Current	2.5 ± 0.1 A
Voltage	11.9 ± 0.1 V
Time	300.0 ± 0.2 s
Temperature rise	22 ± 1 °C
Mass	1.000 ± 0.002 g

Table A1

A2 Express the experiment's final value of the specific heat capacity of copper with a suitable level of precision, and comment on how near the true value it is (as listed earlier in Table 1).

A3 Suggest at least two ways in which the accuracy of the experiment could be Improved.

As seen in Worked example 1 and in Assignment 1, transfer of heat to bodies other than the one under investigation is always a complication in problems involving energy transfer. So when considering the thermal energy transferred to a liquid and its resultant increase in temperature, it is necessary to take into account the thermal energy transferred to and the specific heat capacity of its container, as well as of the liquid itself.

Worked example 2

Determine the energy needed to boil 1.0 kg of water in an aluminium pan of mass 0.35 kg from 16 °C to 100 °C, assuming no energy is transferred to the surroundings. [Specific heat capacity of water is $4190\,\text{J kg}^{-1}\,°\text{C}^{-1}$; specific heat capacity of aluminium is $910\,\text{J kg}^{-1}\,°\text{C}^{-1}$]

The temperature rise is 84 °C. The energy needed is the sum of the energy required to raise the temperature of the water and the energy required to raise the temperature of the aluminium pan. Applying $Q = mc\Delta\theta$ to both the water and the pan gives the energy needed as

$$Q_{total} = Q_{water} + Q_{pan} = (1 \times 4190 \times 84) + (0.35 \times 910 \times 84) = 380\,000\,\text{J}$$

QUESTIONS

1. Describe the changes to the internal energy of the brass block in Worked example 1 while it is being heated.
2. An induction hob (Figure 5) heats a pan by electromagnetic induction, instead of by thermal conduction from a flame or an electrical heating element. An electromagnet underneath the hob's ceramic surface induces circles of electric current in the body of the pan. Because of the electrical resistance of

the metal of the pan, it is the pan itself that generates the heat required to cook the food. For electromagnetic induction to take place, the pan must be made of cast iron or stainless steel.

Figure 5 *An induction hob achieves very rapid temperature rises*

a. A 2.0 kg cast iron pan on an induction hob heats 1 pint (5.7×10^{-4} m^3) of milk from 18 °C to 100 °C in 120 s. Determine the power being supplied from the induction hob's electromagnet to the pan and the milk as their temperature is raised to 100 °C. [Density of milk is 1033 kg m^{-3}; specific heat capacity of milk is 3930 J kg^{-1} °C^{-1}; specific heat capacity of cast iron is 461 J kg^{-1} °C^{-1}]

b. Suggest why an electric induction hob is much more energy efficient than either a gas hob or a conventional electric hob.

3. A 2.2 kW stainless steel electric kettle of mass 0.50 kg is used to heat 1.0 kg of water. Assuming that no energy is transferred from the kettle to the surroundings, determine the time taken for water at room temperature (18 °C) to reach boiling point. [For stainless steel $c = 502$ J kg^{-1} °C^{-1}; and for water $c = 4190$ J kg^{-1} °C^{-1}]

Determination of specific heat capacity by the method of mixtures

The specific heat capacity of a solid object can be determined by an alternative method – by 'mixing' it with a liquid, usually water. This 'method of mixtures' involves heating the solid object in boiling water to get its temperature to 100 °C. The object is then quickly transferred to an insulated beaker containing water at room temperature, and the temperature of the 'mixture' is monitored with a thermometer.

This method can be used to determine the specific heat capacity of brass, c_{brass} (Figure 6). The data required are the mass m_{brass} of the piece of brass, the volume V of the water in the beaker, the initial temperature θ_1 of the water and the final maximum temperature θ_2 of the mixture.

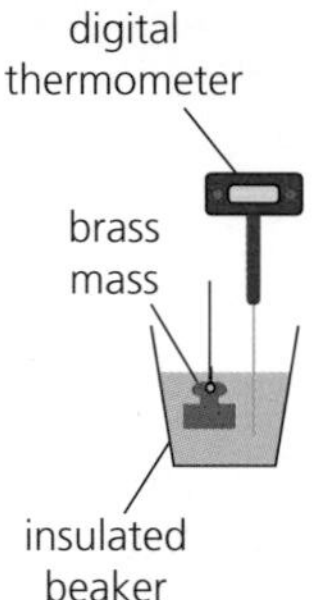

Figure 6 *Method of mixtures for finding the specific heat capacity of brass*

The energy transferred from the brass to the cool water is

$$Q = m_{brass}c_{brass} \times (100 - \theta_2)$$

Assuming that all the internal energy lost by the brass is transferred to the water, and that all the energy gained by the water is used to raise its temperature, the energy gained by the water is

$$Q = m_{water}c_{water} \times (\theta_2 - \theta_1)$$

So equating these two expressions for Q, we obtain

$$m_{brass}c_{brass} \times (100 - \theta_2) = m_{water}c_{water} \times (\theta_2 - \theta_1)$$

The mass of water is $m_{water} = \rho V$, where ρ is the density of water (1000 kg m^{-3}); and the specific heat capacity of water is $c_{water} = 4190$ J kg^{-1} °C^{-1}. Therefore we have

$$m_{brass}c_{brass} \times (100 - \theta_2) = 1000 \times V \times 4190 \times (\theta_2 - \theta_1)$$

which can be rearranged to determine the specific heat capacity of brass:

$$c_{brass} = \frac{1000 \times V \times 4190 \times (\theta_2 - \theta_1)}{m_{brass} \times (100 - \theta_2)}$$

QUESTIONS

4. a. What would be the main cause of inaccuracy in the value of c_{brass} in the above method?

   b. Would this be a random error or a systematic error?

   c. How could this be minimised?

5. A 50 g stainless steel mass is heated to 100 °C in a beaker of boiling water heated by a Bunsen burner. The mass is then transferred to a polystyrene beaker containing 100 cm^3 of water at room temperature of 18.2 °C. Assuming that there are no unwanted energy transfers, determine the final maximum temperature of the mixture of water and steel, given that the specific heat capacity of stainless steel is 502 $J\,kg^{-1}\,°C^{-1}$, the density of water is 1000 $kg\,m^{-3}$ and the specific heat capacity of water is 4190 $J\,kg^{-1}\,°C^{-1}$.

Increasing internal energy by doing work

If an object of mass m is dropped from a height h and hits the ground, without bouncing, all of its kinetic energy on reaching the ground is transferred as internal energy. The temperature of the object rises, by an amount dependent on its mass and its specific heat capacity. We can assume that the temperature increase of the Earth is negligible, because of its huge mass.

The energy transferred is equal to the final kinetic energy of the object, which is equal to the gravitational potential energy gained, or work done on the object, in raising it to the height h:

$$\text{work done} = mgh$$

Joule's experiments on the equivalence of heating and working with regard to raising the temperature of a substance (see section 3.1) show that the energy transferred to the substance by doing work, resulting in a temperature rise $\Delta\theta$, is equal to $mc\,\Delta\theta$. Therefore $mgh = mc\,\Delta\theta$ and

$$\Delta\theta = \frac{gh}{c}$$

Worked example 3

A 2.0 kg block of copper (c = 385 $J\,kg^{-1}\,°C^{-1}$) is dropped from a height of 10 m and hits the ground. Calculate by how much its temperature increases, stating any assumptions you make.

Assuming it does not bounce and all its initial gravitational potential energy is converted to internal energy of the block's atoms, the resulting temperature rise of the block is

$$\Delta\theta = \frac{gh}{c} = \frac{9.81 \times 10}{385} = 0.25\,°C$$

QUESTIONS

6. What would be the expected temperature difference between the water at the top of Niagara Falls and the water at the bottom, given that the height of the Falls is 51 m? [Specific heat capacity of water is 4190 $J\,kg^{-1}\,°C^{-1}$]

Determination of specific heat capacity by calculating work done

An estimate for the specific heat capacity of lead can be determined using an inversion tube method. Lead shot of total mass m is contained in a cardboard tube (Figure 7), which has a rubber bung at one end and a second rubber bung with a hole through its centre at the other end. The initial temperature of the lead is measured by inserting a thermometer through the hole in the bung. The tube, of length h, is held vertically and then repeatedly turned end to end, so that the lead is continually being lifted and allowed to fall to the other end of the tube. The internal energy of the lead shot increases because work is done on the falling lead shot to bring it to rest as it hits the bung at the bottom of the tube.

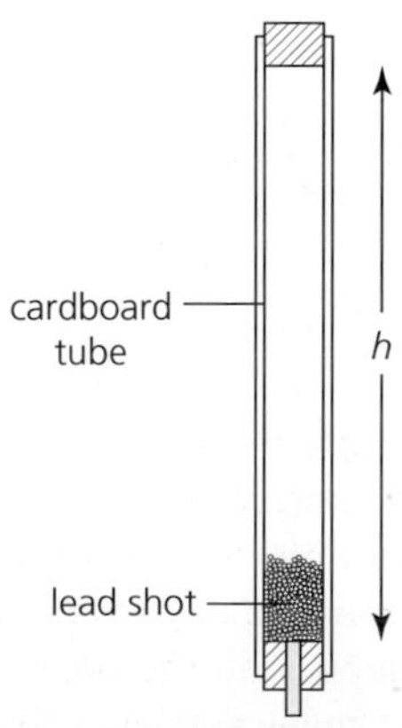

Figure 7 *Cardboard inversion tube containing lead shot*

On inverting the tube, the lead shot gains an additional amount of gravitational potential energy equal to *mgh*. As the lead shot falls, its gravitational potential energy is converted to kinetic energy, and the work done by the bung to bring the lead shot to rest is equal to that initial gravitational potential energy, *mgh*. Therefore, if the total number of inversions is *N*, the total work done on the lead shot is *Nmgh*. The final temperature of the lead is measured by replacing the glass rod with a thermometer, enabling the temperature rise $\Delta\theta$ to be determined.

Assuming that all the work done, *Nmgh*, is converted to the internal energy of the lead shot, then

$$c = \frac{\text{energy transferred}}{m\Delta\theta} = \frac{Nmgh}{m\Delta\theta} = \frac{Ngh}{\Delta\theta}$$

The value obtained for the specific heat capacity of lead is very much an estimate. Some of the lead shot does not fall the full distance *h*, partly because of the depth of the lead shot itself and also because some of the shot slides down the side of the tube before it reaches the vertical position. So the actual work done on the shot will in practice be less than *Nmgh*. Also, some of the internal energy will be lost to the bungs and to the air in the tube.

QUESTIONS

7. An inversion tube of internal length 1.0 m contains 250 g of lead shot. The initial temperature of the lead shot is 17.1 °C. The tube is held vertically and turned end to end, so that the lead is continually being lifted and allowed to fall, hitting the bottom of the tube. After 50 inversions, a thermometer is inserted and the final temperature of the lead recorded as 19.7 °C.

    a. Obtain a value for the specific heat capacity of lead.

    b. Suggest whether the value obtained for the specific heat capacity is likely to be an underestimate or overestimate.

Flow calculations

Because of its high specific heat capacity, water is good for storing internal energy and also for transporting it, for example, around the home using central heating pipes. Water is used as the fluid in the cooling systems of many engines and other machinery (Figure 8) as its high specific heat capacity means that a lot of excess thermal energy can be removed from the engine without the water temperature becoming too high. So energy transfer calculations often need to be done involving a flow of water.

Figure 8 *This water-cooled Aquasar supercomputer consumes up to 40% less energy than a comparable air-cooled machine, and 9 kW of thermal power are fed into the building's heating system.*

Worked example 4

An electric shower heats the water flowing through it from 15 °C to 43 °C when the flow rate is 5.0 litre per minute. Determine the electrical power supplied to the shower, assuming that all the energy supplied is converted to internal energy of the water. [Take the specific heat capacity of water as $4190\,\text{J}\,\text{kg}^{-1}\,°\text{C}^{-1}$ and the density of water as $1000\,\text{kg}\,\text{m}^{-3}$]

Mass of water flowing in 1 minute = water density × volume flow per minute, so

$$m_{\text{water}} = 1000 \times 5.0 \times 10^{-3} = 5.0\,\text{kg}\,\text{min}^{-1}$$

Energy supplied to water in 1 minute is

$$Q = m_{water} c\,\Delta\theta = 5.0 \times 4190 \times (43 - 15)$$
$$= 5.866 \times 10^5\,\mathrm{J\,min^{-1}}$$

Power supplied to shower = energy supplied per second, so

$$\text{power to shower} = \frac{5.866 \times 10^5}{60} = 9777\,\mathrm{W} = 9.8\,\mathrm{kW}$$

QUESTIONS

8. A car engine produces a lot of heat when it is running, and must be cooled continuously to maintain its correct operating temperature. This is done by circulating a liquid coolant, usually water mixed with antifreeze, around the engine. The coolant gains internal energy from the engine, then enters the radiator, where some of its internal energy is transferred to the surrounding air. The coolant leaves the radiator at a lower temperature and is pumped back around the engine. The coolant circulates around the engine at a rate of 80 litre per minute. How much of its internal energy is transferred to the surroundings per second from the radiator if it enters at a temperature of 94 °C and leaves at 81 °C? [Specific heat capacity of coolant is $3900\,\mathrm{J\,kg^{-1}\,{}^\circ C^{-1}}$ and density of water is $1000\,\mathrm{kg\,m^{-3}}$]

9. Determine the power supplied to a shower that heats water flowing through it at 4.0 litre per minute from 14 °C to 39 °C, assuming that all the energy supplied is converted to internal energy of the water. [Take the specific heat capacity of water as $4190\,\mathrm{J\,kg^{-1}\,{}^\circ C^{-1}}$ and the density of water as $1000\,\mathrm{kg\,m^{-3}}$]

10. Determine the output power of a radiator if the water temperature at the input pipe is 80 °C and at the output pipe is 60 °C and the rate at which water flows through the pipe is $0.013\,\mathrm{kg\,s^{-1}}$. [Specific heat capacity of water is $4190\,\mathrm{J\,kg^{-1}\,{}^\circ C^{-1}}$]

KEY IDEAS

- Specific heat capacity c is the energy needed to raise the temperature of 1 kg of a material by 1 °C without any change of state. It is measured in units of $\mathrm{J\,kg^{-1}\,{}^\circ C^{-1}}$.
- Energy transferred = $mc\,\Delta\theta$, where $\Delta\theta$ is the change in temperature.
- In cases of fluid flow, rate of energy transfer = rate of mass flow × $c\,\Delta\theta$.

3.3 SPECIFIC LATENT HEAT

Heating an object does not always lead to an increase in temperature. For example, if water, contained in a beaker, has been heated to its boiling point by a Bunsen burner, any additional heating will not cause a further increase in temperature. When a substance changes from a solid to a liquid, or from a liquid to a gas, energy is needed to do work against the attractive forces holding the solid or liquid together. Therefore, when energy is being supplied to cause a change of state, the potential energy component of the substance's internal energy increases, but, since the temperature is constant, the average kinetic energy of the particles does not increase. During a change of state there is usually an increase in volume as a solid melts or a liquid boils. Energy is therefore also needed to do work against atmospheric pressure as the substance expands. For example, when water turns to steam, about 7% of the energy supplied is needed to do work against atmospheric pressure.

- The **specific latent heat of vaporisation** l of a substance is the energy required to change 1 kg of a liquid into 1 kg of gas with no change in temperature, and is measured in $\mathrm{J\,kg^{-1}}$.
- The **specific latent heat of fusion** l of a substance is the energy required to change 1 kg of a solid into 1 kg of liquid with no change in temperature, and is measured in $\mathrm{J\,kg^{-1}}$.

Values of specific latent heat of vaporisation and specific latent heat of fusion of some substances are shown in Table 2. Note that the values are in $\mathrm{kJ\,kg^{-1}}$.

Material	Specific latent heat of vaporisation / $kJ\,kg^{-1}$	Specific latent heat of fusion / $kJ\,kg^{-1}$
Water	2260	334
Oxygen	243	14
Helium	25	5
Mercury	290	11
Iron	6339	276
Lead	854	25

Table 2 *Some values of specific latent heat*

Suppose thermal energy Q has to be transferred to mass m of a solid substance, already at its melting point, to cause a complete change from solid to liquid. Then the specific latent heat of fusion l is given by

$$l = \frac{Q}{m}$$

which rearranges to give the energy transferred

$$Q = ml$$

Worked example 1

An electric kettle is used to raise the temperature of 1.1 kg of water from 18.5 °C to 100 °C. The automatic off switch on the kettle is faulty and the kettle continues to heat the water until 10% of the water has been converted to steam before switching off. The kettle switch was in the 'on' position for 275 s. Determine

a. the total energy supplied to raise the water temperature to 100 °C and vaporise 10% of the water

b. the total energy transferred to the body of the kettle and to the surroundings during the time that the kettle was switched on.

[The power rating of the kettle is 2.5 kW. Use other data from Tables 1 and 2]

a. Energy supplied to raise the water temperature to 100 °C is

$$Q_1 = mc\Delta\theta = 1.1 \times 4190 \times (100 - 18.5) = 3.756 \times 10^5\,J$$

Energy supplied to cause a change of state from water to steam is

$$Q_2 = ml = 0.1 \times 1.1 \times 2.26 \times 10^6 = 2.486 \times 10^5\,J$$

Total energy supplied $= Q_1 + Q_2 = 6.242 \times 10^5 = 6.2 \times 10^5\,J$

b. Energy supplied by the heating element of the kettle is

$$\text{power} \times \text{time} = 2500 \times 275 = 6.875 \times 10^5\,J$$

Total energy supplied to the body of the kettle and transferred to the surroundings is

$$\text{energy to surroundings} = 6.875 \times 10^5 - 6.242 \times 10^5 = 0.633 \times 10^5 = 6.3 \times 10^4\,J$$

QUESTIONS

11. Determine the energy needed to completely melt 1.0 kg of copper that is initially at room temperature of 20 °C, given the relevant data in Table 3.

Quantity	Value of the quantity
Melting point of copper	1083 °C
Specific heat capacity of copper	385 $J\,kg^{-1}\,°C^{-1}$
Specific latent heat of fusion of copper	$2.07 \times 10^5\,J\,kg^{-1}$

Table 3

12. An ice cube of mass 20 g and temperature −10 °C is added to 200 g of water that is initially at 20 °C. As the ice melts, it has a cooling effect on the water.

a. If all the energy needed to melt the ice comes from the water, calculate the consequent temperature drop of the water.

b. Has the water now reached its final temperature? Explain and justify your answer.

[Specific heat capacity of ice, 2000 $J\,kg^{-1}\,°C^{-1}$; specific heat capacity of water, 4190 $J\,kg^{-1}\,°C^{-1}$; specific latent heat of water, $3.34 \times 10^5\,J\,kg^{-1}$]

If energy is supplied at a constant rate to a solid substance (Figure 9), its temperature will rise until it reaches its melting point. The substance will then melt at constant temperature. When all the solid has turned into liquid, the temperature will rise again until the liquid reaches its boiling point. The substance will then boil at constant temperature until all the liquid has turned into gas, when the temperature will rise again.

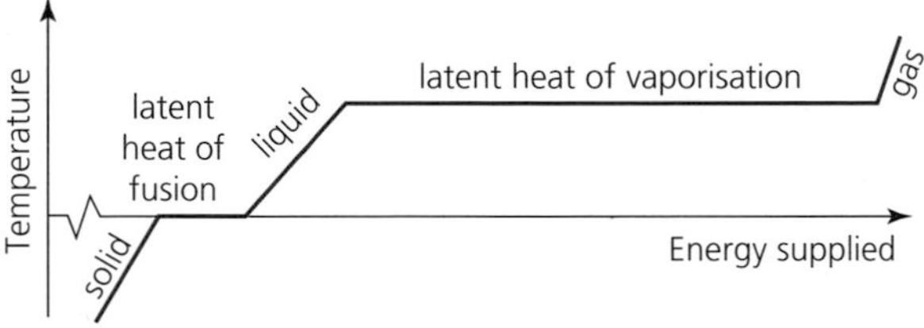

Figure 9 *The temperature change of a substance that is heated at a constant rate*

Producing a heating curve, followed by a cooling curve, showing temperature versus time for a substance undergoing a change of state, is most efficiently done using a temperature sensor and data logging equipment, because the process can take an extended period of time (Figure 10). The data logger can be set to instruct the temperature sensor to take a measurement every 2 s so the sampling rate is 0.5 Hz.

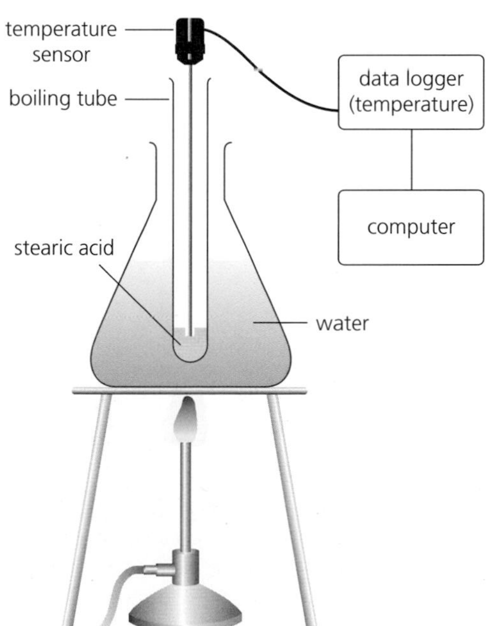

Figure 10 *Using data logging equipment to monitor the change of state of stearic acid*

Stearic acid, which is solid at room temperature, is heated in a water bath that is itself heated by a Bunsen burner. Once the heating is started, the Bunsen is not adjusted, so that the heating takes place at an approximately constant rate. Once the stearic acid has melted and its temperature once again starts to rise, the boiling tube is removed from the water bath and the stearic acid allowed to cool.

The data logger can be connected to a computer to display the temperature versus time graph (Figure 11) for the heating and subsequent cooling of the stearic acid. Although the heating takes place at a constant rate, the cooling does not. The rate of cooling depends on the difference between the object's temperature and the surrounding room temperature.

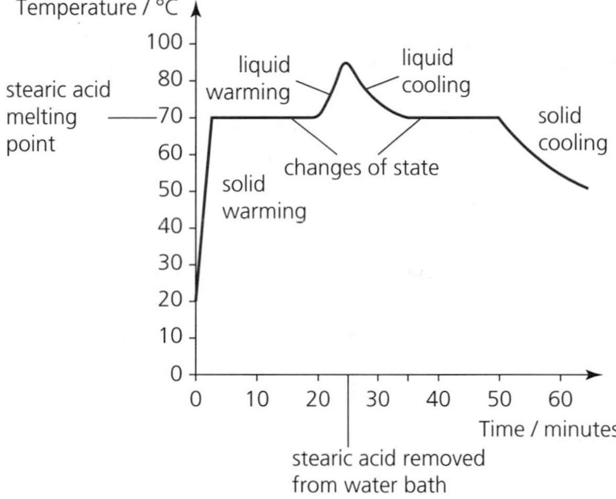

Figure 11 *Experimental heating curve and cooling curve for stearic acid*

QUESTIONS

13. The following questions refer to Figure 11.

a. What does the gradient of the curve at any point represent?

b. How does the motion of the stearic acid molecules change during the first change of state?

c. How do the components of internal energy of the stearic acid change during the second change of state?

The energy needed to change the state of a substance is often used to transfer thermal energy. The cooling towers of a power station transfer energy from the power station by evaporating large amounts of water. Thermal energy is **dissipated** to the surroundings. Dissipation is a thermal energy transfer that cannot easily be reversed – the energy has been randomly spread out.

Refrigerators work on a similar principle on a smaller scale, and we do the same thing ourselves – when we sweat, we are transferring excess thermal energy by evaporating liquid.

Worked example 2

The amount of water that we lose through the evaporation of sweat depends on the temperature of our surroundings, as well as on the air humidity and wind speed. On average, a typical person loses about 0.5 litre of sweat per day. Calculate the average power of this energy transfer.

The energy needed to evaporate 0.5 litre of sweat is $Q = ml$, where m is the mass of 0.5 litre of water, which is 0.5 kg, and l is the specific latent heat of vaporisation of water, $2.260 \times 10^6\ \text{J kg}^{-1}$. So the energy transferred is

$$Q = 0.5 \times 2.260 \times 10^6 = 1.13 \times 10^6\ \text{J}$$

If this is transferred over a period of 24 hours, or $24 \times 60 \times 60 = 86\,400$ s, then

$$\text{average power of this energy transfer} = \frac{1.13 \times 10^6}{86400} = 13\,\text{W}$$

QUESTIONS

14. A runner loses energy at the rate of 500 W by the evaporation of sweat. What mass of sweat is evaporated each minute? [Specific latent heat of vaporisation of water is $2.3 \times 10^6\ \text{J kg}^{-1}$]

15. In a geothermal power station, water at 15 °C is pumped into hot underground rocks and vaporises. Steam at 100 °C is returned to the power station at the rate of $200\,\text{kg s}^{-1}$. Determine the rate at which energy is transferred from the rocks to the power station. [Specific heat capacity of water is $4190\,\text{J kg}^{-1}\,{}^\circ\text{C}^{-1}$; specific latent heat of vaporisation of water is $2.3 \times 10^6\ \text{J kg}^{-1}$]

KEY IDEAS

- The specific latent heat of vaporisation l of a substance is the energy required to change 1 kg of a liquid into 1 kg of gas, with no change in temperature, and is measured in J kg^{-1}.
- The specific latent heat of fusion l of a substance is the energy required to change 1 kg of a solid into 1 kg of liquid, with no change in temperature.
- Energy transferred $Q = ml$.
- On heating or cooling a substance, during a change of state its temperature remains constant.
- In a change of state, the potential energy component of the substance's internal energy changes but the average kinetic energy of the particles does not.

3.4 THE GAS LAWS

Robert Boyle, born in Ireland in 1627, was one of the first prominent scientists to perform controlled experiments and to publish his work detailing his procedures and observations. Boyle's experiments on air (Figure 12) resulted in what is now known as **Boyle's law**, which established the relationship between the pressure and volume of a gas at constant temperature.

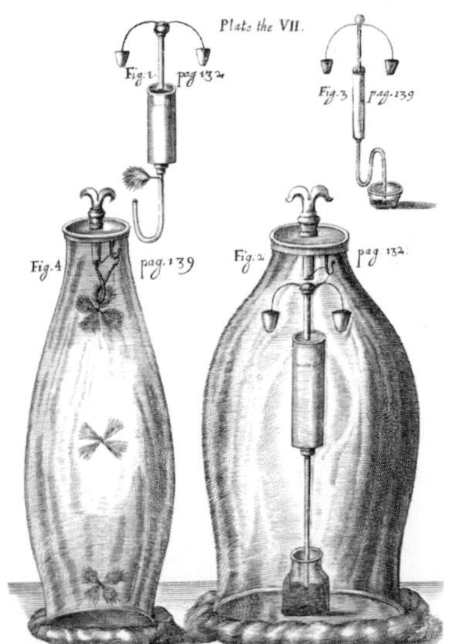

Figure 12 *Working in Oxford with Robert Hooke, Boyle developed an improved version of the vacuum pump and was able to show by experiment that, in the absence of air resistance, bodies of different mass fall at the same rate and that sound requires air for its transmission.*

The pressure of a gas is defined as the force it exerts per unit area and is measured in N m^{-2} or **pascal** (Pa). For example, the pressure exerted by the atmosphere at the surface of the Earth is typically about 100 kPa.

REQUIRED PRACTICAL: APPARATUS AND TECHNIQUES

Part 1: Investigation into the relationship between pressure and volume for a gas at constant temperature (Boyle's law)

The aim of the required practical is to obtain volume and pressure data for air to establish the relationship between these two quantities whilst the air temperature remains constant. The practical gives you the opportunity to:

- use appropriate analogue apparatus to record a range of measurements (to include pressure and volume) and to interpolate between scale markings.

Apparatus

There are different designs of apparatus that can be used to investigate the relationship between pressure and volume. Figure P1 shows a glass tube, closed at the top, containing oil. The air trapped in the tube can be compressed by using the foot pump to force oil to move up the tube. The tube is made of extra-strong glass and has a plastic safety screen secured in front of it.

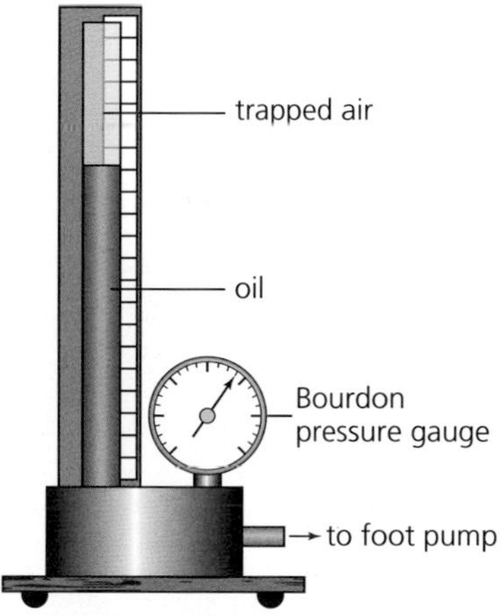

Figure P1 *Apparatus for investigating the pressure and volume of air*

A scale graduated in cm^3 is positioned alongside the glass tube to enable the volume of air to be measured. When the meniscus is stationary, the air pressure in the tube is the same as the oil pressure, which can be read from the pressure gauge, which has a typical precision of ±5 kPa.

Techniques

The volume of the air can be reduced by applying pressure with the foot pump, and a series of volume and pressure measurements can be taken. When measuring the volume of air in the tube, the line of sight of your eye should be at the same level as the meniscus of the oil to avoid parallax error (Figure P2).

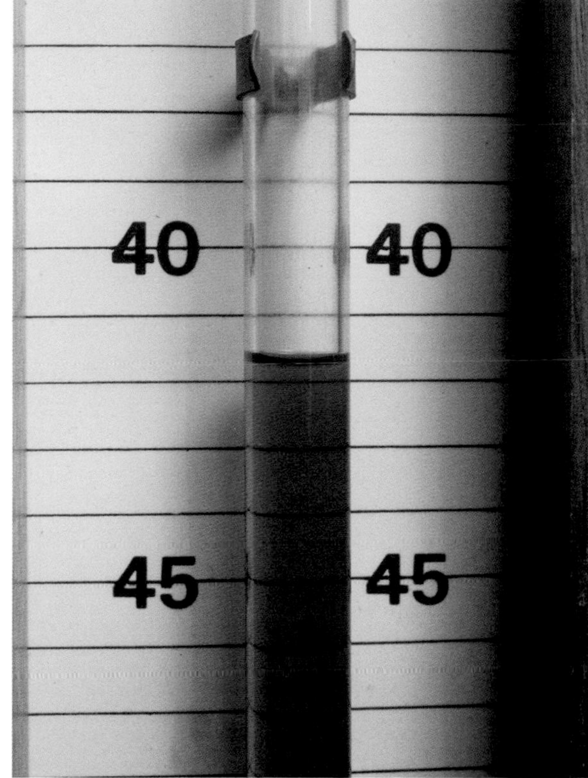

Figure P2 *The volume of the air can reasonably be measured to half a division, which is 0.5 cm^3.*

The pressure gauge has an analogue scale, which also creates the possibility of a parallax error (*see section 1.3 in Chapter 1 of Year 1 Student Book*). Although the smallest division on the pressure gauge is likely to be 5kPa, it is possible to interpolate to one-fifth of a division and to record measurements to 1kPa. To ensure that there is no change in temperature of the air in the tube, it is important to apply pressure to the pump slowly and allow a few minutes between readings. It is unlikely that repeat readings can be taken with this apparatus, because setting the pressure to a specific value is very difficult to achieve.

Analysis

A graph of pressure versus volume is a curve (Figure P3) showing that the volume of a gas decreases as its pressure is increased.

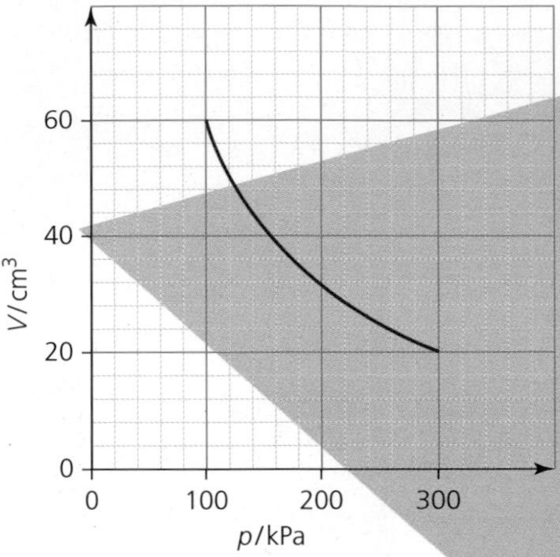

Figure P3

In addition to a plot of volume V versus pressure p, the following two graphs can also be used to show the relationship between the pressure and volume of a gas:

1. A plot of $\frac{1}{V}$ on the y-axis versus p on the x-axis could demonstrate whether V and p are inversely proportional. If they are, then the graph will be a straight line through the origin.
2. A plot of log V on the y-axis versus log p on the x-axis could demonstrate whether V and p are related by a power law such as $V = kp^n$, where k and n are constants. If such a power law holds, then $\log V = \log k + n \log p$.

QUESTIONS

P1 **a.** How should the Bourdon pressure gauge be viewed to minimise parallax error?

b. Some Bourdon pressure gauges have an antii-parallax mirror. Find out and explain how an anti-parallax mirror can be used to improve the accuracy of a measurement.

P2 A student obtained pressure and volume measurements for a gas at a constant temperature of 20 °C (293 K). He then plotted a graph of log(V / m^3) on the y-axis versus log(p / Pa) on the x-axis. This produced a straight line of gradient of −1.00 and an intercept on the y-axis of 0.917.

a. Determine the relationship between V and p.

b. Predict the volume of the gas at a pressure of 300 kPa.

Boyle's law

Boyle's law is quoted as:

> The volume of a fixed mass of gas at constant temperature is inversely proportional to its pressure.

Mathematically, this can be written in a number of different ways:

$$V \propto \frac{1}{p} \quad \text{or} \quad V = \frac{k}{p}$$

where k is a constant. Alternatively, the relationship can be written

$$pV = \text{constant}$$

If the pressure and volume change from initial values p_1 and V_1 to final values p_2 and V_2, at constant temperature, then Boyle's law can be written

$$p_1V_1 = p_2V_2$$

Worked example 1

A volume 20 cm^3 of air at atmospheric pressure, 1×10^5 Pa, is trapped in a bicycle pump when a finger is placed over the end of the pump. If the piston is pushed in until the volume is 5 cm^3, find the new pressure, assuming the temperature is unchanged.

Since the temperature is constant, we can apply Boyle's law, pV = constant. It is helpful to assign the symbols p_1 and V_1 to the initial pressure and volume, and p_2 and V_2 to the final pressure and volume.

So Boyle's law can be written:

$$p_1V_1 = p_2V_2$$

It is not necessary to convert the volume data from cm^3 to m^3, because the conversion factor would appear on both sides of the above equation and would therefore cancel.

Substituting the data:

$$1 \times 10^5 \times 20 = p_2 \times 5$$

which rearranges to give the final pressure, $p_2 = 4 \times 10^5$ Pa.

QUESTIONS

16. An alternative apparatus for testing Boyle's law is shown in Figure 13. This apparatus does not require oil and a foot pump. Instead, there is a screw that can be turned to reduce the volume and increase the pressure of the air in the tube. The volume of the air is initially at 15 cm^3 when the pressure is 1.0×10^5 Pa. Predict what the volume would have to be reduced to in order to increase the pressure to 1.4×10^5 Pa. It can be assumed that the volume reduction takes place slowly and the room temperature is unchanged.

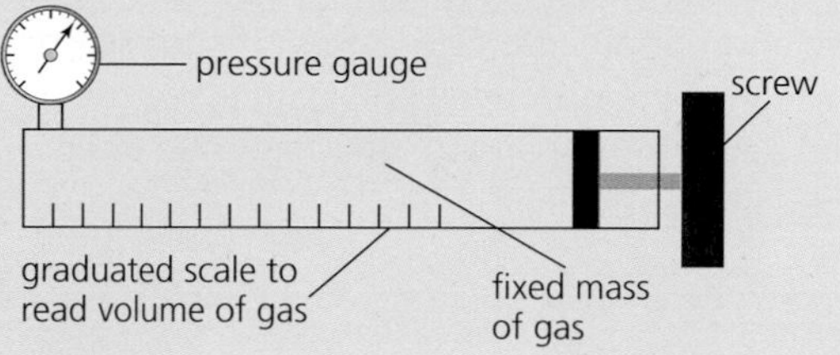

Figure 13 *Alternative Boyle's law apparatus*

The effect of temperature on volume: Charles's law

We have seen that, when a graph of pressure p versus volume V is plotted for a constant mass of gas at constant temperature, a curve is obtained because there is an inverse relation between pressure and volume. If the temperature of the same gas is then raised to a new constant value, and pressure and volume measurements retaken and plotted, the second curve is slightly higher than the first (Figure 14). Since each curve represents a specific temperature, they are called **isotherms**.

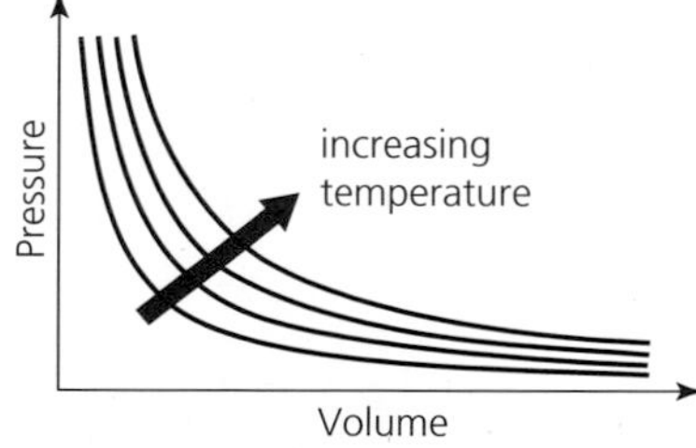

Figure 14 *Some p–V curves for a constant mass of gas*

In the late 18th century, Jacques Charles, a French scientist and balloonist, discovered that different gases, at constant pressure, all expand at the same rate with increasing temperature. For every 1 °C increase in temperature, they expand by about $\frac{1}{273}$ of their volume at 0 °C.

REQUIRED PRACTICAL: APPARATUS AND TECHNIQUES

Part 2: Investigation into the relationship between volume and temperature at constant pressure (Charles's law)

The aim of this required practical is to obtain volume and temperature data for air to establish the relationship between these two quantities whilst the pressure of the air remains constant. The practical gives you the opportunity to:

- use appropriate analogue apparatus to record a range of measurements (to include volume and temperature) and to interpolate between scale markings.

Apparatus

There are different designs of apparatus that can be used to investigate the relationship between the volume and temperature of a gas. Figure P4 shows a capillary tube, which has a graduated millimetre scale. The tube is sealed at one end, and contains a small length of concentrated sulfuric acid about half-way along the tube. The air trapped between the closed end of the tube and the bottom of the sulfuric acid index is the gas under investigation. Since the top of the tube is open, the pressure of the trapped air is equal to atmospheric pressure plus the pressure due to the weight of the sulfuric acid. Since it is reasonable to assume that these do not change during the experiment, it can be assumed that the pressure of the trapped air remains constant.

The capillary tube is attached to a thermometer using rubber bands. Together, they are placed in a beaker of iced water, with the open end of the capillary tube uppermost. The depth of water in the beaker must be sufficient to ensure that the length of trapped air is always below the water level.

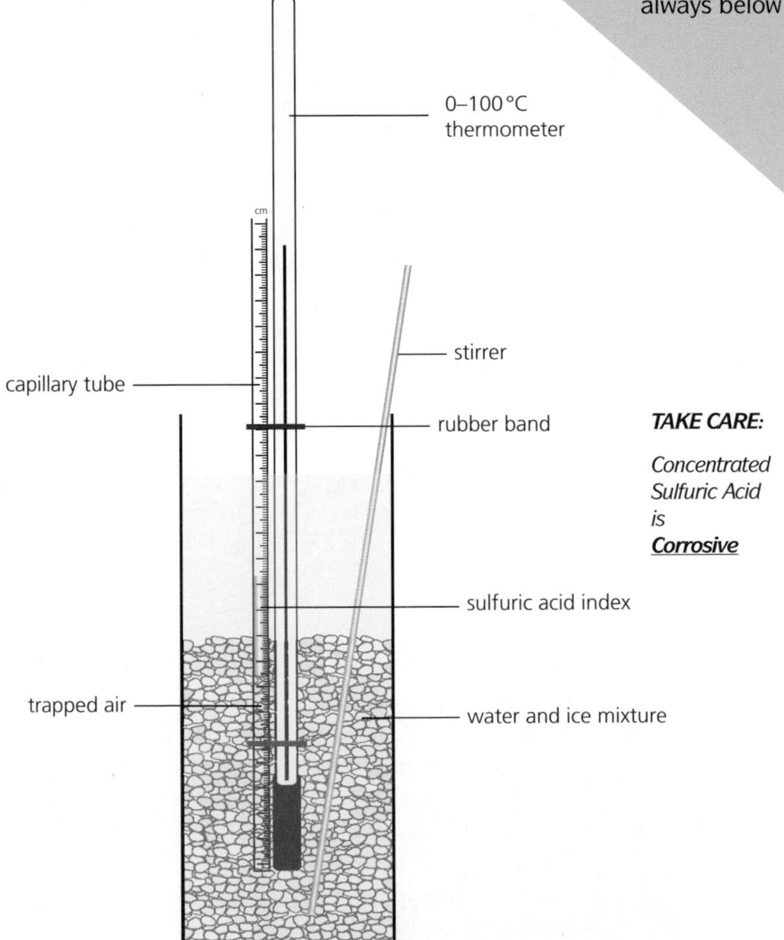

TAKE CARE:

*Concentrated Sulfuric Acid is **Corrosive***

Figure P4 *Apparatus for investigating the temperature and volume of air*

Techniques

The length l of trapped air is measured at different temperatures as the temperature of the ice/water mixture increases with the addition of small amounts of warm water. Since there can be variations in temperature within a liquid, it is important to stir the water before making a temperature measurement. If the thermometer has a precision of 1 °C, it may be possible to interpolate to 0.5 °C. The capillary tube is graduated in millimetres, and it is likely that measuring to the nearest millimetre is the best that can be achieved. Repeat measurements can be difficult to achieve with this apparatus, because precise control of the temperature is not possible.

Analysis

Since the length of the trapped air is directly proportional to its volume, the analysis of the data can be based on the length rather than the volume. A graph of length l on the y-axis versus Celsius temperature θ on the x-axis can be plotted to determine the relationship between the length of the air column and its Celsius temperature. An estimate of the uncertainties in the length and temperature measurements would enable error bars to be added to the points.

The graph should be plotted so that the negative temperature scale extends to −300 °C. Then it is possible to extrapolate the line to the temperature axis (Figure P5), corresponding to zero length, and to determine a value for absolute zero. That is the temperature at which atoms and molecules effectively stop moving. Using the error bars as a guide, a line of maximum gradient and a line of minimum gradient could be added to the graph. They can be used to generate an uncertainty on the value obtained for absolute zero.

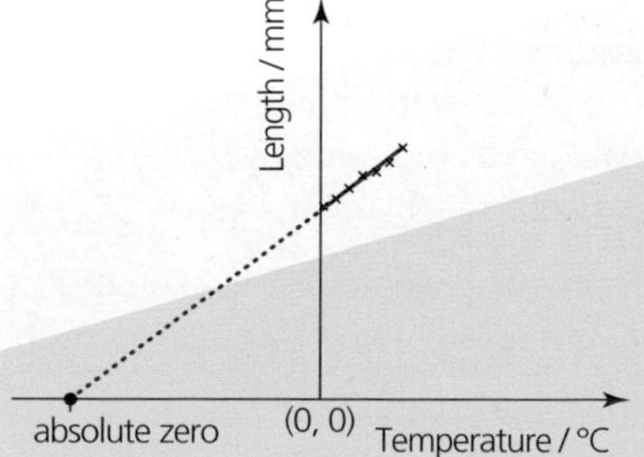

Figure P5 *Extrapolation of the length versus Celsius temperature graph*

QUESTIONS

P3 Suggest how the accuracy of the temperature and the length measurements could be improved.

P4 A student undertakes the experiment and produces the data in Table P1. The uncertainty in each temperature measurements is ±0.5 °C and the uncertainty in each length measurement is ±0.5 mm.

Temperature, θ/ °C	Length of air column, l / mm
5.0	118.0
10.0	121.0
15.0	123.0
20.5	126.0
25.0	127.0
30.0	129.0

Table P1

a. Plot a graph of the student's results, with length l on the y-axis versus Celsius temperature θ on the x-axis, and include error bars. The purpose of the graph is to obtain accurate values for the gradient and the y intercept. The x-axis should therefore start at zero, but the y-axis can start at 110 mm.

b. Determine the gradient of the best-fit line, but also draw maximum and minimum gradient lines using the error bars as a guide to obtain an uncertainty in the value of the gradient. Determine the y intercept, again using the maximum and minimum gradient lines to obtain an uncertainty in the value of the intercept.

c. Write an equation for the line in the form of $y = mx + c$, where $y = l$ and $x = \theta$, using your values for the gradient m and the y intercept c to show the linear relationship between length and Celsius temperature. Rearrange the equation to make temperature θ the subject, so that the equation is of the form

$$\theta = \frac{l - c}{m}$$

d. Substitute $l = 0$ into the equation and determine the corresponding value for θ that gives the temperature of absolute zero.

e. Work out the percentage uncertainties in the gradient m and y intercept c, and add them to determine the percentage uncertainty in the value for absolute zero.

Absolute zero is the lowest possible temperature and is the point at which the atoms in a substance have effectively zero random kinetic energy. By international agreement, absolute zero is defined as 0 K on the Kelvin scale (or absolute temperature scale) and this is equivalent to –273.15 °C on the Celsius scale.

A temperature change of 1 K is identical to a temperature change of 1 °C, so temperature T on the Kelvin scale is related to Celsius temperature θ by the equation

$$T = \theta + 273.15$$

The relationship between the volume of a gas and its temperature is usually expressed with regard to its kelvin temperature rather than its Celsius temperature. A graph of the volume of a gas versus its kelvin temperature (Figure 16) is a straight line through the origin, so it can be deduced that the two quantities are directly proportional.

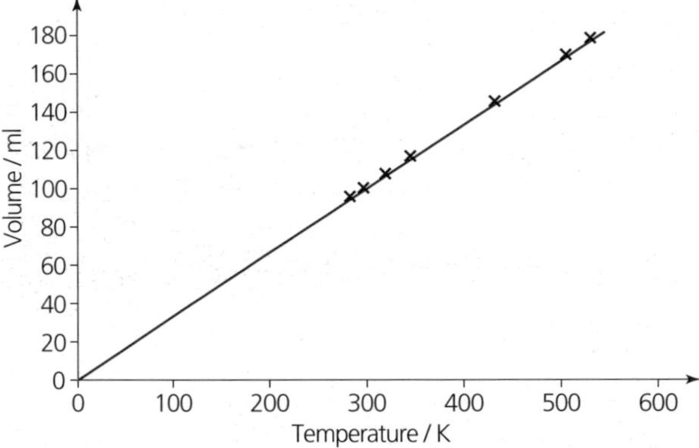

Figure 15 *Gas volume versus kelvin temperature*

Charles's law is expressed as:

The volume of a fixed mass of gas at constant pressure is directly proportional to its kelvin temperature.

Mathematically, Charles's law can therefore be written:

$$V \propto T \qquad \text{or} \qquad V = kT$$

where k is a constant, or

$$\frac{V}{T} = \text{constant}$$

If the volume and temperature change from initial values V_1 and T_1 to final values V_2 and T_2, at constant pressure, then Charles's law can be written

$$\frac{V_1}{T_1} = \frac{V_2}{T_2}$$

Worked example 2

A gas is contained in a cylinder sealed by a movable piston so that the gas pressure does not change. At a temperature of 350 K the volume of the gas is 5.0×10^{-3} m^3. If the gas is now heated to 500 K, what will its volume become?

Since the pressure is constant and no gas can escape so the mass is constant, Charles's law can be applied:

$$\frac{V_1}{T_1} = \frac{V_2}{T_2}$$

Substituting data:

$$\frac{5 \times 10^{-3}}{350} = \frac{V_2}{500}$$

gives $V_2 = 7.1 \times 10^{-3}\,\text{m}^3$.

QUESTIONS

17. A gas is contained in a cylinder sealed by a movable piston so that the gas pressure does not change. At a temperature of 100 °C the volume of the gas is 3.5×10^{-3} m^3. If the gas is now heated to 250 °C, what is its new volume?

The pressure law

The relation between pressure and temperature at constant volume was established in 1802 by Joseph Louis Gay-Lussac, a French chemist and physicist. It can be demonstrated using a copper flask, containing air, connected to a pressure gauge (Figure 16). The flask is heated by a water bath, and it is assumed that a thermometer recording the temperature of the water gives a temperature value that is very close to the temperature of the air in the copper flask.

A graph of pressure versus kelvin temperature (Figure 17) is a straight line through the origin.

The ***pressure law*** states that:

For a fixed mass of gas at constant volume, the pressure is directly proportional to the kelvin temperature.

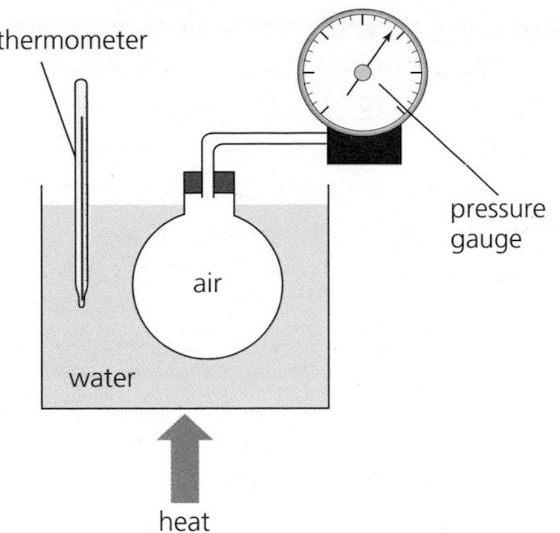

Figure 16 *Apparatus for investigating the temperature and pressure of a constant volume of air*

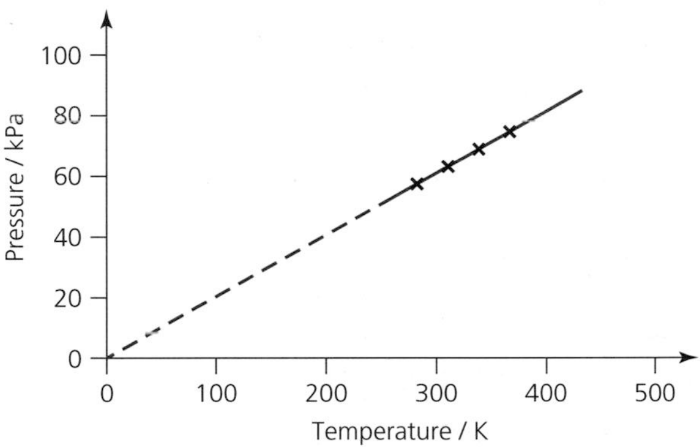

Figure 17 *Pressure law graph*

Mathematically, this can be written:

$$p \propto T \qquad \text{or} \qquad p = kT$$

where k is a constant, or

$$\frac{p}{T} = \text{constant}$$

If the pressure and temperature change from initial values p_1 and T_1 to final values p_2 and T_2, at constant volume, then the pressure law can be written

$$\frac{p_1}{T_1} = \frac{p_2}{T_2}$$

QUESTIONS

18. A gas is at a pressure of 1.1×10^5 Pa and temperature of 300 K in a rigid container. If the gas is heated to a temperature of 400 K, determine the new pressure, assuming no gas escapes.

The gas laws

In 1811, the Italian scientist Amedeo Avogadro published an article in which he put forward an explanation of experimental observations made by Gay-Lussac in 1808 involving reactions between specific volumes of gases. Avogadro hypothesised that 'equal volumes of gases at the same temperature and pressure contain equal numbers of molecules'. Although largely ignored for half a century, the hypothesis, now known as **Avogadro's law**, was a major factor in advances in chemistry during the second half of the 19th century. Avogadro's law can be stated mathematically as $V \propto N$ at constant pressure and temperature, where V represents the volume and N the number of molecules of the gas.

Boyle's law, Charles's law, the pressure law and Avogadro's law are referred to as the 'gas laws' and, being based on experiments, are described as being **empirical**.

KEY IDEAS

- Boyle's law states that the volume of a fixed mass of gas at constant temperature is inversely proportional to its pressure:

$$V \propto \frac{1}{p} \qquad \text{or} \qquad pV = \text{constant}$$

- Charles's law states that the volume of a fixed mass of gas at constant pressure is directly proportional to its kelvin temperature:

$$V \propto T \qquad \text{or} \qquad \frac{V}{T} = \text{constant}$$

- The pressure law states that the pressure of a fixed mass of gas at constant volume is directly proportional to the kelvin temperature:

$$p \propto T \qquad \text{or} \qquad \frac{p}{T} = \text{constant}$$

3.5 THE IDEAL GAS

Gases at moderate pressures and densities obey the gas laws (section 3.4) relatively closely, but at higher pressures and densities, the gas laws are much less successful. The reason for this is to do with attractive electrostatic forces, called van der Waals forces, between the molecules of a gas. At low to moderate pressures and densities, the average separation of the molecules is much greater than the molecular diameter, and the van der Waals forces are so weak that they have a negligible effect. However, at higher pressures and densities, the average separation of the molecules is much reduced, and the van der Waals forces become significant and affect the large-scale properties of the gas. The attractive forces between the gas molecules also increase significantly if a gas is cooled to low enough temperatures that it approaches liquefaction.

It is helpful to introduce the concept of an **ideal gas** as a theoretical gas that obeys the gas laws at all pressures and temperatures. The effect of attractive forces between its molecules can always be regarded as negligible. Real gases are compared with an ideal gas, and the extent to which a real gas approaches ideal behaviour depends on its physical conditions.

The ideal gas equation

Boyle's law, Charles's law, the pressure law and Avogadro's law can be combined to give

$$V \propto \frac{NT}{p}$$

where N is the number of molecules of the gas. Replacing the proportional sign with an equals sign and a constant of proportionality k gives

$$V = \frac{NkT}{p}$$

The equation is usually written in the form

$$pV = NkT$$

where $k = 1.38 \times 10^{-23}$ J K^{-1} and is known as the **Boltzmann constant**.

It can be useful to work in moles rather than in numbers of molecules.

1 mole (usually abbreviated to 1 mol) is defined as the number of atoms in 12 grams of carbon–12 and is equal to 6.02×10^{23}.

The number 6.02×10^{23} is known as the **Avogadro constant** and has the symbol N_A.

Hence, if a gas contains N molecules, then the number n of moles of gas present is

$$n = \frac{N}{N_A} \quad \text{and so} \quad N = nN_A$$

Then $pV = NkT$ can now be written as

$$pV = nN_AkT$$

where N_Ak is a constant that can be replaced by a single constant R, giving

$$pV = nRT$$

Here $R = 8.31$ J K^{-1} mol^{-1} and is known as the **molar gas constant**.

The equations $pV = NkT$, with N = number of molecules, and $pV = nRT$, with n = number of moles, are both forms of the **ideal gas equation**. The ideal gas equation can be applied to real gases to a reasonable degree of accuracy provided the pressure and density values are relatively moderate.

Note that the mass of one mole is called the **molar mass**, and so

$$\text{molecular mass} = \frac{\text{molar mass}}{N_A}$$

Worked example 1

Gas, in a cylinder of fixed volume 7.1×10^{-2} m^3, is maintained at a temperature of 20 °C and pressure 500 kPa. It can be assumed that the gas behaves like an ideal gas.

a. Calculate

- **i.** the number of moles of gas in the cylinder
- **ii.** the mass of gas in the cylinder
- **iii.** the number of molecules in the cylinder, given that the molar mass of the gas is 0.029 kg.

[Avogadro constant is 6.02×10^{23}]

b. Some of the gas is released and the gas pressure in the cylinder falls to 400 kPa. The temperature of the gas is unchanged. Determine the number of moles of gas that were released from the cylinder.

a. i. Rearranging $pV = nRT$ gives the number of moles:

$$n = \frac{pV}{RT} = \frac{500 \times 10^3 \times 7.1 \times 10^{-2}}{8.31 \times 293}$$

$$= 14.58 = 15 \text{mol}$$

ii. Mass of gas in the cylinder = $n \times$ molar mass = $14.58 \times 0.029 = 0.4228 = 0.42$ kg

iii. Number of molecules = $n \times N_A$ = $14.58 \times 6.02 \times 10^{23} = 8.777 \times 10^{24}$ = 8.8×10^{24}

b. The number of moles remaining in the cylinder can be found from

$$n = \frac{pV}{RT} = \frac{400 \times 10^3 \times 7.1 \times 10^{-2}}{8.31 \times 293} = 11.66 \text{ mol}$$

The number of moles released from the cylinder = 14.58 – 11.66 = 2.9 mol

QUESTIONS

19. A diver's air cylinder has a volume of 2.2×10^{-2} m^3.

a. Calculate the mass of air in the cylinder if the air is at a pressure of 2.0×10^7 Pa and at a temperature of 10 °C, given that the molar mass of air is 0.029 kg.

b. Determine the number of molecules of air that are in the cylinder.

20. A cylinder of volume 7.2×10^{-2} m^3 contains gas at a pressure of 500 kPa and temperature 290 K. A quantity of gas is released from the cylinder and the pressure drops to 410 kPa with no change in temperature. Determine the number of moles of gas released from the cylinder, assuming that the gas behaves as an ideal gas.

The ideal gas equation $pV = nRT$ can be rearranged to give $\frac{pV}{T} = nR$, which if applied to a *constant number of moles* of gas can be rewritten as

$$\frac{p_1V_1}{T_1} = \frac{p_2V_2}{T_2}$$

where 1 indicates the initial state and 2 indicates the final state after a change in conditions. This can be helpful for solving problems involving changes in pressure, volume and temperature, provided the number of moles does not change.

Worked example 2

A cylinder with a gas-tight piston contains gas at a temperature of 20.5 °C and pressure of 101 kPa. The volume of the gas is 4.51×10^{-3} m^3. An external force is suddenly applied to the piston, reducing the gas volume to 3.25×10^{-3} m^3 and causing the gas temperature to increase to 35.5 °C. Determine the new pressure, assuming the gas behaves like an ideal gas.

Rearranging

$$\frac{p_1V_1}{T_1} = \frac{p_2V_2}{T_2}$$

to make P_2 the subject gives

$$p_2 = \frac{p_1V_1T_2}{T_1V_2} = \frac{101 \times 10^3 \times 4.51 \times 10^{-3} \times (273.15 + 35.5)}{(273.15 + 20.5) \times 3.25 \times 10^{-3}}$$

$$= 1.47 \times 10^5 \text{Pa} = 147 \text{kPa}$$

QUESTIONS

21. At sea level, atmospheric pressure is 101 kPa and 1.0 m^3 of air at 20 °C contains 41 moles. At an altitude of 2000 m, the atmospheric pressure is 79 kPa and the temperature falls to 4 °C. Determine the volume occupied by 41 moles of air at an altitude of 2000 m.

22. A bubble of air escapes from a diver's breathing apparatus at a depth of 45 m. The bubble has a volume of 2.0×10^{-5} m^3. The pressure due to the water at a depth of 45 m is 450 kPa and the water temperature is 5 °C. What is the volume of the bubble when it has risen to the surface, where the temperature is 10 °C? Take atmospheric pressure as 100 kPa and assume that the air in the bubble behaves as an ideal gas.

Work done by an expanding gas

When a gas expands, unless it expands into a vacuum, it does work. Consider a gas contained in a cylinder sealed by a frictionless piston (Figure 18). Both the gas inside and the atmosphere outside are exerting pressure on the frictionless piston. Since the piston is stationary and is therefore in equilibrium, the pressure of the gas is equal to atmospheric pressure. Now consider the gas being gently warmed. The gas pressure briefly increases and the piston starts to move slowly to the right, as it is no longer in equilibrium. Once the piston has started to move, it will continue to move steadily provided the gas continues to be warmed steadily. During this steady expansion, the gas pressure remains equal to atmospheric pressure, and the gas does work in pushing back the atmosphere.

At the instant that the piston reaches the position indicated by the dotted lines in Figure 18, the work

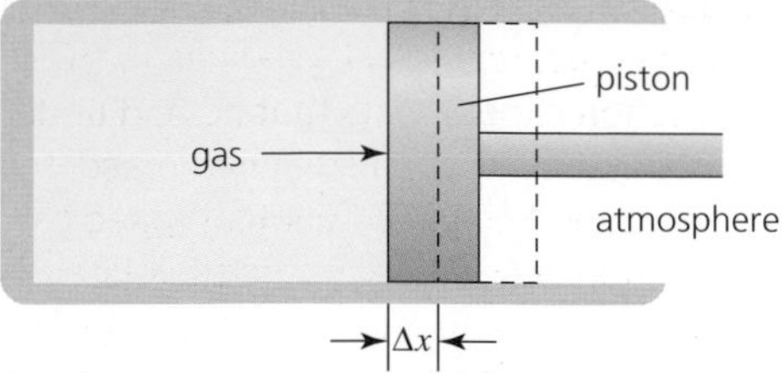

Figure 18 *A gas can do work when it expands.*

W done by the gas in expanding at constant pressure is given by $W = F\Delta x$, where F is the force exerted by the gas on the piston and Δx is the distance moved by the piston (*see section 11.5 in Chapter 11 of Year 1 Student Book*).

The force exerted by the gas on the piston is given by $F = pA$, where p is the gas pressure in the cylinder and A is the cross-sectional area of the piston. Therefore, the work done by the gas is given by $W = F\Delta x = pA\Delta x$. However, $A\Delta x$ is the increase in volume of the gas, ΔV. Therefore, the work W done by a gas expanding at constant pressure is given by the equation

$$W = p\Delta V$$

Worked example 3

A gas is contained in a cylinder similar to that in Figure 18. It is warmed gently so that it expands slowly, pushing the piston out a distance of 12 cm against atmospheric pressure of 1.0×10^5 Pa. Determine the work done by the gas in expanding at constant pressure if the cross-sectional area of the piston is 80 cm^2.

Since the expansion takes place slowly, it can be assumed that the gas pressure is equal to atmospheric pressure. Therefore the work done is

$$W = p\Delta V = 1.0 \times 10^5 \times 0.12 \times 80 \times 10^{-4} = 96\,\text{J}$$

QUESTIONS

23. When an amount of water is heated and converted to steam, its volume increases by a factor of 1600. How much work is done against the atmosphere when 1.5 litre of water is converted to steam, if atmospheric pressure is 1.0×10^5 Pa?
[1.0 litre = 1.0×10^{-3} m^3]

KEY IDEAS

- An ideal gas is a gas that obeys Boyle's law, Charles's law and the pressure law at all pressures and temperatures, since the effect of attractive forces between its molecules can always be regarded as negligible.
- The ideal gas equation is

  $pV = NKT$ for N molecules, where k is the Boltzmann constant

  or

  $pV = nRT$ for n moles, where R is the molar gas constant.
- The work done by a gas when it expands by ΔV at constant pressure p is

  $$W = p\Delta V$$

3.6 THE MOLECULAR KINETIC THEORY MODEL

In the late 17th century, based on his observation that gases could be compressed, Robert Boyle suggested that a gas could be made up of tiny invisible particles. In 1803, John Dalton, the English chemist and physicist, following on from the work of Boyle, Charles and Gay-Lussac, and based on his own experiments with gases, put forward his atomic theory. He proposed that elements are made up of extremely small particles called 'atoms' and that the atoms of a particular element are identical.

> "Matter, though divisible in an extreme degree, is nevertheless not infinitely divisible. That is, there must be some point beyond which we cannot go in the division of matter. ... I have chosen the word 'atom' to signify these ultimate particles."
>
> *John Dalton, 1810*

Dalton's atomic theory was the subject of much debate, because there was a lack of direct evidence for the existence of atoms. In 1827, when the Scottish botanist Robert Brown was studying pollen suspended in water under a microscope, he observed that the tiny particles of pollen had a random jittery motion. Brown thought at first that he was observing something that was alive, so he repeated his experiment with tiny particles of glass and also with dust from a stone, and observed the same random jittery motion. Smoke particles suspended in air demonstrate the same jittery motion observable under a microscope (Figure 19).

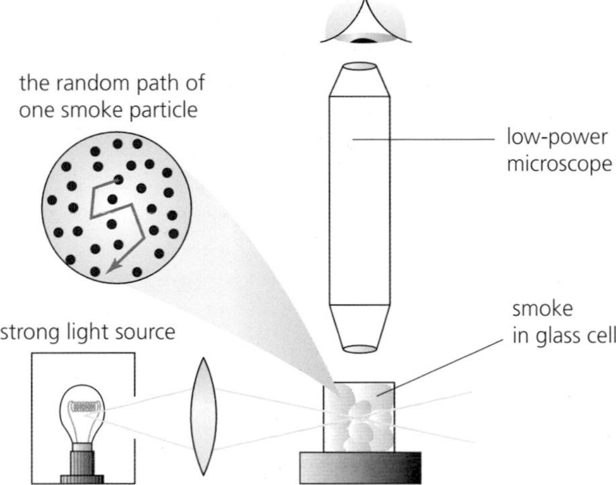

Figure 19 *Observing the Brownian motion of smoke particles*

Brown thought that the effect may be due to incident light or possibly evaporation currents or even some source of vibration, but he was unable to reach any conclusions. However, Brown's observations would, many decades later, provide crucial evidence leading to the acceptance of the existence of atoms. Meanwhile, the debate about the existence of atoms continued. In 1833, the eminent English physicist, Michael Faraday commented "though it is very easy to talk of atoms, it is very difficult to form a clear idea of their nature ...".

During the 1860s and 1870s, James Maxwell, Ludwig Boltzmann and others were developing the **kinetic theory** in which liquids and gases were considered to be made up of small particles (atoms or molecules) that are in constant random motion. The theory attempted to explain the gas properties of pressure and temperature in terms of the movement of these particles. In 1905, Albert Einstein published his theoretical analysis of Brownian motion in what was essentially seen as a crucial test of Maxwell and Boltzmann's kinetic theory. He showed, in precise mathematical detail, that microscopic particles, suspended in a liquid or gas, would perform random movements observable under a microscope, caused by the high-speed thermal motion of the liquid or gas molecules. This is now known as **Brownian motion**. Einstein's paper was seen as providing evidence in support of the kinetic theory and of the existence of atoms.

In 1909, Jean Perrin, the French physicist and future Nobel Prize winner, reported the results of the Brownian motion experiments that he had undertaken to test the equations that Einstein had presented in his 1905 paper. Perrin had photographed the Brownian motion of spherical particles of latex with a diameter of 0.53 μm in a liquid consisting of water and methanol, and recorded the positions of particles every 30 s on a grid (Figure 20). Perrin's analysis of his experimental results was in agreement with Einstein's equations, providing confirmation of the kinetic theory and the physical reality of atoms and molecules.

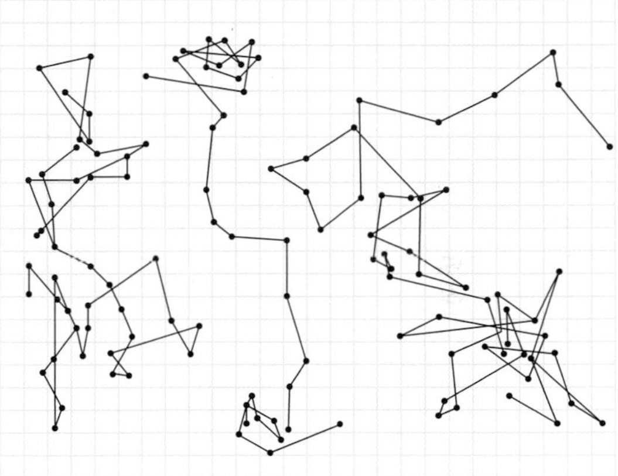

Figure 20 *Grid showing changing positions of latex particles undergoing Brownian motion*

The observed jittery motion of the smoke particles is due to bombardment by air molecules moving randomly. Smoke particles are small, but air molecules are very much smaller. The air molecules must be moving randomly at very high speeds to have sufficient momentum to cause the heavier smoke particles to exhibit Brownian motion. At any instant, the imbalance in the number of molecules hitting a smoke particle on one side (Figure 22) give the smoke particle a push in a particular direction.

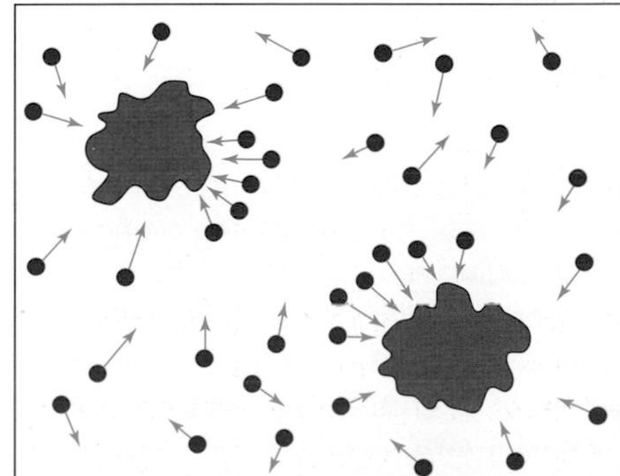

Figure 21 *A resultant force on a smoke particle pushes it in a particular direction.*

QUESTIONS

24. a. Suggest why particles larger than about 1 μm do not display Brownian motion.

b. Experiments showed that a rise in temperature caused an increase in the rapid erratic motion of the particles undergoing Brownian motion. Suggest a possible reason for this.

c. Suggest why the debate about the possible existence of atoms continued for over 200 years.

The **molecular kinetic theory model** of a gas considers the molecules to be moving around colliding with each other and with the walls of their container, exerting pressure and moving faster if the temperature of the gas is raised. Apparatus illustrating a simple model of a gas exerting pressure is shown in Figure 22, in which a vibrating plate causes small plastic spheres to move around quickly, exerting forces and therefore supporting a section of polystyrene.

Figure 22 Kinetic theory model apparatus

The kinetic theory describes the pressure exerted by a gas in terms of the momentum change of the molecules colliding with the walls of the container. When a molecule collides with the container wall, it bounces off, changing its direction. The molecule's momentum, being a vector quantity, must therefore have changed. Newton's second law states that 'resultant force = rate of change of momentum' (*see section 11.1 in Chapter 11 of Year 1 Student Book*). Therefore, the momentum change of the molecule shows that the wall exerted a force on the molecule. Since the wall exerts a force on the molecule, then Newton's third law states that the molecule must exert an equal but opposite force on the wall.

In 1920, Otto Stern, a German-born American physicist, showed, by experiment, that, in a gas, there is a distribution of molecular speeds (Figure 23). This confirmed Maxwell and Boltzmann's kinetic theory predictions regarding the speeds and energies of molecules in a gas. Stern's experiments demonstrated that, if the gas is heated to a higher temperature, the peak in the distribution shifts to a higher speed, showing that:

An increase in temperature results in an increase in the average speed of the gas molecules.

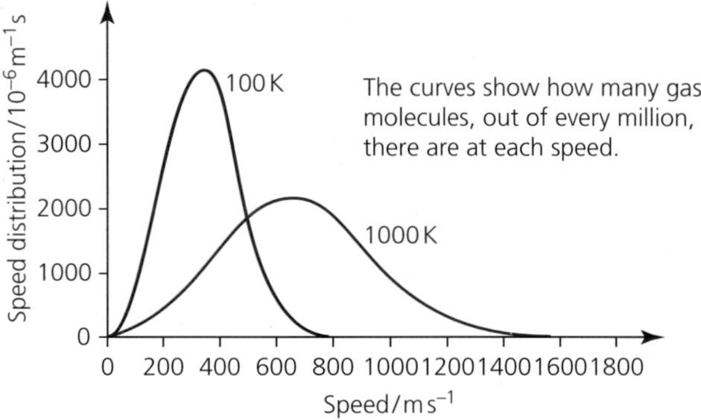

Figure 23 Distribution of molecule speeds in a gas

QUESTIONS

25. Boyle's law states that $p \propto \frac{1}{V}$ at constant temperature. Consider what happens to the molecules of a gas if its volume is reduced at constant temperature (Figure 24). Answer the following questions, giving an explanation of your answers.

a. Does the average speed of the molecules change?

b. Does the frequency of collisions of molecules with the container walls change?

c. Suggest why the pressure exerted by the molecules on the cylinder walls has increased.

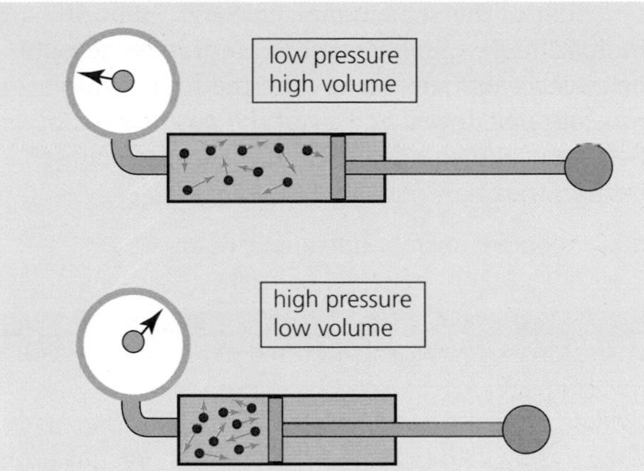

Figure 24 Changing the volume of a gas at constant temperature and constant number of molecules

26. Charles's law states that $V \propto T$ at constant pressure. Consider what happens to the molecules of a gas if it is heated, raising its temperature, but the gas pressure remains constant (Figure 25). Answer the following questions, giving an explanation of your answers.

a. Does the average speed of the molecules change?

b. Does the average force exerted by a molecule on the piston change?

c. The gas expands until the pressure of the gas is once again balanced by the pressure exerted by the atmosphere and the weight of the piston. What can you conclude regarding the frequency of collisions of molecules with the piston?

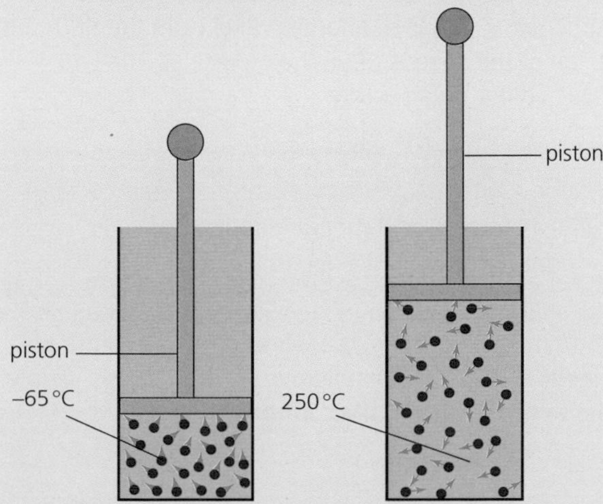

Figure 25 Changing the temperature of a gas at constant pressure and constant number of molecules

27. The pressure law states that $p \propto T$ at constant volume. Consider what happens to the molecules of a gas if it is heated, raising its temperature, but the volume remains constant (Figure 27). Answer the following questions, giving an explanation of your answers.

a. Does the average speed of the molecules change?

b. Does the average force exerted by a molecule on the container walls change?

c. The volume is constant, so the distance travelled by a molecule in between collisions with the container walls is unchanged. Would you expect the frequency of collisions to change?

d. Comment on how your answers to parts **b** and **c** relate to the pressure exerted by the gas.

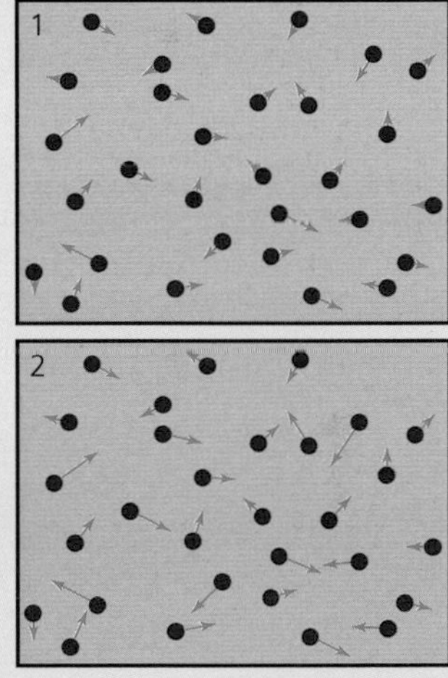

Figure 26 Increasing the temperature of a gas at constant volume and constant number of molecules

Deriving a term for the molecular kinetic energy

The random kinetic energy of gas molecules consists of three different types. Although most of the random kinetic energy of gas molecules is translational kinetic energy due to the movement of molecules around the container, the molecules also have some rotational and vibrational kinetic energy (Figure 27). However, a gas made up of single atoms (a monatomic gas), such as helium, has no vibrational kinetic energy and very little rotational energy.

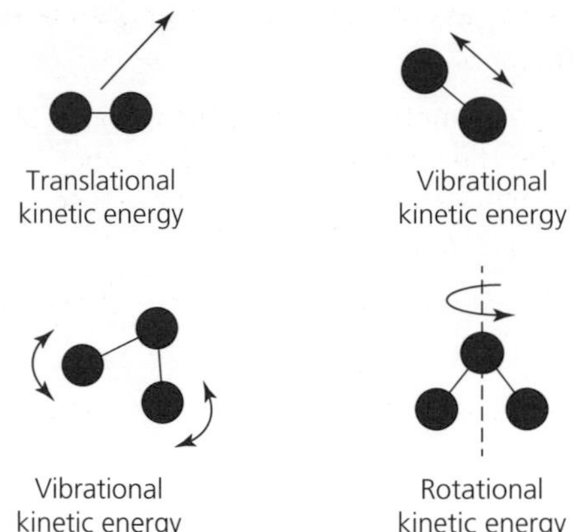

Figure 27 *Types of molecular kinetic energy*

When the distribution of molecular speeds for different gases at the same temperature (Figure 28) was investigated experimentally, it showed that the average speed of gas molecules depended on the particular gas considered. On average, heavier molecules such as oxygen travelled more slowly than lighter atoms such as helium. This suggested that the temperature of a gas may be linked to the average kinetic energy of the molecules rather than just their speed.

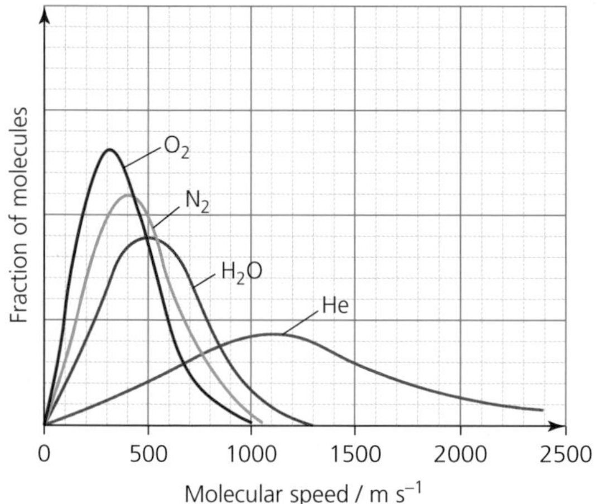

Figure 28 *Comparing the distribution of molecular speeds of different gases*

To obtain an expression for the random *translational* kinetic energy per molecule of a gas, it is first necessary to consider the total random translational kinetic energy of a gas composed of N molecules. The gas must be in a closed container, so that N is constant. Using c to represent the speed of molecules, the speed of the first molecule is c_1, that of the second molecule is c_2, that of the third molecule is c_3 and so on. If m represents the mass of one molecule, then the total random translational kinetic energy of the gas is given by the sum of the kinetic energies of each individual particle, that is,

total random translational kinetic energy

$$= \frac{1}{2}mc_1^2 + \frac{1}{2}mc_2^2 + \frac{1}{2}mc_3^2 + \ldots + \frac{1}{2}mc_N^2$$

Dividing this total random translational kinetic energy of the gas by the number of molecules in the gas gives the average kinetic energy per molecule:

average translational kinetic energy per molecule

$$= \frac{\frac{1}{2}mc_1^2 + \frac{1}{2}mc_2^2 + \frac{1}{2}mc_3^2 + \ldots + \frac{1}{2}mc_N^2}{N}$$

This can be rewritten as

$$\text{average translational kinetic energy per molecule} = \frac{\frac{1}{2}m(c_1^2 + c_2^2 + c_3^2 + \ldots + c_N^2)}{N}$$

To make these equations more manageable, the **mean square speed**, symbol $(c_{rms})^2$, is defined as

$$(c_{rms})^2 = \frac{c_1^2 + c_2^2 + c_3^2 + \ldots + c_N^2}{N}$$

which is 'the sum of the squares of the speeds of all the molecules of a gas divided by the number of molecules'. The speed c_{rms} is called the **root mean square speed** of the gas molecules in the container. Although c_{rms} has a different value from the arithmetic mean of the speeds of all the molecules, the two values differ only slightly.

Then we obtain the expression:

$$\text{average translational kinetic energy per molecule} = \frac{1}{2}m(c_{rms})^2$$

For a monatomic gas, the vibrational and rotational kinetic energies are negligible, so $\frac{1}{2}m(c_{rms})^2$ is simply the **average molecular kinetic energy**.

QUESTIONS

28. Consider 10 gas molecules, each with the following speed in m s^{-1}: 330, 245, 425, 399, 343, 601, 515, 287, 454, 312.

a. Determine the root mean square speed, c_{rms}, for these 10 molecules using

$$(c_{rms})^2 = \frac{c_1^2 + c_2^2 + c_3^2 + \ldots + c_N^2}{N}$$

b. Determine the average speed of these 10 molecules by simply adding the speeds and dividing by the number of molecules.

c. How do your answers to parts **a** and **b** compare?

Deriving the kinetic theory equation

The molecular kinetic theory model of a gas is a theoretical approach to the study of gases. It aims to link the macroscopic quantities of pressure, volume and temperature with microscopic quantities, such as the mass, speed and kinetic energy of the gas molecules. The molecules of a real gas exert attractive (van der Waals) forces on each other, and, during collisions, exert repulsive forces on each other. In order to make the mathematics of the kinetic theory model more straightforward, it is necessary to treat the gas as an ideal gas and its molecules as objects that are moving around independently of each other. On this basis, the molecular kinetic theory model makes the following assumptions about an ideal gas:

1. The gas molecules move in random directions.
2. The volume of the gas molecules is negligible compared with the volume of the gas (so the molecules are relatively far apart and the volume of the actual molecules is considered negligible).
3. The gas molecules do not exert any force on each other, except during collisions. There are no intermolecular forces of attraction.
4. The time that a gas molecule spends travelling in-between collisions is significantly longer than the duration of a collision.
5. The gas molecules have elastic collisions. There is no loss of kinetic energy when a gas molecule collides with the walls of its container.
6. Newton's laws of motion can be applied to the gas molecules.

The above assumptions simplify the theoretical approach to the study of a gas and are not unreasonable assumptions to make because of the following observations:

- Brownian motion provides evidence of the random motion of gas molecules.
- Gases can be compressed so the gas atoms are, on average, far apart.
- Gas molecules in a container at the same temperature are not observed to lose kinetic energy and slow down, so their collisions must be elastic.

The molecular kinetic theory model considers the molecules of a gas moving around independently in a container of side L (Figure 29). The pressure p that the gas exerts on the walls of the container is due to the collisions that the gas molecules, each of mass m, make with the container walls. One molecule is chosen (referred to as 'molecule 1), and its speed denoted by c_1.

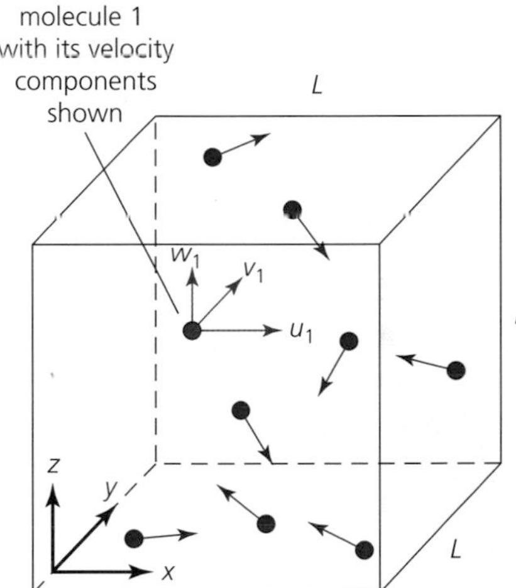

Figure 29 Gas molecules moving randomly in a cubical container

Since molecule 1 is moving in three dimensions, it has three components of velocity, which we shall call u_1, v_1 and w_1, along the x, y and z directions, respectively. These can be combined using Pythagoras's theorem to give

$$c_1^2 = u_1^2 + v_1^2 + w_1^2$$

Now consider molecule 1 colliding with one of the walls of the container (Figure 30). The molecule collides with the wall, bounces back and travels in the opposite direction. The molecule has therefore experienced a change of momentum, which means, according to Newton's second law, that the wall has exerted a resultant force on the molecule. According to Newton's third law, it can be concluded that the molecule exerted an equal but opposite force on the wall (*see section 11.1 in Chapter 11 of Year 1 Student Book*).

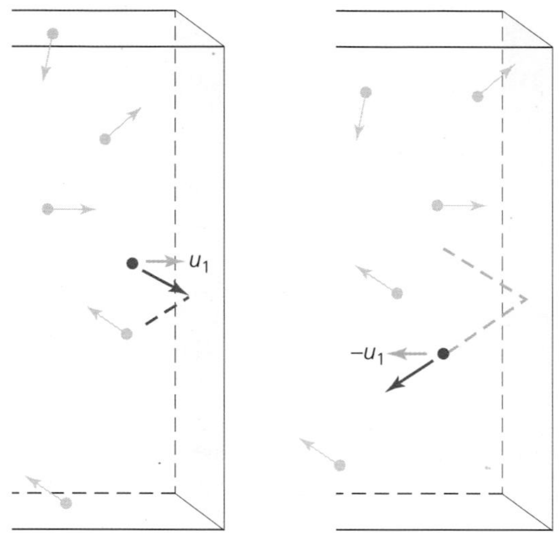

Figure 30 *A molecule making an elastic collision with the container wall*

If an expression for the force exerted by one molecule can be determined, the total force on the wall exerted by all the molecules in the box can be found by adding up all the forces due to the individual molecules. An expression for the pressure of the gas can then be found using the defining formula for pressure $p = \frac{F}{A}$.

If molecule 1 collides with the right hand wall of the box, it is only necessary to consider the x component of the velocity of the molecule, u_1, which takes the molecule towards this wall, with which it collides and bounces back:

$$\begin{aligned}&\text{momentum change of molecule 1 on hitting the wall}\\&= mu_1 - (-mu_1)\\&= 2mu_1\end{aligned}$$

The molecule will hit this right hand wall again after travelling to the other side of the box and back again. So

$$\begin{aligned}&\text{average time between consecutive collisions of molecule 1 with right hand wall}\\&= \frac{\text{distance}}{\text{speed}}\\&= \frac{2L}{u_1}\end{aligned}$$

Then

$$\begin{aligned}&\text{number of collisions of molecule 1 with right hand wall per second}\\&= \frac{1}{\text{time between collisions}}\\&= \frac{u_1}{2L}\end{aligned}$$

The momentum change per second of molecule 1 is equal to the momentum change per collision multiplied by the number of collisions per second. Therefore

$$\text{momentum change of molecule 1 per second} = 2mu_1 \times \frac{u_1}{2L} = \frac{mu_1^2}{L}$$

Since Newton's second law can be written as 'resultant force = rate of change of momentum', the force F that the right hand wall of the container exerts on the molecule is

$$F = \frac{mu_1^2}{L}$$

Since the force F exerted by the wall on molecule 1 is equal in size to the force exerted by molecule 1 on the wall (Newton's third law), the above equation for F gives the equation for the force exerted by molecule 1 on the container wall.

Now consider all N molecules in the box. Since we are assuming that the gas molecules are moving around independently of each other, to get the total force due to all the molecules we just have to add up the forces due to each one. Then the total force exerted on the right hand wall is

$$F_{\text{total}} = \frac{mu_1^2}{L} + \frac{mu_2^2}{L} + \frac{mu_3^2}{L} + \ldots + \frac{mu_N^2}{L}$$

which can be rewritten as

$$F_{\text{total}} = \frac{m}{L}(u_1^2 + u_2^2 + u_3^2 + \ldots + u_N^2)$$

This equation can be simplified using the expression for the mean square speed in the x direction:

$$(u_{\text{rms}})^2 = \frac{u_1^2 + u_2^2 + u_3^2 + \ldots + u_N^2}{N}$$

which gives the total force on the right hand wall as

$$F_{\text{total}} = \frac{Nm(u_{\text{rms}})^2}{L}$$

The area of the right hand wall is L^2 and pressure $p = \frac{F}{\text{area}}$, so the pressure p exerted by the molecules on the right hand wall is

$$p = \frac{Nm(u_{rms})^2}{L \times L^2} = \frac{Nm(u_{rms})^2}{L^3} = \frac{Nm(u_{rms})^2}{V}$$

since the container volume $V = L^3$.

Given that the molecules are moving in all directions, the mean square speeds in the *x*, *y* and *z* directions can be combined using Pythagoras's theorem:

$$(c_{rms})^2 = (u_{rms})^2 + (v_{rms})^2 + (w_{rms})^2$$

Also, since the molecules will be distributed randomly throughout the container, we can assume that the mean square speed in each of the three different directions must be equal:

$$(u_{rms})^2 = (v_{rms})^2 = (w_{rms})^2$$

Therefore, we obtain

$$(u_{rms})^2 = \frac{1}{3}(c_{rms})^2$$

which gives the gas pressure as

$$p = \frac{1}{3} \times \frac{Nm(c_{rms})^2}{V}$$

This is usually written as

$$pV = \frac{1}{3}Nm(c_{rms})^2$$

and is referred to as the **kinetic theory equation**.

The derivation of the kinetic theory equation was based on a number of assumptions, which were made to enable a simple application of Newton's laws. The equation can however be applied to real gases with a reasonable degree of accuracy provided the gases involved do not have high pressures and high densities.

Worked example 1

Calculate the root mean square speed of oxygen molecules in a room at 20 °C if the density of oxygen at 20 °C and atmospheric pressure of 101 kPa is 1.33 kg m^{-3}.

A density of 1.33 kg m^{-3} means that a volume of 1 m^3 has a mass of 1.33 kg. The mass of the gas is equal to the number of molecules *N* multiplied by the mass *m* of one molecule.

The kinetic theory equation $pV = \frac{1}{3}Nm(c_{rms})^2$ can be rearranged to give

$$c_{rms} = \sqrt{\frac{3pV}{Nm}} = \sqrt{\frac{3 \times 101 \times 10^3 \times 1}{1.33}} = 477\,\text{m s}^{-1}$$

QUESTIONS

29. Calculate the root mean square speed of carbon dioxide molecules in a room at 20 °C if the density of carbon dioxide at 20 °Cand atmospheric pressure of 101 kPa is 1.84 kg m^{-3}.

Stretch and challenge

30. In the derivation of the kinetic theory equation, it was assumed that Pythagoras's theorem could be applied to the mean square speeds in the *x*, *y* and *z* directions:

$$(c_{rms})^2 = (u_{rms})^2 + (v_{rms})^2 + (w_{rms})^2$$

When Pythagoras's theorem is applied to the speed components of just molecule 1, we get $c_1^2 = u_1^2 + v_1^2 + w_1^2$, for molecule 2 we get $c_2^2 = u_2^2 + v_2^2 + w_2^2$ and for molecule *N* we get $c_N^2 = u_N^2 + v_N^2 + w_N^2$. Use the application of Pythagoras's theorem to individual molecules to show that $(c_{rms})^2 = (u_{rms})^2 + (v_{rms})^2 + (w_{rms})^2$ is correct.

Dependence of molecular kinetic energy on temperature

Experiments have shown that raising the temperature of a gas results in the peak in the distribution of the speeds of the gas molecules (see Figure 23) shifting to correspond to a higher speed and therefore a higher average translational kinetic energy per molecule. The kinetic theory equation and the ideal gas equation can be combined to predict an equation showing the relationship between temperature and the average molecular kinetic energy (of translation) of an ideal gas.

Equating $pV = \frac{1}{3}Nm(c_{rms})^2$ and $pV = NkT$ gives

$$\frac{1}{3}Nm(c_{rms})^2 = NkT$$

which can be rearranged to give

$$m(c_{rms})^2 = \frac{3NkT}{N}$$

Multiplying both sides of the equation by $\frac{1}{2}$ and cancelling N gives

$$\frac{1}{2}m(c_{rms})^2 = \frac{3kT}{2}$$

This shows that the average molecular kinetic energy (of translation) of an ideal gas is directly proportional to the kelvin temperature of the ideal gas. Again, this equation can be applied to real gases with a reasonable degree of accuracy provided that the gas pressures and densities are low or moderate.

Since the Boltzmann constant $k = \frac{R}{N_A}$, where R is the molar gas constant and N_A is the Avogadro constant, the equation above can be rewritten as

$$\frac{1}{2}m(c_{rms})^2 = \frac{3RT}{2N_A}$$

Note that the total kinetic energy of translation of a gas can be found by multiplying the average molecular kinetic energy (of translation) by the number of gas molecules present.

Worked example 2

Determine the total kinetic energy of 2.0 mol of a monatomic gas at a temperature of 20 °C and at atmospheric pressure.

The total kinetic energy is equal to the average molecular kinetic energy multiplied by the number of molecules. So

$$\text{total kinetic energy} = \text{number of molecules} \times \frac{3kT}{2}$$

Substituting the values for number of molecules in 2.0 mol = $2N_A = 2 \times 6.02 \times 10^{23}$, temperature $T = (20 + 273.15)$ K = 293.15 K and $k = 1.38 \times 10^{-23}$ J K^{-1}, this gives

total kinetic energy

$$= 2 \times 6.02 \times 10^{23} \times \frac{3 \times 1.38 \times 10^{-23} \times 293.15}{2}$$

$$= 7.3 \times 10^3 \text{ J}$$

QUESTIONS

31. A monatomic gas, which can be assumed to be ideal, is at a pressure of 210 kPa and a temperature of 290 K in a rigid container of volume 0.51×10^{-3} m^3. Determine

a. the number of gas atoms present

b. the total random kinetic energy of the gas.

KEY IDEAS

- The observation of Brownian motion provided direct evidence for the existence of atoms.
- The assumptions of the molecular kinetic theory model of gases led to the derivation of the kinetic theory equation showing the relation between the macroscopic quantities of pressure, volume and temperature and the microscopic quantities of molecular mass and speed:

$$pV = \frac{1}{3}Nm(c_{rms})^2$$

where $(c_{rms})^2$ is the mean square speed of the molecules.

- The average molecular kinetic energy (of translation) is directly proportional to the kelvin temperature:

$$\frac{1}{2}m(c_{rms})^2 = \frac{3kT}{2} = \frac{3RT}{2N_A}$$

PRACTICE QUESTIONS

1. An ideal gas at a temperature of 22 °C is trapped in a metal cylinder of volume 0.20 m^3 at a pressure of 1.6×10^6 Pa.
   a. Outline what is meant by an *ideal gas*.
   b. i. Calculate the number of moles of gas contained in the cylinder.
      ii. The gas has a molar mass of 4.3×10^{-2} kg mol^{-1}. Calculate the density of the gas in the cylinder. State an appropriate unit for your answer.
      iii. The cylinder is taken to high altitude where the temperature is –50 °C and the pressure is 3.6×10^4 Pa. A valve on the cylinder is opened to allow gas to escape. Calculate the mass of gas remaining in the cylinder when it reaches equilibrium with its surroundings. Give your answer to an appropriate number of significant figures.

   AQA Unit 5 Section 1 June 2013 Q4

2. a. Define the Avogadro constant.
   b. i. Calculate the mean kinetic energy of krypton atoms in a sample of gas at a temperature of 22 °C.
      ii. Calculate the mean square speed $(c_{rms})^2$ of krypton atoms in a sample of gas at a temperature of 22 °C. [Mass of 1 mole of krypton = 0.084 kg]
   c. A sample of gas consists of a mixture of krypton and argon atoms. The mass of a krypton atom is greater than that of an argon atom. State and explain how the mean square speed of krypton atoms in the gas compares with that of the argon atoms at the same temperature.

   AQA Unit 5 Section 1 June 2014 Q3

3. An electrical immersion heater supplies 8.5 kJ of energy every second. Water flows through the heater at a rate of 0.12 kg s^{-1} as shown in Figure Q1.

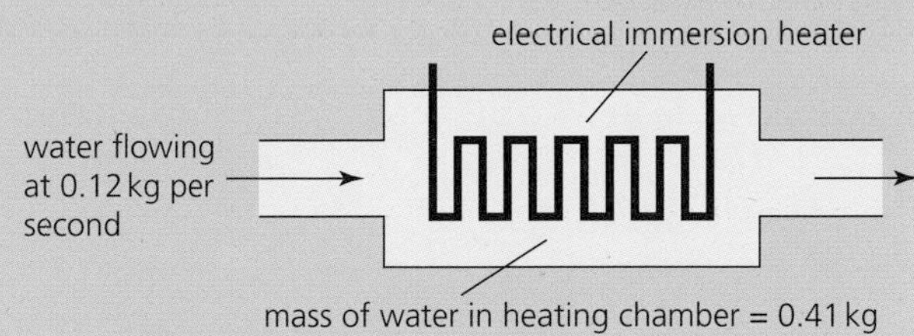

Figure Q1 *Electrical immersion heater*

   a. Assuming all the energy is transferred to the water, calculate the rise in temperature of the water as it flows through the heater. [Specific heat capacity of water 4200 J kg^{-1} °C^{-1}]
   b. The water suddenly stops flowing at the instant when its average temperature is 26 °C. The mass of the water trapped in the heater is 0.41 kg. Calculate the time taken for the water to reach 100 °C if the immersion heater continues supplying energy at the same rate.

   AQA Unit 5 Section 1 June 2012 Q1

4. An electrical heater is placed in an insulated container holding 100 g of ice at a temperature of –14 °C. The heater supplies energy at a rate of 98 J per second.
   a. After an interval of 30 s, all the ice has reached a temperature of 0 °C. Calculate the specific heat capacity of ice.
   b. Show that the final temperature of the water formed when the heater is left on for a further 500 s is about 40 °C. [Specific heat capacity of water = 4200 J kg^{-1} °C^{-1}. Specific latent heat of fusion of water = 3.3×10^5 J kg^{-1}]
   c. The whole procedure is repeated in an uninsulated container in a room at a temperature of 25 °C. State and explain whether the final temperature of the water formed would be higher or lower than that calculated in part **b**.

   AQA Unit 5 Section 1 June 2011 Q4

5. Which of the following relationships does Boyle's law describe?
   - **A** The relationship between volume and kelvin temperature at constant pressure
   - **B** The relationship between volume and Celsius temperature at constant pressure
   - **C** The relationship between volume and pressure at constant temperature
   - **D** The relationship between pressure and temperature at constant volume

6. Which of the following statements about an ideal gas is **incorrect**?
   - **A** The volume of the molecules is negligible.
   - **B** Forces of attraction between the molecules are negligible.
   - **C** The distance between the molecules is negligible.
   - **D** The collisions between the molecules are elastic.

7. A fixed mass of gas is heated in a container of constant volume. Which of the following statements about the gas is ***incorrect***?
   - **A** The gas pressure increases because the collisions of the molecules with the container walls occur more frequently.
   - **B** The gas pressure increases because the molecules experience a bigger momentum change on colliding with the container walls.
   - **C** The average random potential energy of the molecules increases.
   - **D** The pressure increases in direct proportion with the kelvin temperature.

8. A gas is heated in a cylinder with a gas-tight frictionless piston, resulting in the gas expanding and its temperature increasing. Which of the following statements about the gas is **incorrect**?
   - **A** The average kinetic energy of the molecules increases.
   - **B** The average random potential energy of the molecules increases.
   - **C** The average speed of the molecules increases.
   - **D** The distance travelled by the molecules between collisions with the cylinder walls decreases.

9. A monatomic gas, which can be assumed to behave as an ideal gas, is maintained at a constant temperature of 50 °C. What is the average kinetic energy per molecule of the gas, in joules?

   A 1.0×10^{-21} **B** 6.7×10^{-21}

   C 2.0×10^{-21} **D** 1.3×10^{-20}

10. An ideal gas has a volume *V*, pressure *p* and kelvin temperature *T*. If the pressure is doubled and the temperature increased by a factor of 4, what would be the new volume, assuming the mass of the gas is unchanged?

   A $\frac{1}{4}V$ **B** $\frac{1}{2}V$

   C V **D** $2V$

4 GRAVITY

PRIOR KNOWLEDGE

You will be familiar with acceleration due to gravity, kinetic energy and potential energy. You will have an understanding of centripetal force and centripetal acceleration from Chapter 1.

LEARNING OBJECTIVES

In this chapter you will learn about Newton's law of gravity and how it can be applied to satellites, moons and planets. You will learn how to calculate the gravitational field strength and gravitational potential of a planet. You will also gain experience in solving problems related to the orbits of satellites and astronomical bodies.

(Specification 3.7.1 part, 3.7.2.1 to 3.7.2.4)

In solving physics problems, you have probably used a value for the gravitational field strength, g, at the Earth's surface of $9.81\,\mathrm{N\,kg^{-1}}$, a value assumed constant. In fact, the value of g varies from place to place, because it is affected by the Earth's rotation, the presence of mountains, ocean trenches and variations in the density of the Earth's interior.

The *Gravity Field and Steady-State Ocean Circulation Explorer* (*GOCE*) satellite (Figure 1) was launched in 2009, its mission being to measure the gravitational field strength at the Earth's surface to an accuracy of $1 \times 10^{-5}\,\mathrm{N\,kg^{-1}}$.

To obtain the best possible measurements, *GOCE* flew in a very low orbit, just 255 km above the Earth's surface. This meant that the satellite would be affected by the Earth's atmosphere, so, in order to minimise drag, *GOCE* was given a sleek aerodynamic shape.

Figure 1 *The GOCE satellite, dubbed the Ferrari of space*

The satellite's mapping of the Earth's gravity field is shown in Figure 2. Although *GOCE* was only designed to last 20 months, the satellite operated for 56 months (4 years and 8 months) before disintegrating in November 2013. High-resolution data showed that ice loss in the west of Antarctica between 2009 and 2012 caused a measurable reduction in the gravitational field.

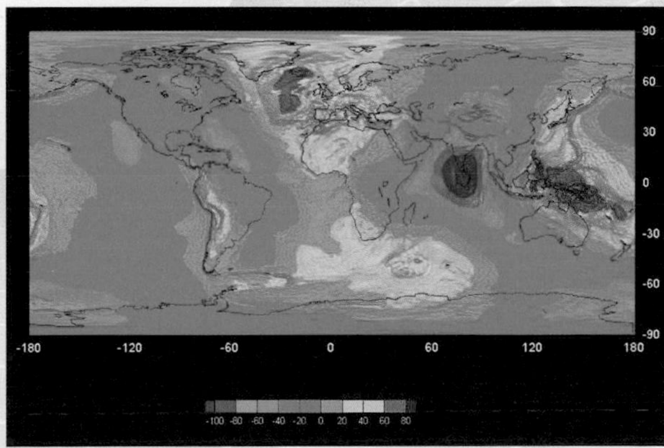

Figure 2 *Earth's gravity map: the red and orange areas have a higher-than-average gravitational field strength, while the blue areas have lower-than-average field strength.*

4.1 NEWTON'S LAW OF GRAVITY

In 1684, three members of the Royal Society, Christopher Wren, Robert Hooke and Edmond Halley, debated the idea that a gravitational force that varied inversely with the square of distance could explain the way that planets in the solar system orbited the Sun and also the Moon's orbit around the Earth. Unable to find evidence to support the idea, Halley decided to approach Isaac Newton for help in the matter.

Newton was familiar with Galileo's experiments on projectiles, in which it was established that an object projected horizontally followed a parabolic path to the ground. Newton had the idea that the motion of the Moon around the Earth could be explained by considering the Moon as a projectile. He devised a thought experiment in which he considered a cannon firing cannon balls horizontally from the top of a very high mountain (Figure 3) with increasing amounts of gun powder so that successive shots would fire the cannon balls faster and faster. Newton realised that a speed would be reached at which the curvature of the Earth would have to be taken into consideration in predicting where the falling cannon ball would land and that eventually, at a high enough speed, the falling cannon ball would never hit the ground but would continue in orbit around the Earth. He thought that the Moon could be regarded as a body that was also falling towards the Earth, but, because of the Earth's curvature, never reached the surface, just like the high-speed cannon ball. By comparing the motion of a falling object at the Earth's surface and the motion of the Moon falling towards the Earth, Newton was to able demonstrate mathematically that the acceleration caused by the Earth's gravitational force varied inversely with the square of the distance from the Earth's centre.

$$\text{gravitational force} \propto \frac{1}{(\text{distance})^2}$$

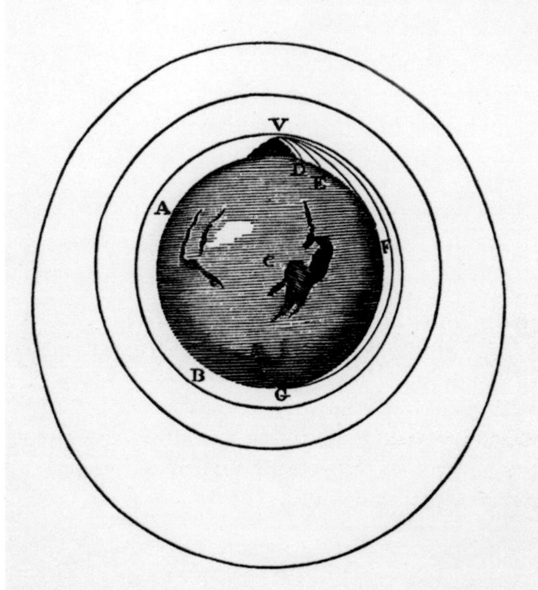

Figure 3 Newton's drawing of his cannon ball thought experiment

ASSIGNMENT 1: DEMONSTRATING THAT THE FORCE OF GRAVITY VARIES INVERSELY WITH THE SQUARE OF DISTANCE

(MS 0.3, MS 1.1, MS 2.3)

The aim of this assignment is for you to compare the force of gravity exerted by the Earth on an apple at the Earth's surface with the force of gravity exerted by the Earth on the Moon, in order to determine how the force of gravity weakens with distance. Obviously, the apple and the Moon have very different masses, so, instead of comparing the sizes of the gravitational forces on the apple and the Moon, you will compare the *acceleration* of each caused by the Earth's gravitation (since, from Newton's second law of motion, the acceleration is the force per unit mass).

To compare the acceleration due to gravity of both the apple and the Moon requires the data shown on Figure A1, along with the (average) value of the acceleration due to gravity at the Earth's surface, $9.81\,\text{m s}^{-2}$, and the time for the Moon to complete one orbit around the Earth, which is 27.3 days.

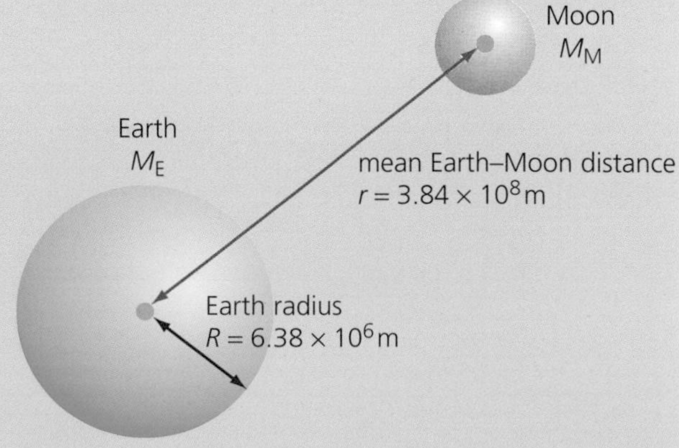

Figure A1 The Moon in orbit around the Earth

Questions

A1 Use your knowledge of circular motion (see section 1.2 of Chapter 1) to determine the centripetal acceleration of the Moon in its orbit around the Earth.

The Moon is kept in its orbit by the force of gravity, so the centripetal acceleration you have calculated in question **A1** is equal to the acceleration due to the Earth's gravitation at a distance of 3.84×10^{8} m from the Earth's centre.

A2 How many times greater is the acceleration due to gravity at the Earth's surface, $9.81 \, \text{m s}^{-2}$, than your answer to question **A1**?

A3 Determine the number of times greater the Earth–Moon distance is compared with the distance from the Earth's surface to its centre (its radius).

A4 Use your answers to questions **A2** and **A3** to demonstrate that the acceleration caused by the Earth's gravitation decreases inversely with the square of distance from the Earth's centre.

Newton concluded that gravity was an attractive force that acted between all masses. It was a universal force that not only made objects fall towards the Earth's surface, but also kept the Moon in orbit around the Earth, and the Earth and the other planets in orbit around the Sun. He deduced that the force of gravity between two masses would depend not only on the distance between them but also on the mass of each of the two bodies. **Newton's law of universal gravitation** states that:

> Any two point masses attract each with a force F that is directly proportional to the product of their masses $m_1 m_2$ and inversely proportional to the square of their separation r.

Note that Newton's law refers to *point masses*, which means spherically symmetrical objects of uniform density whose gravitational effect is as if all their mass is concentrated at their centre (Figure 4).

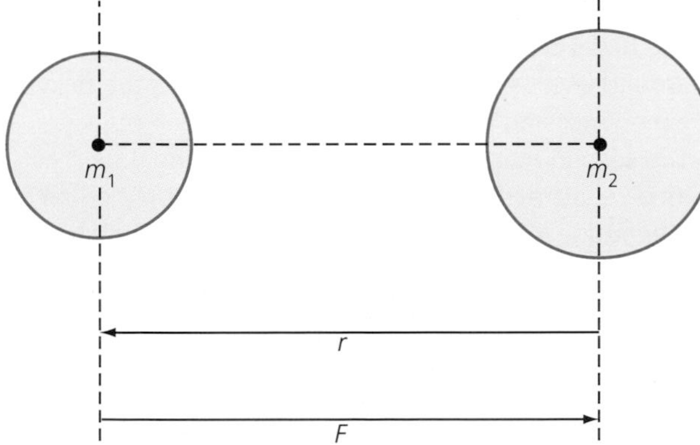

Figure 4 *Representation of m_1, m_2, r and F*

Newton's law is summarised by the equation:

$$F = \frac{Gm_1m_2}{r^2}$$

where G is a constant of proportionality, now known as the **gravitational constant**, which has the value $6.67 \times 10^{-11} \, \text{N m}^2 \, \text{kg}^{-2}$.

Newton presented his law of gravity along with his laws of motion in his book *Philosophiae Naturalis Principia Mathematica* first published in 1687.

The value of the gravitational constant G was determined as a consequence of an experiment to measure the density of the Earth undertaken by the British scientist Henry Cavendish. He constructed an arrangement consisting of two small lead spheres stuck to the ends of a six-foot-long wooden rod that was suspended from a fibre and free to rotate (Figure 5). He then brought a second dumbbell consisting of two much larger lead spheres and positioned it so that the gravitational attraction between the large and small spheres caused the suspended rod to oscillate. From his measurements of the oscillation, Cavendish determined the gravitational force between the large and small spheres. Knowing the gravitational force between the Earth and the spheres, he calculated the average density of the Earth, obtaining a value to within 1% of the currently accepted value. The value for the gravitational constant G was determined from Cavendish's experimental data.

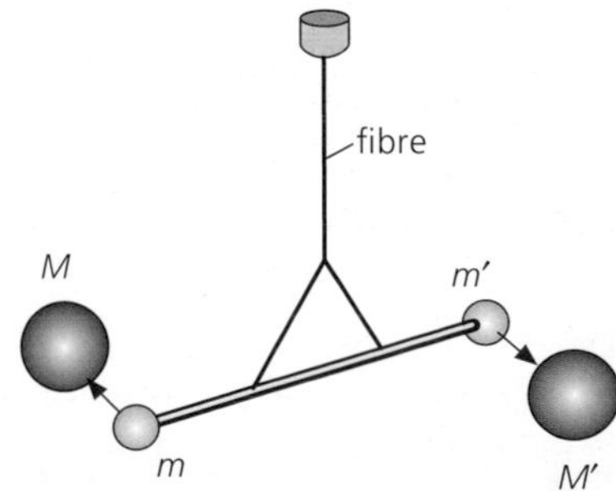

Figure 5 *Cavendish's experiment to measure the density of the Earth*

Worked example

Calculate the gravitational force between the Earth and a GPS satellite with a mass of 1600 kg orbiting at an altitude of 20 200 km. [Earth mass $= 5.97 \times 10^{24}$ kg ; Earth radius $= 6.37 \times 10^{6}$ m; gravitational constant $= 6.67 \times 10^{-11}$ N m^2 kg^{-2}]

Distance from centre of Earth to GPS satellite is

$$6.37 \times 10^{6} + 20200 \times 10^{3} = 2.657 \times 10^{7}\,\text{m}$$

Gravitational force $F = \dfrac{Gm_1m_2}{r^2}$

$$= \frac{6.67 \times 10^{-11} \times 5.97 \times 10^{24} \times 1600}{(2.657 \times 10^{7})^2}$$

$$= 902\,\text{N}$$

QUESTIONS

1. Calculate the gravitational force between the dwarf planet Pluto and its largest moon Charon (Figure 6). [Pluto's mass $= 1.31 \times 10^{22}$ kg; Charon's mass $= 1.52 \times 10^{21}$ kg; distance between Pluto and Charon = 19 640 km; gravitational constant $= 6.67 \times 10^{-11}$ N m^2 kg^{-2}]

Figure 6 Pluto and Charon, imaged by the New Horizons spacecraft during its fly-by in July 2015

2. NASA's *SMAP* satellite (see Figure 19) is in low Earth orbit, monitoring the water in the top layer of the Earth's soil. Calculate the gravitational force between the Earth and the *SMAP* satellite, which has a mass of 1123 kg and orbits at a height of 685 km. [Earth mass $= 5.97 \times 10^{24}$ kg; Earth radius $= 6.37 \times 10^{6}$ m , gravitational constant $= 6.67 \times 10^{-11}$ N m^2 kg^{-2}]
3. Estimate the gravitational force between you and a person near to you.

KEY IDEAS

- Newton's law of universal gravitation states that the gravitational force of attraction, F, between two point masses, m_1 and m_2, whose centres of mass are separated by distance r, is given by

$$F = \frac{Gm_1m_2}{r^2}$$

where G is the gravitational constant, 6.67 N m^2 kg^{-2}.

- A 'point mass' means a uniform spherical object whose gravitational effect acts as if all its mass is concentrated at its centre.

4.2 GRAVITATIONAL FIELDS

The concept of a field

Physicists introduced the concept of a **field** of force to describe non-contact interactions, such as an object falling towards the Earth, one magnet attracting or repelling another magnet, and electrically charged materials attracting or repelling each other. A 'source', for example, a mass, a magnet, or a static or moving charge, creates a field in space. Other objects experience a force dependent on the strength of that field at their own location. The field strength, whether gravitational, magnetic or electric, is represented by a vector, which has, at any point, both a magnitude and a direction.

Gravitational field strength

Gravity is a non-contact force, so it is helpful to use the concept of a force field to describe its effects. The mass of an object creates a **gravitational field** around itself. Any other mass positioned in the field experiences an attractive force towards the object. The second mass also creates a gravitational field of its own, attracting the first object with a force equal in magnitude (Gm_1m_2/r^2). For example, a person exerts an attractive gravitational force on the Earth equal in size to the gravitational force the Earth exerts on the person.

The **gravitational field strength**, g, at a point is defined as the gravitational force per unit mass at that point. So, if a small mass m experiences a force F when placed at a point in a gravitational field of strength g, the gravitational field strength at that point is given by

$$g = \frac{F}{m}$$

The unit of gravitational field strength is $N\,kg^{-1}$.

Gravitational field strength is a vector quantity, and the field can be represented by lines with arrows showing the direction of the force on a mass placed in the field. If the Earth's field is considered near to its surface over a small area, the field lines can be drawn almost parallel to each other, since the force on an object acts vertically downwards (Figure 7a). On this scale, the Earth's field is approximately uniform. However, if planet Earth is viewed as a whole, then the field lines are drawn inwards towards the centre of the Earth and the field is described as a **radial field** (Figure 7b). An increase in the separation of the field lines represents a reduction in the strength of the field.

Newton's second law of motion, showing the resultant force F causing a mass m to have acceleration a, can be represented by the equation $F = ma$, which can be rearranged to give $a = \frac{F}{m}$. A comparison between $a = \frac{F}{m}$ and $g = \frac{F}{m}$ shows that the gravitational field strength is equivalent to the acceleration due to gravity. For example, at the Earth's surface, the acceleration due to gravity is $9.81\,m\,s^{-2}$ and the gravitational field strength is $9.81\,N\,kg^{-1}$. Both quantities are given the symbol g. (It can be seen that the units are equivalent if you remember than one newton, N, is defined as the force that produces an acceleration of $1\,m\,s^{-2}$ in a mass of 1 kg.)

The magnitude of g in a radial field

An equation for the magnitude of the gravitational field strength in a radial field can be found using Newton's law of gravity. Consider a small mass m in the gravitational field of a planet of mass M. If the distance from the centre of the planet to the small mass is r, then the gravitational force between the planet and the small mass is given by

$$F = \frac{GMm}{r^2}$$

The magnitude of the gravitational field strength g of the planet at the position of the small mass is equal to $\frac{F}{m}$ and therefore

$$g = \frac{F}{m} = \frac{GM}{r^2}$$

This shows that the Earth's gravitational field strength varies inversely with the square of the distance from its centre. A sketch graph of the Earth's gravitational field strength versus distance in units of R, where R represents the Earth's radius, is shown in Figure 8.

QUESTIONS

4. Determine the magnitude of the Earth's gravitational field strength at the height above the Earth's surface corresponding to the orbit of NASA's *SMAP* satellite (see question **2** for the required data).
5. At what height above the Earth's surface isthe gravitational field strength half of its value at the surface? [Earth mass $= 5.97 \times 10^{24}$ kg; Earth radius $= 6.37 \times 10^{6}$ m; gravitational constant $= 6.67 \times 10^{-11}\,N\,m^2\,kg^{-2}$]

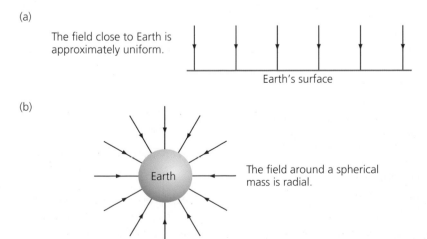

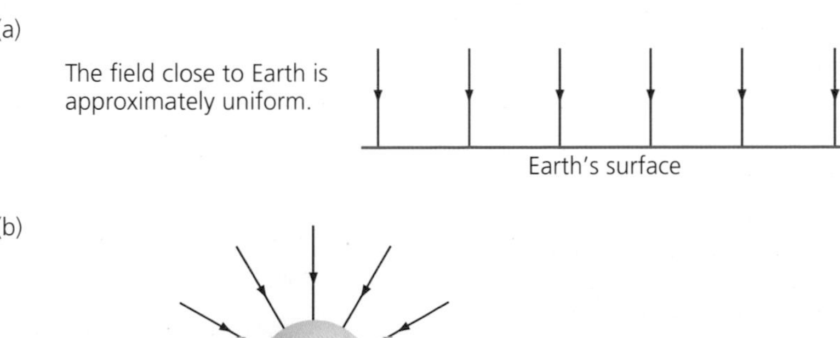

The field around a spherical mass is radial.

Figure 7 *The Earth's gravitational field lines*

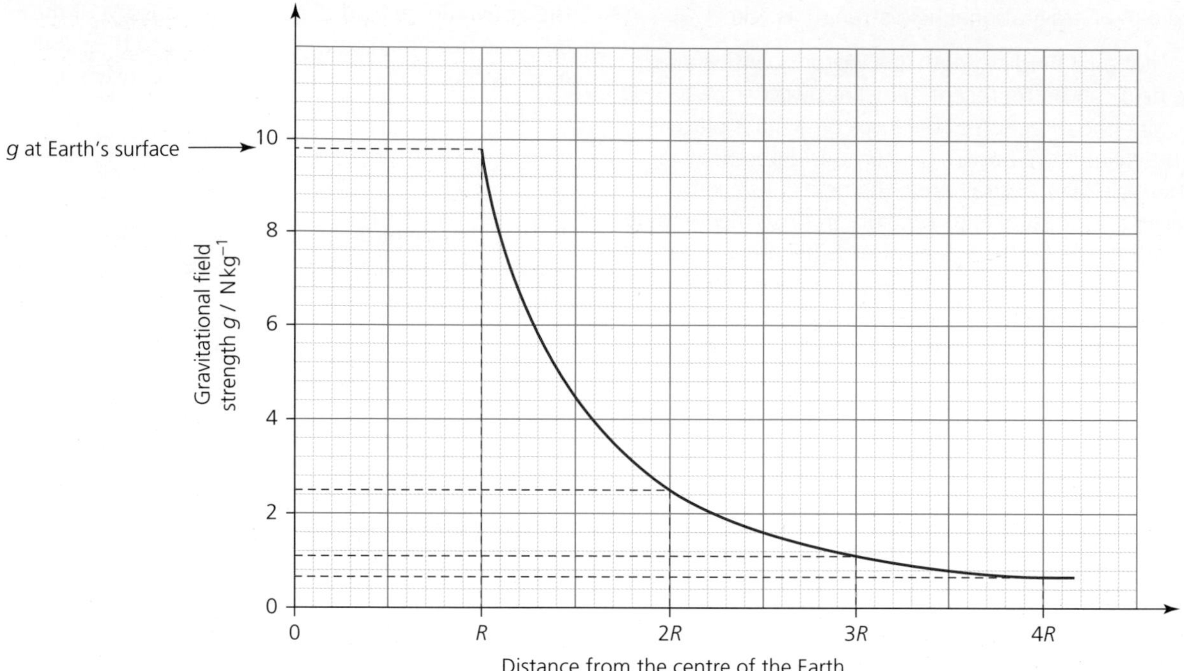

Figure 8 *Variation in magnitude of the gravitational field strength of the Earth with distance from the Earth's centre: an inverse square relationship*

Between the Earth and the Moon is a neutral point (Figure 9), where the gravitational field strength of the Moon is equal, but in the opposite direction, to the Earth's gravitational field strength. The greater mass of the Earth means that the neutral point is much closer to the Moon than to the Earth. Once a spacecraft passes through the neutral point on its way to the Moon, it can continue its journey without having to expend any additional fuel to provide thrust from its engines.

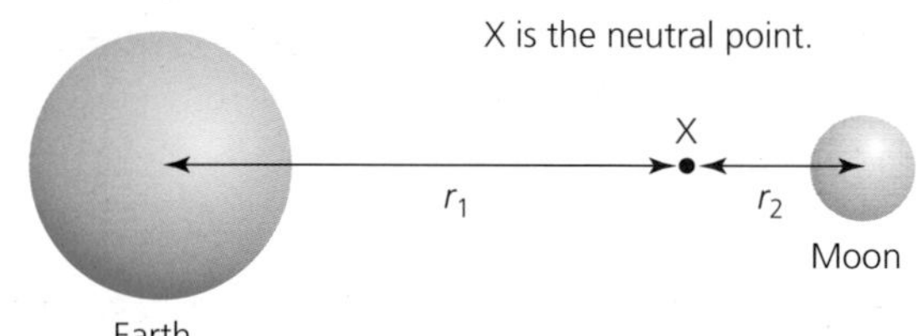

Figure 9 *The neutral point between the Earth and the Moon*

At the neutral point, the magnitudes of the Earth's and the Moon's gravitational field strengths are equal:

$$\frac{GM_E}{r_1^2} = \frac{GM_M}{r_2^2}$$

which rearranges to give:

$$\frac{r_1^2}{r_2^2} = \frac{M_E}{M_M}$$

and therefore

$$\frac{r_1}{r_2} = \sqrt{\frac{M_E}{M_M}}$$

Since $r_1 + r_2$ is the distance from the centre of the Earth to the centre of the Moon, the values of r_1 and r_2 can be determined.

Worked example

The distance from the centre of the Earth to the centre of the Moon is 3.84×10^8 m. Determine the distance of the neutral point from the Earth given that the mass of the Earth is 5.97×10^{24} kg and the mass of the Moon is 7.35×10^{22} kg.

$$\frac{r_1}{r_2} = \sqrt{\frac{M_E}{M_M}} = \sqrt{\frac{5.97 \times 10^{24}}{7.35 \times 10^{22}}}$$

which gives $r_1 = 9.01 r_2$. Since $r_1 + r_2 = 3.84 \times 10^8$, we can write:

$$9.01 r_2 + r_2 = 3.84 \times 10^8$$

$$10.01 r_2 = 3.84 \times 10^8$$

which gives $r_2 = 3.84 \times 10^7$ m. Distance of neutral point from (centre of) Earth is thus

$$r_1 = 3.84 \times 10^8 - 3.84 \times 10^7 = 3.46 \times 10^8 \text{ m}$$

QUESTIONS

6. The average distance from the centre of the Earth to the centre of the Sun is 1.4960×10^{11} m. A neutral point exists at the point where the gravitational fields of the Sun and the Earth have the same magnitude. Determine the distance of the neutral point from Earth, given that the mass of the Sun is 1.989×10^{30} kg and the mass of the Earth is 5.972×10^{24} kg.

KEY IDEAS

- A field is a region in which an object experiences a non-contact force.
- Gravitational field strength g at a point is defined as the gravitational force per unit mass at that point, in $\mathrm{N\,kg^{-1}}$. This is equivalent to the acceleration due to gravity, in $\mathrm{m\,s^{-2}}$:

$$g = \frac{F}{m}$$

- Gravitational field strength is a vector quantity and can be represented by field lines labelled with arrows to show the field's direction. The closer together the field lines, the stronger the field.
- For a radial field due to a point mass M, the magnitude of the gravitational field strength g at a distance r is given by

$$g = \frac{GM}{r^2}$$

- Close to the Earth's surface over a small area, the gravitational field is uniform.

4.3 GRAVITATIONAL POTENTIAL

The rocket carrying the *New Horizons* spacecraft that was to fly by Pluto in July 2015 (see Figure 6) was launched from Earth in January 2006. A crucial calculation that NASA scientists had to make when planning the launch was how much fuel would be required by the rocket (Figure 10) to take the space probe to the required height above the Earth's surface to begin its trajectory to Pluto. To answer this type of question, scientists need to work out how great an increase in gravitational potential energy would be required.

Figure 10 *The Lockheed Martin Atlas V launch rocket carrying the New Horizons spacecraft at Cape Canaveral Air Force Station, Florida, January 2006*

Gravitational potential energy is the energy an object has because of its position in a gravitational field (*see section 11.5 in Chapter 11 in Year 1 Student Book*).

The equation for the change in gravitational potential energy ΔE_p of a mass m experiencing a change in height Δh should be familiar: $\Delta E_p = mg\Delta h$. However, this equation only applies to situations where the gravitational field strength g is constant, for example close to the Earth's surface. For situations involving moving a mass through a distance over which the gravitational field strength varies, a different approach is required.

It can be shown using calculus that the work done W (energy transferred) in moving a mass m from a point at distance r from the Earth's centre to a point beyond the influence of the Earth's gravitational field is given by

$$W = \frac{GMm}{r}$$

where M is the mass of the Earth. The work done in moving *unit mass* is then equal to $\frac{GM}{r}$.

We define the **gravitational potential** V at a point in a gravitational field as follows:

The gravitational potential V at a point is the work done (or gain in potential energy) per unit mass to move a mass from infinity to that point. The unit is J kg^{-1}.

(Note that 'a point beyond the influence of the Earth's gravitational field' is simply described as infinity.)

This definition of V refers to a mass being brought *into* the field rather than being removed from the field, so it is necessary to introduce a minus sign, which gives

$$V = -\frac{GM}{r}$$

A graph of gravitational potential versus distance from the Earth (Figure 11) shows that the potential approaches zero at infinity and becomes increasingly negative as the distance from the Earth is decreased.

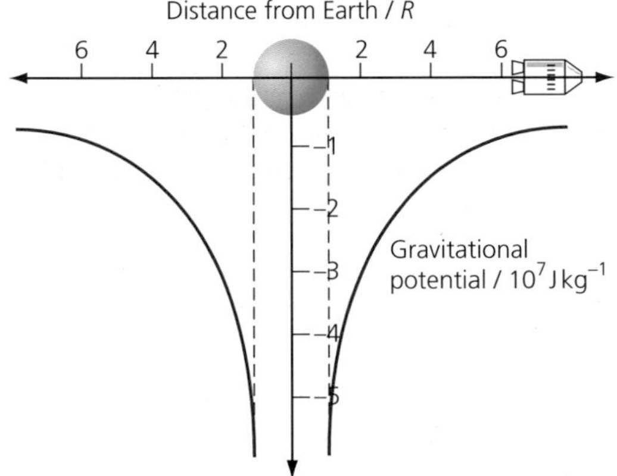

Figure 11 *Gravitational potential versus distance from the Earth*

The equation for V (or its graph, Figure 11) can be used to determine values for the gravitational potential at the Earth's surface and at a specific height above the Earth's surface. The difference between these two values is the **gravitational potential difference** ΔV. The work done ΔW (energy needed) to move a mass m from the Earth's surface to the height specified can then be found from

$$\Delta W = m\Delta V$$

Worked example

a. Determine the gravitational potential at **(i)** the Earth's surface and **(ii)** a height of 650 km above the Earth's surface.

b. To launch a satellite into orbit requires a launch vehicle consisting of several rocket stages, which are discarded one by one as the vehicle gains altitude. Estimate the minimum energy needed to transfer an average mass of 50 000 kg from the Earth's surface to a height of 650 km. [Earth's mass is ; 5.97×10^{24} kg; Earth's radius is 6.37×10^6 m ; gravitational constant is $6.67 \times 10^{-11}\text{ N m}^2\text{ kg}^{-2}$]

a. i. Gravitational potential at the Earth's surface:

$$V = -\frac{GM}{R} = -\frac{6.67 \times 10^{-11} \times 5.97 \times 10^{24}}{6.37 \times 10^6}$$
$$= -6.25 \times 10^7\text{ J kg}^{-1}$$

ii. Gravitational potential at a height of 650 km above the Earth's surface:

$$V = -\frac{GM}{R+h} = -\frac{6.67 \times 10^{-11} \times 5.97 \times 10^{24}}{6.37 \times 10^6 + 650 \times 10^3}$$
$$= -5.67 \times 10^7\text{ J kg}^{-1}$$

b. To determine the energy needed to transfer an average mass of 50 000 kg from the Earth's surface to a height of 650 km:

$$\text{work done } \Delta W = m\Delta V$$
$$= 50000 \times (6.25 \times 10^7 - 5.67 \times 10^7)$$
$$= 2.9 \times 10^{11}\text{ J}$$

QUESTIONS

7. A small space probe of mass 510 kg has been landed on the surface of a small asteroid. Determine the output energy of the probe's onboard rockets for the probe to escape the gravitational field of the asteroid, given that the gravitational potential at the surface of the asteroid is -40 kJ kg^{-1}.

8. The gravitational potential at the surface of a planet of radius r is V. What would be the gravitational potential at a height of r above the planet's surface?

A $\frac{V}{2}$ **B** V **C** $2V$ **D** $4V$

Equipotentials

The equation $V = -\frac{GM}{r}$ shows that, at a specific distance from the Earth, the gravitational potential has the same value. This means that surfaces of equal potential can be used to represent a gravitational field, as well as field lines. In a two-dimensional diagram (Figure 12), an **equipotential surface** around a sphere is represented by a circle. The difference in potential between adjacent circles is equal, so the increasing separation of the circles with increasing distance from the Earth represents the gravitational field becoming weaker. Work would have to be done to move a satellite from an inner equipotential surface to an outer equipotential surface, but no work is done as the satellite moves in its orbit along an equipotential surface.

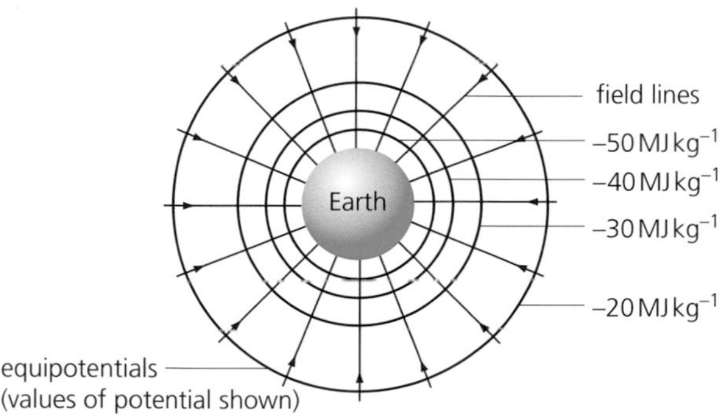

Figure 12 *Field lines and equipotential surfaces around the Earth*

QUESTIONS

9. A satellite of mass 800 kg needs to change its orbit using its own onboard propulsion system. The satellite is currently in an orbit of gravitational potential −34.5 MJ kg^{-1}. Determine the output energy of the satellite's propulsion system needed to move the satellite to an orbit of gravitational potential −33.9 MJ kg^{-1}.

10. Close to the Earth's surface, the gravitational field strength can be considered uniform. Describe the equipotential surfaces and sketch them in two dimensions.

Unsurprisingly, the two quantities, gravitational potential V and gravitational field strength g, are mathematically related. Consider a small mass m experiencing a gravitational force F towards the Earth. The mass is then moved away from the Earth by a very small distance Δr. Over such a small distance, the gravitational force on the mass would hardly have changed at all, so the force that was needed to move the mass away from the Earth would be equal to $-F$.

The work done in moving the mass is $\Delta W = -F \times \Delta r$, which must also equal the gain in potential energy of the mass, $\Delta W = m\Delta V$. Therefore

$$-F\Delta r = m\Delta V$$

which rearranges to give

$$-\frac{F}{m} = \frac{\Delta V}{\Delta r}$$

But $\frac{F}{m}$ is equal to the gravitational field strength g, so it can be concluded that

$$g = -\frac{\Delta V}{\Delta r}$$

$\frac{\Delta V}{\Delta r}$ represents the gradient of the graph of gravitational potential versus distance (Figure 13). Therefore the relationship between V and g can be summarised as:

> Gravitational field strength is equal to '–gradient' of the graph of gravitational potential versus distance.

This gradient is sometimes referred to simply as the 'potential gradient'.

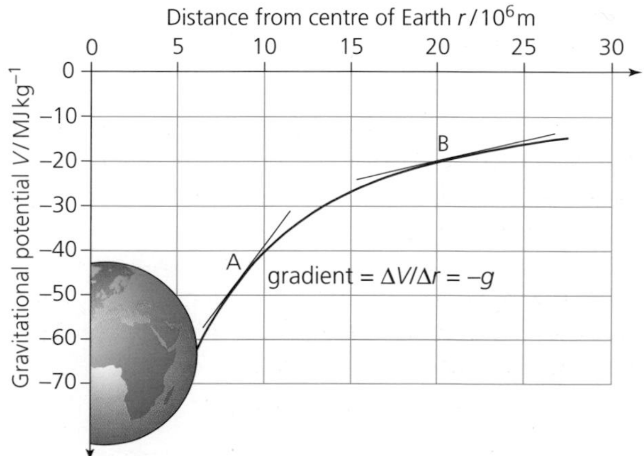

Figure 13 *Gravitational potential versus distance in the Earth's gravitational field*

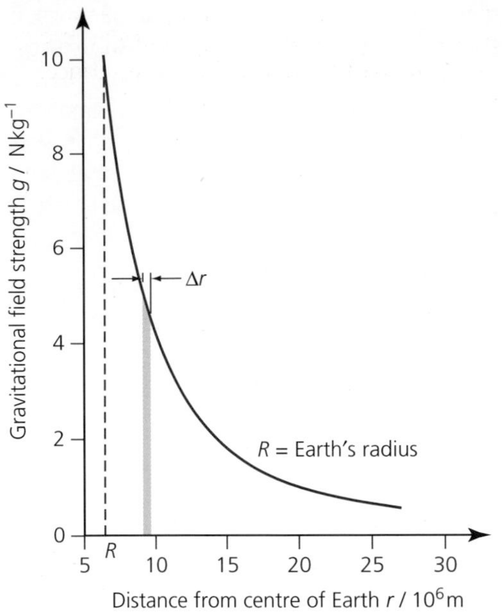

Figure 14 *Gravitational field strength versus distance in the Earth's gravitational field*

There is an alternative way of describing the relationship between g and V. Again, consider a small mass m experiencing a gravitational force F towards the Earth. The mass is then moved away from the Earth by a very short distance Δr and the work done is $\Delta W = -F \times \Delta r$. Now suppose the mass is 1 kg. Force F is now equal to g and the work done is $g \times \Delta r$, which corresponds to the blue shaded area of the graph of g versus distance in Figure 14. The total area between the curve of the graph and the distance axis must then correspond to the total work done in moving 1 kg through the distance indicated by the curve. But the total work done in moving 1 kg between two points in a gravitational field is also equal to the potential difference ΔV. Therefore, the relationship between V and g can also be summarised as follows:

Potential difference ΔV is equal to the area between the curve and the distance axis of a gravitational field strength versus distance graph.

QUESTIONS

11. What would be the gravitational force on a 900 kg satellite orbiting at a height in the Earth's gravitational field where the potential gradient is 7.9 J m^{-1}?
12. Figure 15 is a graph of gravitational field strength versus distance (in Earth radii) from the Earth's centre. Use the graph to estimate the work done in moving a mass of 1 kg from the Earth's surface to a height of three Earth radii above the Earth's surface. [Earth's radius $R = 6.37 \times 10^6$ m]

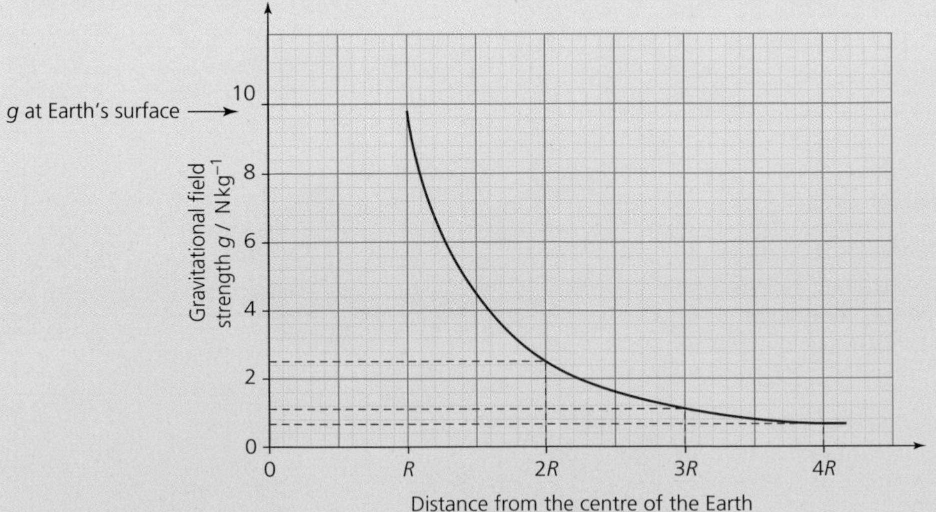

Figure 15 *The g–r graph for the Earth*

13. Figure 16 shows a graph of the Earth's gravitational potential versus distance from the Earth's centre in units of the Earth radius, R. Use the graph to estimate the Earth's gravitational field strength at a height of 1.5 times the Earth's radius above the surface. Earth's radius $R = 6.37 \times 10^6$ m

Figure 16 *The V–r graph for the Earth*

Stretch and challenge

14. Consider a 1 kg mass at a distance r from the centre of the Earth, of mass M. In Figure 14, the shaded area of the graph corresponds to the work done in moving a 1 kg mass through a distance Δr and is equal to $g\Delta r$. The total area between the curve of the graph and the distance axis from r to infinity corresponds to the total work done in moving 1 kg from r to infinity. Show, by integration of the expression for gravitational field strength from r to ∞, that the total work done moving 1 kg from r to infinity is equal to $\frac{GM}{r}$.

KEY IDEAS

- The gravitational potential V at a point in a gravitational field is defined as the work done (or gain in potential energy) per unit mass to move a mass from infinity to that point.
- The gravitational potential at a point in a radial field is given by $V = -\frac{GM}{r}$.
- A gravitational field can be represented by a series of equipotential surfaces.
- No work is done when a mass moves along an equipotential surface.
- The work done W in moving mass m through a potential difference ΔV is given by

$$\Delta W = m\Delta V$$

- Gravitational field strength is related to potential gradient by

$$g = -\frac{\Delta V}{\Delta r}$$

- The area under a graph of g versus r is equal to the potential difference ΔV.

4.4 ORBITS OF PLANETS, MOONS AND SATELLITES

The planets of the solar system, with the exception of Mercury and Venus, are orbited, in total, by more than 180 moons. The Earth has only one moon, Mars has two, Jupiter and Saturn both have in excess of 60 moons with known orbits (Figure 17), Uranus has 27 and Neptune has 13.

Figure 17 *An image of Jupiter and its moon Io obtained by the New Horizons spacecraft that flew by in February 2007*

If the radius and time of the orbit, T, of a moon are known, then the mass of its planet can be determined. Using Newton's law of gravity,

$$F = \frac{Gm_1m_2}{r^2}$$

where F is the force of gravity between the moon and its planet, m_1 is the mass of the planet, m_2 is the mass of the moon and r is the radius of the moon's orbit. This can be equated to the centripetal force

$$F = m_2r\omega^2$$

(see section 1.2 of Chapter 1) keeping the moon in its orbit, where ω is its angular speed which is related to the time of orbit by $\omega = \frac{2\pi}{T}$. Thus

$$\frac{Gm_1m_2}{r^2} = m_2r\omega^2$$

which after cancelling m_2 on both sides and rearranging gives an equation for the mass m_1 of the planet:

$$m_1 = \frac{r^3\omega^2}{G}$$

Worked example 1

The radius and time of the orbit for Io, one of Jupiter's moons, are 4.22×10^8 m and 42 h, respectively. Determine the mass of Jupiter. [Gravitational constant = 6.67×10^{-11} N m^2 kg^{-2}]

The angular speed of Io is

$$\omega = \frac{2\pi}{T} = \frac{2\pi}{42 \times 3600} = 4.16 \times 10^{-5}\ \text{rad s}^{-1}$$

The mass of Jupiter is thus

$$m_1 = \frac{r^3\omega^2}{G} = \frac{(4.22 \times 10^8)^3 \times (4.16 \times 10^{-5})^2}{6.67 \times 10^{-11}}$$
$$= 1.9 \times 10^{27}\ \text{kg}$$

QUESTIONS

15. Determine the mass of the Sun, given that the time of orbit of the Earth is 365 days and the radius of the Earth's orbit is 1.5×10^{11} m.

16. It is estimated that the solar system completes an orbit around the Galactic Centre of the Milky Way in 240 million years, with a radius of orbit of 27 000 light years. Determine the effective mass within the orbit that has its centre of mass at the Galactic Centre. [1 light year = 9.46×10^{15} m]

During the early 17th century, Johannes Kepler, a German mathematician, worked as an assistant to Tycho Brahe, the Danish astronomer, who studied the solar system for many years, making observations of the orbital time periods and orbital radii of the planets. At the time, the scale of the solar system was not known, so the orbital radii of the planets were expressed in terms of their size compared with the Earth's radius of orbit, referred to as an astronomical unit (AU).

Kepler's analysis of Brahe's solar system data revealed a relationship between the time period T of orbit and the radius of orbit r of a planet. He established that, for planets in the solar system,

$$T^2 \propto r^3$$

Newton later used this relationship to provide evidence to support his own law of gravity. Newton showed that his law of gravity predicted the $T^2 \propto r^3$ relationship for the planets of the solar system, as observed.

Newton equated his equation for gravitational force to the equation for centripetal force:

$$\frac{Gm_1m_2}{r^2} = m_2r\omega^2$$

where m_1 represents the mass of the Sun, m_2 is the mass of a planet, r is the radius of orbit of the planet and ω is the angular speed of the planet in its orbit. Substituting $\omega = \frac{2\pi}{T}$, where T is the orbital time period, into the equation gives

$$\frac{Gm_1}{r^2} = r\left(\frac{2\pi}{T}\right)^2$$

which rearranges to give

$$T^2 = \frac{4\pi^2}{Gm_1}r^3$$

Since the mass m_1 of the Sun can be assumed to be constant, the equation shows that $T^2 \propto r^3$.

ASSIGNMENT 2: STUDYING THE MOONS OF JUPITER

(MS 0.2, MS 0.5, MS 3.1, MS 3.2, MS 3.3, MS 3.4, MS 3.11)

The aim of this assignment is to determine to what extent the orbital time periods and orbital radii of some of the larger moons of Jupiter (Table A1) follow the relationship $T^2 \propto r^3$ derived from Newton's law of gravitation.

The relation $T^2 \propto r^3$ can be expressed as $T = kr^{1.5}$, where k is a constant of proportionality. This power law can be analysed using logarithms. Taking logarithms of both sides of the equation and applying logarithmic rules (see Assignment 1 of Chapter 2) gives

$$\log T = 1.5\log r + \log k$$

Comparing this equation with $y = mx + c$ suggests that a graph of $\log T$ versus $\log r$ should be a straight line with a gradient of 1.5.

Use the data in Table A1 to answer the questions that follow.

Name of moon	Radius of orbit / m	Orbital time period / Earth days
Himalia	11.5×10^9	250
Leda	11.2×10^9	241
Themisto	7.39×10^9	130
Callisto	1.88×10^9	16.7
Ganymede	1.07×10^9	7.15
Europa	6.71×10^8	3.55
Io	4.22×10^8	1.77

Table A1 *Data for some of the larger moons of Jupiter*

Questions

A1 Plot a graph of $\log T$ on the y-axis versus $\log r$ on the x-axis, using data from Table A1. The graph does not require an origin, so choose a suitable scale so that the points take up over half of the graph paper. Draw a best-fitting line through the points.

A2 Determine the gradient of the graph using as large a gradient triangle as possible. Determine the percentage difference between your gradient and the gradient predicted by Newton's law.

Satellites

There are over 1200 operational satellites in orbit around the Earth, and many more that no longer work. Communication satellites orbit at an altitude of over 35 000 km and relay live television, telephone and radio transmissions all over the Earth. Global positioning system (GPS) satellites essential for navigation orbit at an altitude of over 20 000 km in a medium Earth orbit (MEO) circling the Earth twice a day. Other satellites monitor weather systems and environmental changes, and some are space satellites used to study astronomical objects and phenomena. These tend to have low Earth orbits (LEO), with altitudes between 160 and 2000 km. The *International Space Station* orbits at an average altitude of 400 km.

According to Newton's law, the force of gravity F exerted by a planet on a satellite is

$$F = \frac{GMm}{r^2}$$

where M is the mass of the planet, m is the satellite's mass and r is the distance from the planet's centre to the satellite.

For a satellite orbiting at speed v around a planet, the required centripetal force is provided by the planet's gravitational field, so

$$\frac{mv^2}{r} = \frac{GMm}{r^2}$$

which rearranges to give the orbital speed v as

$$v = \sqrt{\frac{GM}{r}}$$

This shows that satellites in lower orbits require a higher orbital speed.

QUESTIONS

17. Calculate the orbital speed of the *Hubble* space telescope given that it orbits at a height of 547 km above the Earth's surface. [Earth's mass is 5.97×10^{24} kg; Earth's radius is 6.37×10^6 m; gravitational constant is 6.67×10^{-11} N m^2 kg^{-2}]

A satellite following a **synchronous orbit** has a time period equal to the rotational period of the planet being orbited. A satellite following a synchronous orbit around the Earth is described as having a geosynchronous orbit. A communication satellite requires a special type of geosynchronous orbit. The satellite needs always to be above the same position on the Earth, so that the receiving dishes can have a fixed position pointing to the same spot in the sky and can maintain continuous contact with the satellite. To achieve this, the communication satellite, in addition to having a 24 hour orbital time period, must orbit in the same direction as the Earth rotates and orbit in the equatorial plane. This type of orbit is called a **geostationary orbit** (GEO). Although each satellite communicates with a restricted area of the Earth's surface, this area is still considerable (Figure 18). To achieve global communications coverage requires a system of several geostationary satellites.

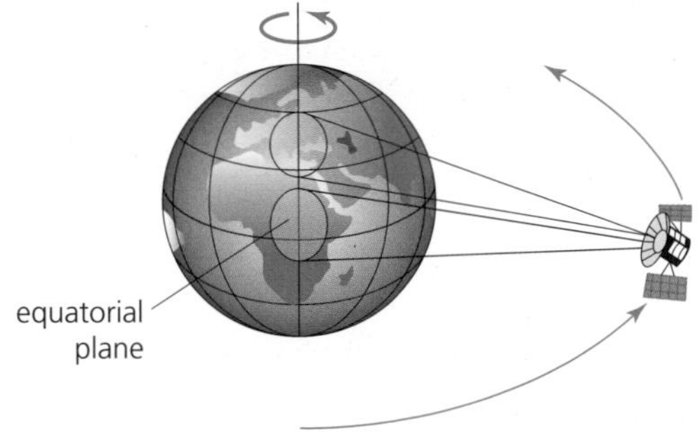

Figure 18 *Geostationary satellite use for communications*

Worked example 2

Determine the radius of orbit and altitude of a satellite in a geostationary orbit. [Earth's mass is 5.97×10^{24} kg; Earth's radius is 6.37×10^6 m; gravitational constant: 6.67×10^{-11} N m^2 kg^{-2}]

Equating Newton's law to the equation for the required centripetal force for the satellite's orbit:

$$\frac{GMm}{r^2} = mr\omega^2$$

where M is the mass of the Earth, m is the satellite's mass, r is the distance from the Earth's centre to the satellite and ω is the satellite's angular speed. Rearranging gives

$$r^3 = \frac{GM}{\omega^2}$$

Since a geostationary satellite has a time period T of 24 h (24 × 60 × 60 s), the angular speed is given by

$$\omega = \frac{2\pi}{T} = \frac{2\pi}{24 \times 3600} = 7.272 \times 10^{-5} \text{rad s}^{-1}$$

Substituting into the previous equation gives

$$r^3 = \frac{GM}{\omega^2} = \frac{6.67 \times 10^{-11} \times 5.97 \times 10^{24}}{(7.272 \times 10^{-5})^2}$$
$$= 7.530 \times 10^{22}\,\text{m}^3$$

which gives the radius of the orbit as

$$r = 4.22 \times 10^7\,\text{m}$$

and the altitude as

$$h = 4.22 \times 10^7 - 6.37 \times 10^6 = 3.58 \times 10^7\,\text{m}$$

Figure 19 NASA's SMAP satellite, launched in January 2015, is in a polar orbit, covering the globe every few days. It measures soil moisture in the top 5 cm of the Earth's surface, allowing changes over both short time scales, due to weather conditions, and longer time scales, due to seasonal changes, to be observed. SMAP's measurements enable drought conditions to be monitored and floods to be predicted, assisting with attempts to improve crop productivity.

A satellite that is put into a low Earth **polar orbit** passes above both the North and South Poles of the Earth during each complete orbit. The plane of a polar orbit is at 90° to the equatorial plane. The satellite is much closer to the Earth than a geostationary satellite, travels at a much greater speed and makes several complete orbits in one day. As the Earth rotates beneath it, the satellite can scan the whole of the Earth's surface in just a few days, making it suitable for wide-ranging purposes:

- mapping land features (Figure 19)
- monitoring the extent of polar ice caps
- monitoring ocean currents
- tracking cloud coverage
- observing short-term environmental changes, such as drifting oil spills or air pollution (Figure 20)
- military surveillance.

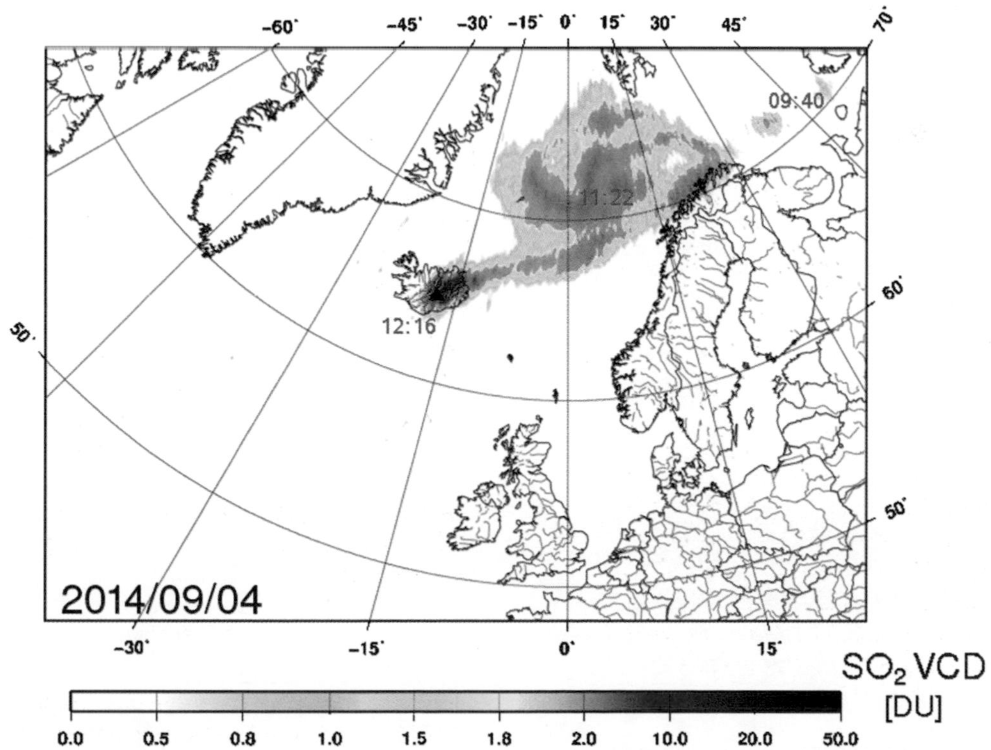

Figure 20 A map of the sulfur dioxide emissions from Iceland's Bardarbunga volcano as detected by the European Space Agency's MetOp-B weather satellite in September 2014

QUESTIONS

18. Determine the radius of orbit and altitude of a GPS satellite, which orbits the Earth twice a day. [Earth's mass is 5.97×10^{24} kg; Earth's radius is 6.37×10^{6} m; gravitational constant is 6.67×10^{-11} N m^2 kg^{-2}]

19. The European Space Agency weather satellite *MetOp-B* has a polar orbit at an altitude of 810 km.

Determine

a. the time period of the orbit

b. the number of orbits it completes in one day

c. the satellite's orbital speed.

[Earth's mass = 5.97×10^{24} kg; Earth's radius = 6.37×10^{6} m; gravitational constant = 6.67×10^{-11} N m^2 kg^{-2}]

ASSIGNMENT 3: RESEARCHING A SATELLITE MISSION

Your task in this assignment is to research and present information about a satellite mission of your choice that is currently in operation. Present the mission details under the following headings.

Observations

Include details of the type of observations and measurements that the satellite makes.

Why it matters

Describe the benefits that are derived from the satellite's observations and measurements.

Satellite orbit

Include the type of orbit and its altitude.

Timeline

Give details of the launch date, the launch location and the satellite's predicted operational lifetime.

The energy of an orbiting satellite

The total energy of a satellite is the sum of its gravitational potential energy and its kinetic energy. The gravitational potential energy E_p of a satellite moving in an orbit of radius r around the Earth is equal to the gravitational potential at distance r from the Earth's centre multiplied by the satellite's mass, m:

$$E_p = mV = (m)\left(-\frac{GM}{r}\right) = -\frac{GMm}{r}$$

where M is the mass of the Earth.

The satellite's kinetic energy is $E_k = \frac{1}{2}mv^2$, where v is the satellite's orbital speed. Equating the centripetal force equation to Newton's law gives:

$$\frac{mv^2}{r} = \frac{GMm}{r^2}$$

so

$$E_k = \frac{1}{2}mv^2 = \frac{GMm}{2r}$$

The satellite's total energy E_{total} is therefore

$$E_{total} = E_k + E_p = \frac{GMm}{2r} - \frac{GMm}{r} = -\frac{GMm}{2r}$$

Note that the expression of the satellite's total energy as a negative number occurs as a consequence of assigning the zero of gravitational potential at infinity.

The variation of a satellite's energy with its distance from the Earth can be shown graphically (Figure 21).

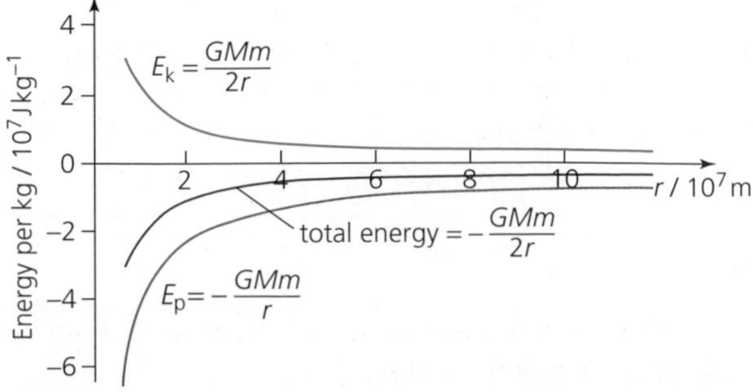

Figure 21 *A satellite's potential energy, kinetic energy and total orbital energy*

Satellites in a low Earth orbit (LEO) are affected by the Earth's atmosphere. Although the atmosphere is very thin at these heights, over time it can still create sufficient air resistance to reduce the satellite's speed, causing it to be pulled towards the Earth. Figure 21 shows that, as a consequence of moving towards the Earth, the satellite's total energy decreases but its kinetic energy actually increases, making the satellite speed up. The satellite spirals lower and faster into the atmosphere and gets hot due to friction caused by the air molecules. Intense heat is generated, causing the satellite to burn up.

QUESTIONS

20. Determine the kinetic energy of

a. the *SMAP* satellite (Figure 19), which has a mass of 1123 kg and orbits the Earth at a height of 685 km

b. the Earth in its orbit around the Sun.

[Earth's mass = 5.97×10^{24} kg; Earth's radius = 6.37×10^6 m; radius of Earth's orbit = 1.50×10^{11} m; Sun's mass = 1.99×10^{30} kg; gravitational constant = 6.67×10^{-11} N m^2 kg^{-2}]

21. Estimate the kinetic energy of a gas giant exoplanet orbiting a star similar to the Sun at an orbital distance about 10 times greater than the Earth's orbital radius. Use data from this chapter to make your estimates.

Escape velocity

An object's **escape velocity** is the speed at which it needs to be travelling away from an astronomical body in order to break free from the gravitational field, without the use of any further propulsion. The energy needed for a mass m to completely escape the gravitational field of a planet of mass M and radius R is equal to the work done, ΔW, in moving the mass from the planet's surface to a point where the planet's gravitational field no longer has any influence (infinity):

$$\Delta W = m\Delta V = m\frac{GM}{R}$$

If the energy needed to do this work is provided by the object's initial kinetic energy, then

$$\frac{1}{2}mv^2 = m\frac{GM}{R}$$

which rearranges to give escape velocity

$$v = \sqrt{\frac{2GM}{R}}$$

Worked example 3

Determine the escape velocity of an object at the Earth's surface. [Earth's mass = 5.97×10^{24} kg; Earth's radius = 6.37×10^6 m; gravitational constant = 6.67×10^{-11} N m^2 kg^{-2}]

Escape velocity is

$$v = \sqrt{\frac{2GM}{R}} = \sqrt{\frac{2 \times 6.67 \times 10^{-11} \times 5.97 \times 10^{24}}{6.37 \times 10^6}}$$

$$= 1.12 \times 10^4 \text{ m s}^{-1}$$

QUESTIONS

22. Determine the escape velocity at the surface of the Moon. [Moon's mass = 7.35×10^{22} kg; Moon's radius = 1.74×10^6 m; gravitational constant = 6.67×10^{-11} N m^2 kg^{-2}]

KEY IDEAS

- Kepler established from observational data that the relationship between the radius of orbit r and the time period of orbit T for planets in the solar system is

$$T^2 \propto r^3$$

This is predicted mathematically from Newton's law of gravitation, by equating gravitational force to centripetal force.

- The plane and height of orbit of a satellite is chosen based on the particular application of the satellite.
- A geostationary orbit is one in an equatorial plane that is synchronous with the rotation of the Earth.
- The total energy of a satellite is

$$E_{total} = E_p + E_k = -\frac{GMm}{r} + \frac{GMm}{2r} = -\frac{GMm}{2r}$$

where M is the mass of the Earth, m is the mass of the satellite and r is the distance from the centre of the Earth to the satellite.

- The escape velocity needed to escape the gravitational field at the surface of a mass M with radius R is

$$v = \sqrt{\frac{2GM}{R}}$$

PRACTICE QUESTIONS

1. a. i. Define *gravitational field strength* and state whether it is a scalar or vector quantity.

   ii. A mass m is at a height h above the surface of a planet of mass M and radius R. The gravitational field strength at height h is g. By considering the gravitational force acting on mass m, derive an equation from Newton's law of gravitation to express g in terms of M, R, h and the gravitational constant G.

   b. i. A satellite of mass 2520 kg is at a height of 1.39×10^7 m above the surface of the Earth. Calculate the gravitational force of the Earth attracting the satellite. Give your answer to an appropriate number of significant figures.

   ii. The satellite in part **b i** is in a circular polar orbit. Show that the satellite would travel around the Earth three times every 24 hours.

   c. State and explain one possible use for the satellite travelling in the orbit in part **b ii**.

AQA Unit 4 Section B June 2014 Q1

2. a. State Newton's law of gravitation.

   b. In 1798, Cavendish investigated Newton's law by measuring the gravitational force between two unequal uniform lead spheres. The radius of the larger sphere was 100 mm and that of the smaller sphere was 25 mm.

   i. The mass of the smaller sphere was 0.74 kg. Show that the mass of the larger sphere was about 47 kg. [Density of lead = 11.3×10^3 kg m^{-3}]

   ii. Calculate the gravitational force between the spheres when their surfaces were in contact.

   c. Modifications, such as increasing the size of each sphere to produce a greater force between them, were considered in order to improve the accuracy of Cavendish's experiment. Describe and explain the effect on the calculations in part **b** of doubling the radius of both spheres.

AQA Unit 4 Section B June 2010 Q1

3. Figure Q1 shows the orbits of two satellites, a communications satellite in a geosynchronous orbit and a monitoring satellite in a low orbit that passes over the poles.

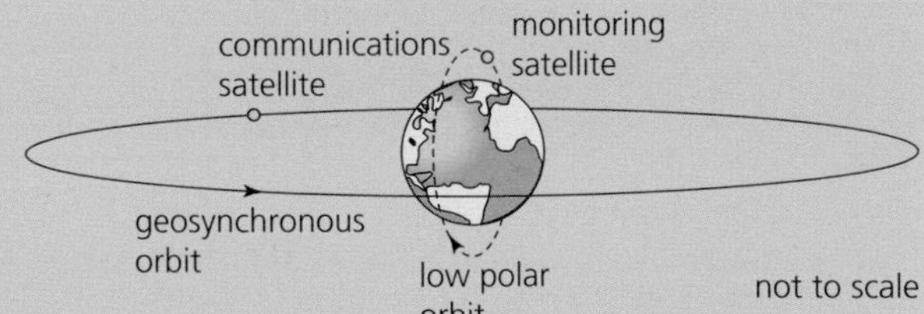

Figure Q1

   a. The time period, T, of any satellite in a circular orbit around a planet is proportional to $r^{3/2}$, where r is the radius of its orbit measured from the centre of the planet. For a satellite in a low orbit that passes over the poles of the Earth, T is 105 minutes when r is 7370 km.

i. Calculate the height above the surface of the Earth, in km, of a satellite in a geosynchronous circular orbit. Give your answer to an appropriate number of significant figures.

ii. Calculate the centripetal force acting on the polar orbiting satellite if its mass is 650 kg.

b. These geosynchronous and polar satellites have different applications because of their different orbits in relation to the rotation of the Earth. Compare the principal features of the geosynchronous and polar orbits and explain the possible uses of satellites in these orbits. In your answer you should explain why

- a low polar orbit is suitable for a satellite used to monitor conditions on the Earth
- a geosynchronous orbit above the equator is especially suitable for a satellite used in communications.

The quality of your written communication will be assessed in your answer.

AQA Unit 4 Section B January 2013 Q2

4. A small mass is situated at a point on a line joining two large masses m_1 and m_2 such that it experiences no resultant gravitational force. Its distance from the centre of m_1 is r_1 and its distance from the centre of m_2 is r_2. What is the value of the ratio $\frac{r_1}{r_2}$?

A $\frac{m_1^2}{m_2^2}$ **B** $\frac{m_2^2}{m_1^2}$ **C** $\sqrt{\frac{m_1}{m_2}}$ **D** $\sqrt{\frac{m_2}{m_1}}$

AQA Unit 4 Section A January 2013 Q10

5. The gravitational potential difference between the surface of a planet and a point P, 10 m above the surface, is 8.0 J kg^{-1}. Assuming a uniform field, what is the value of the gravitational field strength in the region between the planet's surface and P?

A 0.80 N kg^{-1} **B** 1.25 N kg^{-1}

C 8.0 N kg^{-1} **D** 80 N kg^{-1}

AQA Unit 4 Section A June 2010 Q10

6. What is the angular speed, in rad s^{-1}, of a satellite in a geostationary orbit around the Earth?

A 4.4×10^{-3} **B** 4.4×10^{-5}

C 7.3×10^{-3} **D** 7.3×10^{-5}

5 ELECTRIC FIELDS

PRIOR KNOWLEDGE

You will already be familiar with charge and its unit, the coulomb, and know that there are two types of charge, positive and negative. You will understand the difference between an electrical conductor and an insulator. You will have an understanding that a potential difference can accelerate charged particles, and have experience of expressing the energy of a particle in electronvolts. It will help if you have already studied the physics of gravitational fields (Chapter 4).

LEARNING OBJECTIVES

In this chapter you will learn how to represent electric fields with field lines and with equipotential surfaces. You will learn how to calculate electric force, electric field strength and electric potential, and about the trajectories of moving charged particles in an electric field.

(Specification 3.7.1 part, 3.7.3.1 to 3.7.3.3)

The first indication that a proton may have internal structure came in the autumn of 1967 in a deep inelastic scattering experiment using what was then the biggest particle accelerator in the world, the Stanford Linear Accelerator (SLAC), affectionately known as 'the monster' (Figure 1). SLAC accelerated electrons to energies up to 20 GeV through many thousands of electric fields along its 2 mile length and targeted the electron beam at protons. The numbers of electrons scattered at specific angles was measured using a 20 GeV spectrometer. Experiments and analysis of the data continued into 1968 and identified point-like constituents of the proton, originally named 'partons' but subsequently renamed as 'quarks'. By 1973 a coherent picture of the quark structure of both the proton and the neutron had been established.

Figure 1 *The SLAC accelerator tunnel in Stanford, California*

Linear accelerators (linacs) on a smaller scale are used extensively in hospitals to create the high-energy X-rays used for radiotherapy treatments, and are also used at synchrotron facilities. A synchrotron is a particle accelerator that uses magnetic fields to deflect charged particles around a massive ring, while they are repeatedly accelerated by an electric field. A linac accelerates the particles prior to them being injected into the ring of the synchrotron (Figure 2). The accelerated beam from a synchrotron has many applications: in colliders such as the Large Hadron Collider (LHC) at CERN, Geneva, for studies into fundamental particles, for production of radionuclides, for crystallography and chemical analysis, for medical imaging and for cancer therapy, to name a few.

Figure 2 *The linac at the Australian Synchrotron*

5.1 COULOMB'S LAW

Static electricity

Around 585 BC, Thales of Miletus, a Greek philosopher and mathematician, discovered that, if he rubbed amber with a piece of fur, the amber could attract feathers. We now know that this is an effect of **static electricity**, or 'electrostatics'. The same effect can be achieved with a plastic comb, which, after being run through your hair, will attract small pieces of paper (Figure 3a). The comb is electrically charged by friction when run through your hair – it gains electrons from the hair and becomes negatively charged. When the comb is brought near to the pieces of paper, the negative charge on the comb causes a slight redistribution of charge within the atoms in the paper (Figure 3b). The atoms become slightly **polarised** – effectively having a positive end and a negative end. The positive ends of the paper atoms are drawn towards the comb and there will be a net attractive force between the paper and the comb.

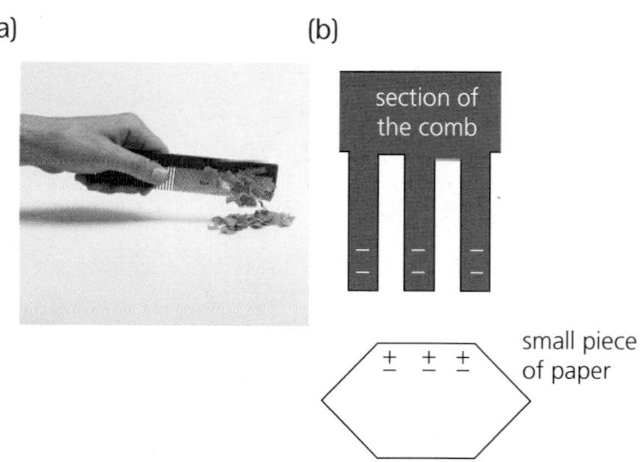

Figure 3 A charged comb attracts uncharged pieces of paper.

A balloon rubbed on your hair also becomes charged by friction and will attract a stream of water (Figure 4). Water molecules are permanently polarised, with one side of a water molecule slightly positive and the other side slightly negative. The presence of negative charge on the balloon causes the water molecules to align so that their positive side is nearest to the balloon, hence the attraction.

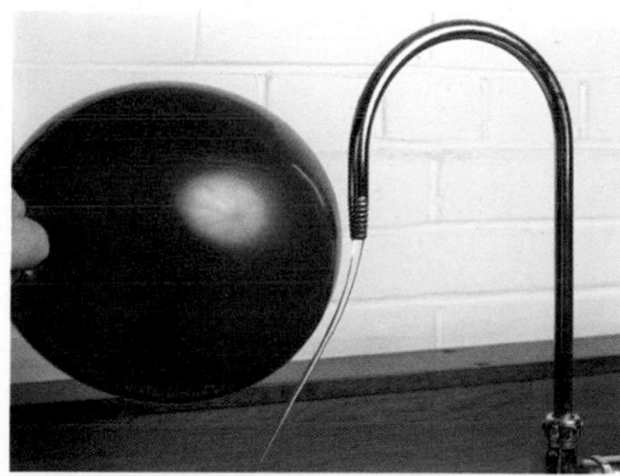

Figure 4 Bending water with a balloon

Two balloons that have been charged by friction repel each other because they are both negatively charged (Figure 5).

Figure 5 Negatively charged balloons repelling each other

The first electrical measuring instrument was the electroscope, invented around 1600, which was able to detect and compare the magnitude of charges on different objects. A modern gold-leaf electroscope is shown in Figure 6. If a polythene rod is rubbed with a cloth, electrons are transferred from the cloth to the rod, which becomes negatively charged. If the negatively charged rod is then scraped across the metal disc of the electroscope, electrons will be transferred to the electroscope's metal stem, plate and gold leaf, causing the gold leaf to diverge away from the plate.

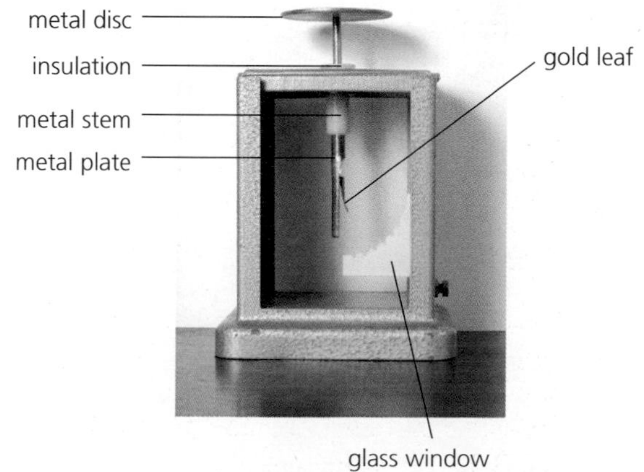

Figure 6 Gold-leaf electroscope

It took until the 18th century, however, for the various observed effects of static electricity to be explained. Benjamin Franklin, the American statesman, scientist and inventor, realised that, in order to explain all the effects, two types of charge were required, which he named 'positive' and 'negative'. He observed that a positive charge would attract a negative charge, but two positive charges would repel each other, and similarly with two negative charges.

QUESTIONS

1. a. A charged object is brought towards the metal disc of a negatively charged electroscope. The gold leaf diverges further. What can be concluded about the type of charge on the object?
   b. What would happen if, instead of the object in part **a**, an oppositely charged object were brought towards the electroscope's disc? Explain this.

Quantifying the effect

In 1785, Charles-Augustin de Coulomb (Figure 7), a French physicist, published the results of his experiments with electricity in which he demonstrated that the force between two charges varied inversely with the square of their distance of separation. Coulomb measured the force between two charged pith balls using a torsional balance that he had designed.

Figure 7 *Charles-Augustin de Coulomb, after whom the unit of electric charge is named*

A modern version of the experiment in which the force between two charged spheres is measured using a sensitive electronic balance is illustrated in Figure 8. Before the spheres are charged, the balance is zeroed. The spheres, which have a metal coating, can be charged by **induction**. For example, if a negatively charged polythene rod is brought close to, but not in contact with, a sphere, and the sphere is then momentarily earthed (connected by a conducting pathway to earth) and the rod is removed, the sphere is left positively charged. After charging both spheres, the force between the spheres at different separations can be determined by converting the mass reading to kg and multiplying by the gravitational field strength, 9.81 N kg^{-1}. The charge on each sphere can be measured using a coulombmeter (Figure 9).

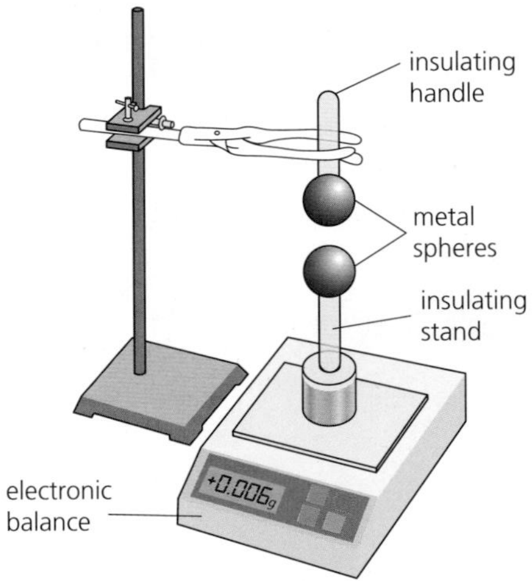

Figure 8 *Measuring the force between two charged spheres*

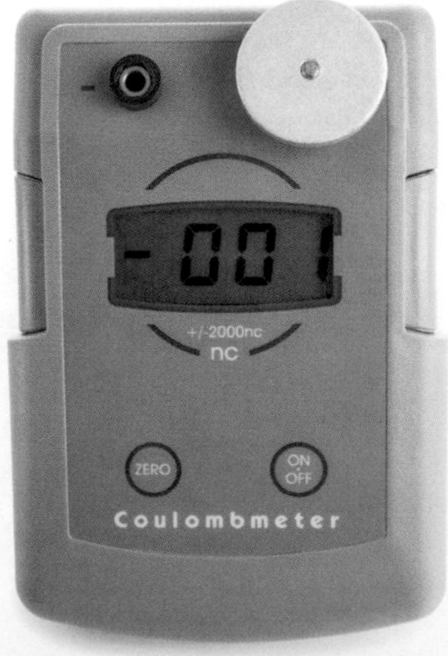

Figure 9 *Coulombmeter for charge measurement*

QUESTIONS

2. Explain how the charging of the metal sphere by induction, as described in the text, leaves it with a positive charge.

Such experiments involving force measurements between charges established what is now known as **Coulomb's law**, which states that:

The force between two point charges separated by a distance r in a vacuum is directly proportional to the product of the two charges and inversely proportional to the square of their separation.

Coulomb's law can be written as an equation:

$$F = \frac{kQ_1Q_2}{r^2}$$

in which F is the electrostatic force (in N) between point charges Q_1 and Q_2 (in coulomb, C), which are separated by distance r (in m). The constant of proportionality k is

$$k = \frac{1}{4\pi\varepsilon_0}$$

where ε_0 is a fundamental constant known as the **permittivity of free space**, so Coulomb's law can be written:

$$F = \frac{1}{4\pi\varepsilon_0}\frac{Q_1Q_2}{r^2} = \frac{Q_1Q_2}{4\pi\varepsilon_0 r^2}$$

The value of ε_0 is $8.85 \times 10^{-12}\,\text{F m}^{-1}$, where F is the unit farad, which is equivalent to one coulomb per volt: $1\,\text{F} = 1\,\text{C V}^{-1}$.

While this equation is only truly valid if the two charges are in 'free space', that is, a vacuum, if they are in air, the equation can still be used because the effect of the air is negligible. The presence of other materials between the two charges can, however, affect the size of the force between them. We will consider this in Chapter 6.

A spherically symmetrical object of uniform charge distribution, such as one of the metal spheres in Figure 8, can be assumed to behave as a point charge at the centre of the sphere.

Worked example

Two small metal spheres are arranged as in Figure 8, and the electronic balance has been zeroed. The spheres are identical and charged initially while in contact, so that they both have the same charge. When the distance between the centres of the spheres is 10 cm, the balance reading is 0.014 g. Determine the charge on each sphere. [Permittivity of free space $= 8.85 \times 10^{-12}\,\text{F m}^{-1}$]

Force between spheres $= mg = 0.014 \times 10^{-3} \times 9.81$

$= 1.373 \times 10^{-4}\,\text{N}$.

Assuming that the charged spheres can be treated as point charges, the electrostatic force F between them is

$$F = \frac{Q_1Q_2}{4\pi\varepsilon_0 r^2}$$

which rearranges to give

$$Q_1Q_2 = F \times 4\pi\varepsilon_0 r^2$$
$$= 1.373 \times 10^{-4} \times 4\pi \times 8.85 \times 10^{-12} \times 0.1^2$$
$$= 1.527 \times 10^{-16}\,\text{C}^2$$

Since the charges are equal, say $Q_1 = Q_2 = Q$, then

$$Q^2 = 1.527 \times 10^{-16}$$
$$Q = \sqrt{1.527 \times 10^{-16}}$$
$$= \sqrt{152.7 \times 10^{-18}}$$
$$= \sqrt{152.7} \times 10^{-9}$$

which gives the charge Q on each sphere as 12 nC.

QUESTIONS

3. Two identical metal spheres are arranged as in Figure 8, with a separation of 12 cm. The electronic balance, which measures to 0.001 g, is zeroed. The spheres are charged while in contact, then separated. Their charges are measured using a coulombmeter and found to be 9 nC. Predict the reading on the balance.

4. Two fixed charges $+Q$ and $-Q$ attract each other with a force F. If an additional charge of $+2Q$ is added to each of them, what is the force between them?

   A $2F$ attractive **B** $2F$ repulsive
   C $3F$ attractive **D** $3F$ repulsive

5. Two identical charges repel each other with a force F. If the size of each charge is doubled and their separation is halved, what is the new force?

ASSIGNMENT 1: ANALYSING COULOMB'S LAW DATA

(MS 0.6, MS 1.1, MS 2.2, MS 2.3, MS 3.1, MS 3.2, MS 3.3, MS 4.1, MS 4.2, MS 4.5, MS 4.6)

In this assignment you are provided with data that will enable you to test Coulomb's relationship between the separation of two charges and the force between them. The data you will use have been obtained using two charged polystyrene balls coated in a metal-based paint. One ball is attached to the end of a clamped polythene rod; the other is suspended on a nylon thread (Figure A1). Each ball is given a positive charge. A half-metre rule clamped and supported just below the polystyrene balls enables the distances r and d shown in the figure to be measured.

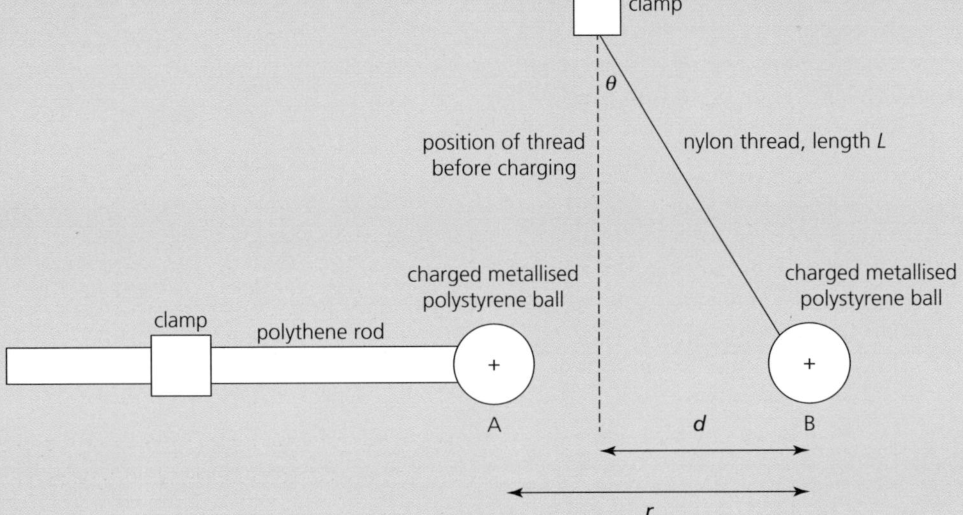

Figure A1 *Experimental set-up*

Prior to being charged, the thread of ball B was vertical and angle θ was zero. Once the balls have been charged, ball A is moved towards ball B and raised slightly to ensure that A and B are on the same horizontal level. This results in B moving, increasing distance d, until B is in equilibrium. The angle θ increases as ball A is moved nearer to ball B. To determine the size of the electrostatic repulsive force between the charged balls, it is helpful to draw a force diagram for ball B (Figure A2), which shows that

$$\tan\theta = \frac{F}{mg}$$

However, we also know that

$$\sin\theta = \frac{d}{L}$$

and provided angle θ is small, the small-angle approximation can be applied: $\sin\theta \approx \tan\theta$. Therefore

$$\frac{F}{mg} = \frac{d}{L}$$

which shows that the electrostatic force F is proportional to distance d.

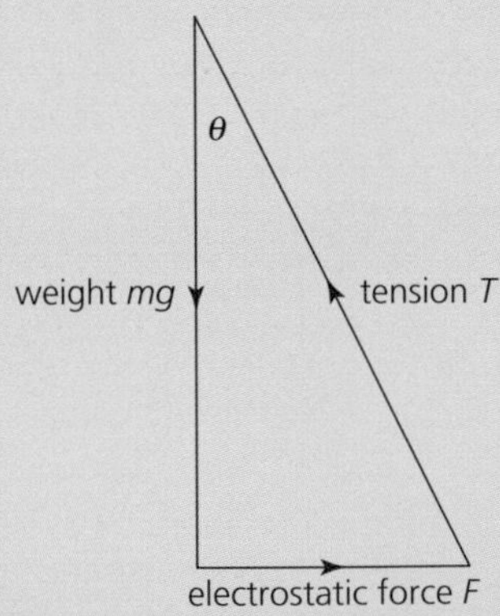

Figure A2 *Force triangle for ball B*

If measurements of d and r (the distance between the centres of the balls) can be made and the relationship between them determined, Coulomb's law stating the relationship between the electrostatic force F and the separation of the charges r

$$F \propto \frac{1}{r^2}$$

can be put to the test.

Questions

A1 Data for distances d and r are provided in Table A1. Since distance d varies in proportion to F, choose and plot a suitable graph to show the relationship between d and r in order to confirm the inverse square nature of Coulomb's law. Explain how your graph confirms Coulomb's law.

r / cm	d / cm
9.0	6.2
9.5	5.5
10.0	5.0
11.1	4.1
11.7	3.7
12.2	3.4
13.0	3.0

Table A1

A2 The small-angle approximation, $\sin\theta \approx \tan\theta$, can be applied for angles up to about 10°. Given that the length of the thread supporting ball B is 40 cm, is it acceptable to apply the small-angle approximation to the experiment set-up that generated the data in Table A1?

A3 Now charged ball A is moved further towards charged ball B, so that, when in equilibrium angle θ is 15°.

a. Draw

i. a free-body force diagram for ball B

ii. a triangle of force vectors for ball B.

b. Given that the mass of each ball is 1.2 g, determine the size of the electrostatic repulsive force between balls A and B.

Forces between charged particles

A hydrogen atom consists of a single electron orbiting a single proton (Figure 10). The centripetal force required for orbital motion is provided by the electrostatic attraction between the positively charged proton and the negatively charged electron. The charge on the proton is $+1.6 \times 10^{-19}$ C and that on the electron is -1.6×10^{-19} C. The diameter of the atom is 0.106 nm.

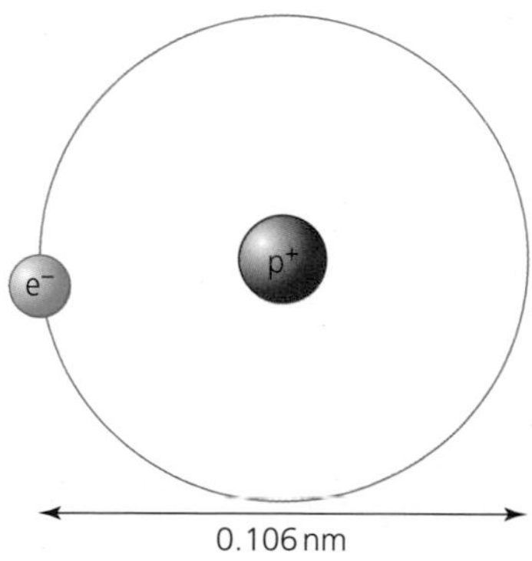

Figure 10 *A hydrogen atom*

The attractive electrostatic force between the electron and the proton can be calculated using Coulomb's law:

$$F_e = \frac{Q_1 Q_2}{4\pi\varepsilon_0 r^2}$$

$$= \frac{(1.6 \times 10^{-19})^2}{4\pi \times 8.85 \times 10^{-12} \times (0.053 \times 10^{-9})^2}$$

$$= 8.20 \times 10^{-8}\,\text{N}$$

It is interesting to compare the size of this electrostatic force between the proton and electron with the size of the gravitational force between them, which can be found using Newton's law (see section 4.1 of Chapter 4). The mass of a proton is 1.67×10^{-27} kg and the mass of an electron is 9.11×10^{-31} kg. The gravitational force is

$$F_g = \frac{Gm_1m_2}{r^2}$$

$$= \frac{6.67 \times 10^{-11} \times 1.67 \times 10^{-27} \times 9.11 \times 10^{-31}}{(0.053 \times 10^{-9})^2}$$

$$= 3.61 \times 10^{-47}\,\text{N}$$

The electrostatic force is over 10^{39} times bigger than the gravitational force. Therefore, when dealing with interactions between charged subatomic particles, the force of gravity can be considered to be negligible.

QUESTIONS

6. Calculate

   a. the electrostatic force

   b. the gravitational force

   between two protons at a separation of 2.0 fm inside a nucleus. Compare the magnitude of your answers and specify if the forces are attractive or repulsive. [1 fm = 1×10^{-15} m; permittivity of free space = $8.85 \times 10^{-12}\,\mathrm{F\,m^{-1}}$; proton charge = $1.6 \times 10^{-19}\,\mathrm{C}$; proton mass = $1.67 \times 10^{-27}\,\mathrm{kg}$; gravitational constant = $6.67 \times 10^{-11}\,\mathrm{N\,m^2\,kg^{-2}}$]

7. In Ernest Rutherford's alpha particle scattering experiment, undertaken in the early 20th century, helium nuclei were fired at gold nuclei. Calculate the repulsive force between a helium nucleus and a gold nucleus when they are at a separation of 50 fm. [The proton numbers of helium and gold are 2 and 79, respectively.]

KEY IDEAS

- Coulomb's law

$$F = \frac{Q_1 Q_2}{4\pi\varepsilon_0 r^2}$$

  shows the force between two point charges in a vacuum, but can also be used for charges in air.

- A uniformly charged sphere may be considered as a point charge at the centre of the sphere.
- The constant ε_0 is the permittivity of free space and has a value of $8.85 \times 10^{-12}\,\mathrm{F\,m^{-1}}$.
- The electrostatic forces between subatomic particles are many orders of magnitude greater than the gravitational forces.

5.2 ELECTRIC FIELDS

Every charged object creates an **electric field** in the region around itself. A charged particle or another charged object in this region experiences a force.

Electric fields can be represented by field lines, with arrows showing their direction. For example, consider an isolated metal sphere that has been charged positively. To determine the direction of the field lines, imagine placing a tiny positive test charge at a particular point in the field and work out the direction of the force on that test charge. The direction of the force gives the direction of the field line at that point (Figure 11a). The field is a radial field. If the metal sphere had been charged negatively, the field lines would be radially inwards (Figure 11b). If two metal spheres, one negative and one positive, are placed near to each other, the field looks like Figure 11c; and if both are positive, the field is as in Figure 11d. Note that the closer together the field lines, the stronger the field.

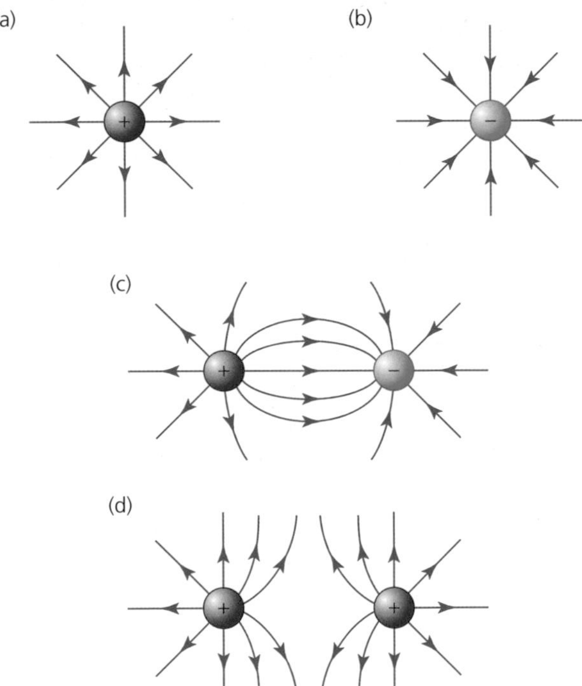

Figure 11 *Electric field patterns around charged spheres*

The definition of the strength of an electric field is based on the size of the force on a tiny positive test charge in the field (Figure 12). The test charge would have to be sufficiently small in both size and charge so as not to distort the electric field it is being used to determine.

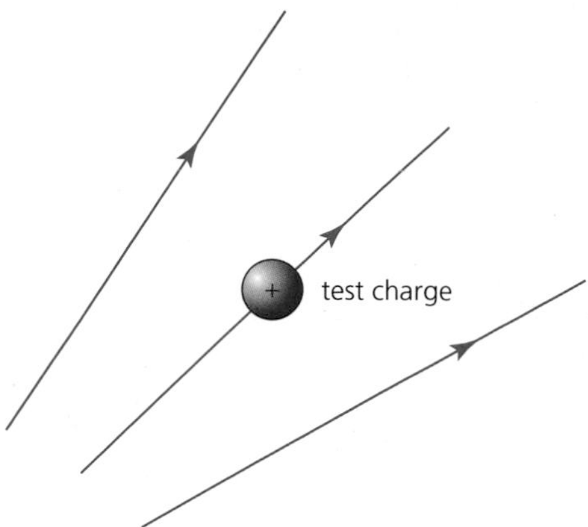

Figure 12 A small positive test charge in an electric field

The **electric field strength** E at a point in the field is defined as the force per unit charge on a positive test charge placed at that point. Electric field strength is a vector quantity and its unit is NC^{-1}. The defining equation for E is

$$E = \frac{F}{Q}$$

where F is the force on test charge Q.

The equation for the electric field strength E in a radial field, at a distance r from a metal sphere carrying charge Q, can be determined using Coulomb's law. Consider a small test charge q placed near to an isolated metal sphere, for example, a metal sphere on an insulating stand (Figure 13). The charge is on the surface of the sphere, but when Coulomb's law is being applied the charge can be considered to act at its centre.

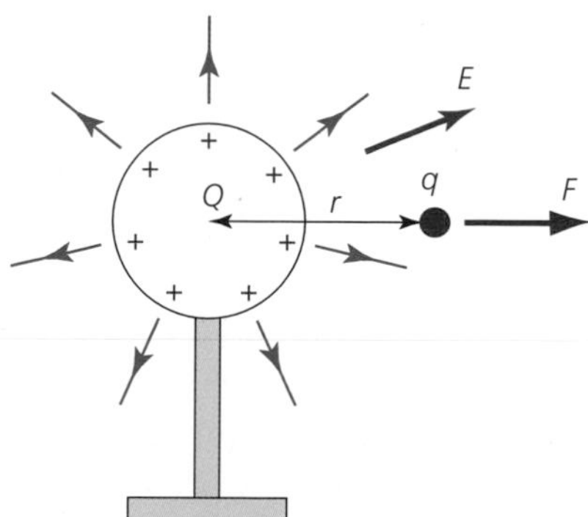

Figure 13 A test charge in the electric field created by a charged metal sphere

The force F between the larger charge Q and the small test charge q is

$$F = \frac{Qq}{4\pi\varepsilon_0 r^2}$$

where r is the distance from the centre of the metal sphere to the test charge. The electric field strength E, due to the charged metal sphere, at the position of the test charge, is equal to $\frac{F}{q}$ and therefore

$$E = \frac{Q}{4\pi\varepsilon_0 r^2}$$

QUESTIONS

8. Determine the electric field strength at a distance of 50 fm from a gold nucleus. [Proton number of gold is 79; permittivity of free space = 8.85×10^{-12} F m^{-1}]

9. An isolated metal sphere is given a positive charge of 20 nC. Determine
   a. the electric field strength at a distance of 20 cm from the centre of the sphere
   b. the magnitude and direction of the force on a molecule of air that has lost one electron and is 20 cm from the charged sphere.

10. Point P in Figure 14 is 40 mm from two +5.0 nC charges, in perpendicular directions. Determine the strength and the direction of the electric field strength at P. [Permittivity of free space = 8.85×10^{-12} F m^{-1}]

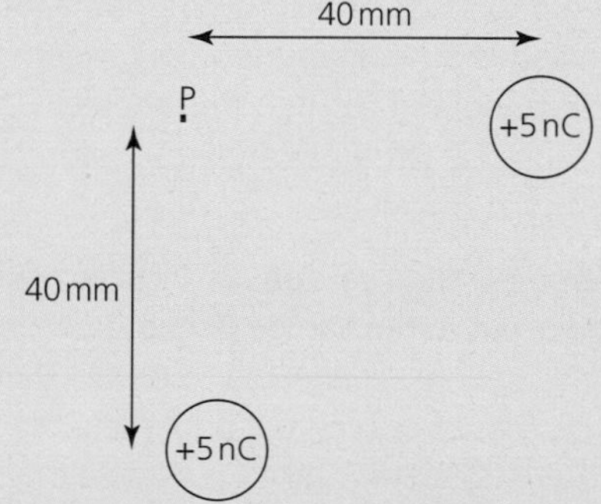

Figure 14

11. Sketch a graph of the electric field strength against distance from a point charge. How is such a graph described?

Generating electric fields

Electrostatic generators are designed to generate large amounts of static electricity and hence high field strengths. Many different designs were developed during the 18th and 19th centuries, including the Wimshurst machine (Figure 15), developed by British inventor James Wimshurst in the 1880s.

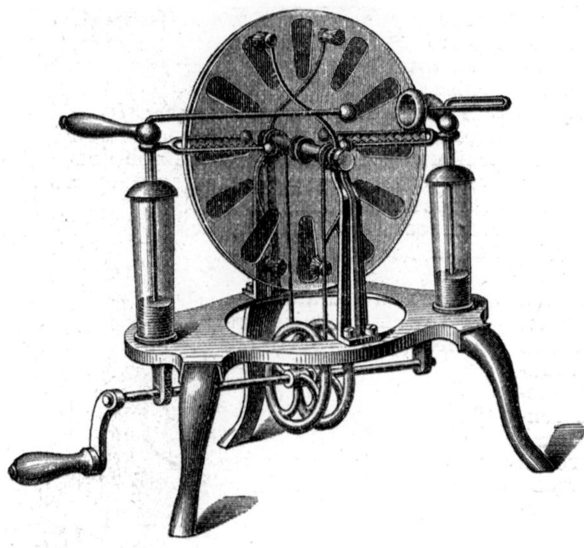

Figure 15 *The Wimshurst machine*

In 1929, the American physicist Robert Van de Graaff invented his electrostatic generator primarily as a particle accelerator. A Van de Graaff generator uses a motor-driven belt to transfer charge onto a hollow metal dome at the top of an insulated column, and can generate up to several million volts. Table-top versions that are used in school laboratories for demonstration can generate up to a few hundred thousand volts (*see section 13.2 and Figure 7 in Chapter 13 in Year 1 Student Book*).

These demonstration versions include a small metal sphere adjacent to the Van de Graaff dome, which is a 'discharging sphere', connected to earth. The electric field in the space between the dome and the discharging sphere is illustrated in Figure 16. Notice that the electric field lines are drawn so that they touch the surface of the charged conductors at 90°. If the electric field between the dome and the sphere is sufficiently large, it can pull electrons out of air molecules, making the air conducting and creating a spark.

If the discharging sphere is removed and the dome of the Van de Graaff is connected to one of two parallel metal plates, the other being earthed, a uniform

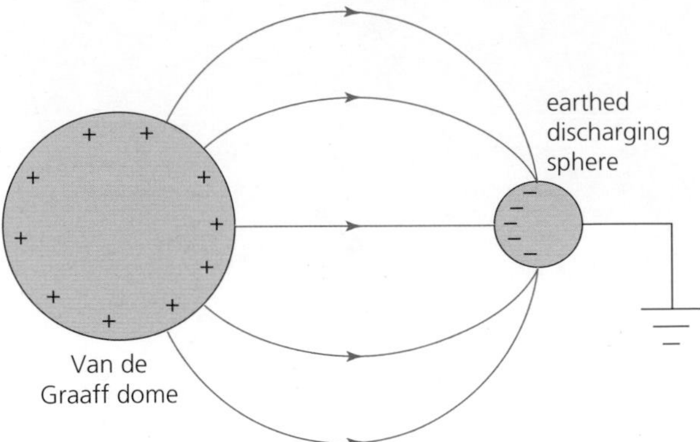

Figure 16 *Electric field between Van de Graaff dome and discharging sphere*

electric field can be created (Figure 17). The electric field lines in a uniform field are equally spaced and are directed from the positive to the negative plate.

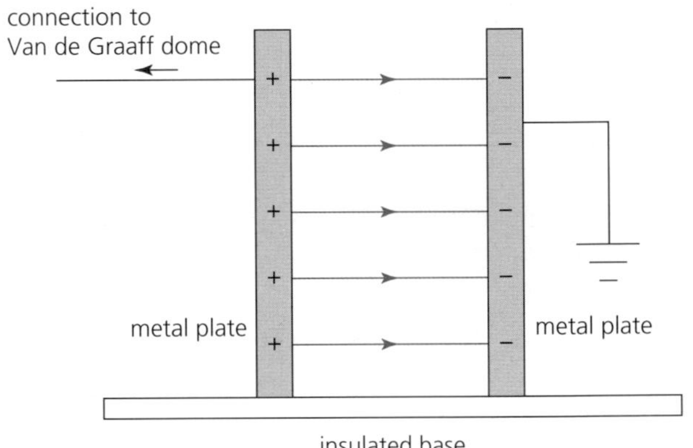

Figure 17 *Uniform electric field*

The uniform field and potential difference

The action of a uniform field can be demonstrated by suspending a metallised polystyrene ball on a nylon thread between the plates (Figure 18). The uncharged ball will be attracted to the plate to which it is nearest due to a redistribution of the electrons in the metal paint caused by the uniform field. On making contact with the plate, the ball becomes charged to the same polarity as the plate and is therefore repelled towards the other plate, where the polarity of the ball changes again on contact. The result is that the polystyrene ball moves rapidly to and fro between the two plates.

Within a uniform field, the electric field strength is constant, which means that a charged particle in the field experiences the same force regardless of its location between the plates. Consider a particle with

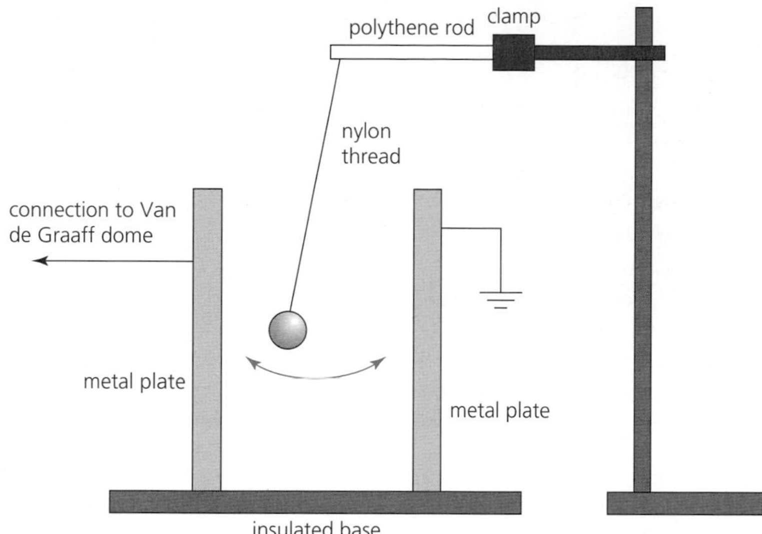

Figure 18 *Charged metallised polystyrene ball oscillating in a uniform field*

positive charge Q that is close to one plate prior to the plate's connection to the Van de Graaff. On making the connection, the electric field E exerts a force F on the particle, pushing the particle a distance d to the other plate.

The work done W by the electric field in moving the particle from one plate to the other is given by $W = Fd$.

However, **electric potential difference** ΔV is defined as the work done per unit charge in moving the charge between two points (*see section 13.3 in Chapter 13 in Year 1 Student Book*):

$$\Delta V = \frac{W}{Q}$$

So the work done can also be written $W = Q\Delta V$ and therefore

$$Fd = Q\Delta V$$

Using $F = EQ$ and substituting into the equation gives $EQd = Q\Delta V$, showing that the electric field strength of a uniform field is given by

$$E = \frac{\Delta V}{d}$$

which is sometimes written as

$$E = \frac{V}{d}$$

where V is the potential difference between the parallel plates giving rise to the field. The equation shows that the unit of electric field strength can be $V m^{-1}$, as well as $N C^{-1}$. The two units are equivalent.

Worked example 1

A uniform electric field is created by a potential difference of 600 V across two metal plates separated by a distance of 10 cm. Determine

a. the force on a singly charged positive air ion that is in the region between the two plates

b. the acceleration of the ion towards the negative plate, assuming its mass is about 5×10^{-26} kg

c. the kinetic energy gained by the ion when it has reached the negative plate, assuming that it was very close to the positive plate when the field was switched on.

a. A singly charged ion has either lost or gained one electron, so its charge is $Q = e = 1.6 \times 10^{-19}$ C. The force F on the ion is given by

$$F = EQ = \frac{V}{d}e$$

$$= \frac{600}{0.1} \times 1.6 \times 10^{-19} = 9.6 \times 10^{-16} \text{N}$$

b. From Newton's second law, $F = ma$, so acceleration is

$$a = \frac{F}{m} = \frac{9.6 \times 10^{-16}}{5 \times 10^{-26}} = 2 \times 10^{10} \text{m s}^{-2}$$

c. The work done W by the electric field in accelerating the ion through a potential difference V can be found using $V = \frac{W}{Q}$ and therefore the work done is

$W = Ve = 600 \times 1.6 \times 10^{-19} = 9.6 \times 10^{-17}$ J.

Since work done = energy transferred, the kinetic energy gained by the ion = 9.6×10^{-17} J.

The electron gun

An electric field can be used to accelerate electrons to form a beam. An arrangement known as an electron gun (Figure 19) consists of two metal electrodes, the cathode and the anode. The cathode is heated, which causes it to emit electrons (when their thermal energy is sufficient to overcome the work function). A high potential difference between the cathode and the anode creates an electric field, which accelerates the electrons towards a hole in the cylindrical anode, to form the beam.

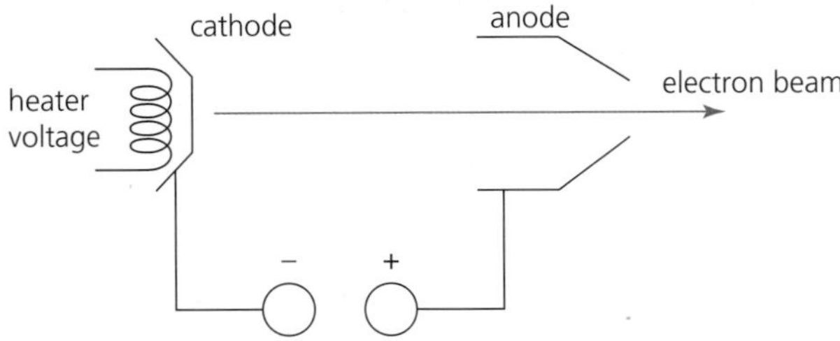

Figure 19 *An electron gun, simplified*

QUESTIONS

12. A free electron is in the air between two conducting plates that are 15 cm apart. Determine

a. the acceleration of the electron in the direction towards the positive plate when a potential difference of 50 V is applied to the plates

b. the kinetic energy gained by the electron if it is very close to the negative plate when the potential difference is applied. [Electron mass = 9.11×10^{-31} kg; magnitude of charge on electron = 1.60×10^{-19} C]

13. In an electron gun arrangement, a potential difference of 2000 V between the cathode and anode creates an electric field of strength 25 kV m^{-1}, which accelerates electrons to form a beam.

Determine

a. the acceleration of the electron in the electric field

b. the kinetic energy gained by an electron.

Moving charged particles in an electric field

Consider an electron fired horizontally from an electron gun and travelling at a constant speed v into a uniform vertical electric field E at 90° to the field lines (Figure 20). The electron continues through the field with a constant horizontal component of velocity v. But it has an increasing vertical component of velocity caused by the downward electrostatic force F due to the electric field.

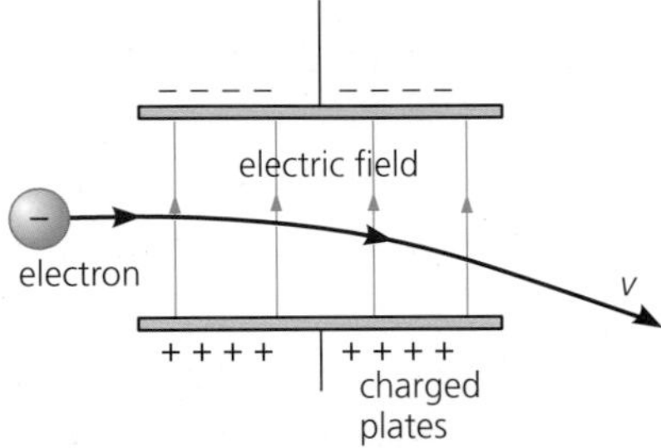

Figure 20 *Deflection of an electron beam by an electric field*

The vertical force on an electron due to the field is

$$F = EQ = Ee$$

since e is the charge on the electron. The vertical acceleration of the electron is

$$a = \frac{F}{m} = \frac{Ee}{m}$$

Suppose the electron has travelled a horizontal distance x within the field in time t. Then

$$t = \frac{x}{v}$$

The vertical displacement y of the electron beam can be found using an equation of motion (*see section 9.5 in Chapter 9 in Year 1 Student Book*), $s = ut + \frac{1}{2}at^2$, which gives

$$y = ut + \frac{1}{2}at^2$$

Since the initial vertical velocity u is zero

$$y = \frac{1}{2}at^2$$

And therefore, substituting for a and t in this equation, the vertical displacement of the electron beam is given by

$$y = \frac{1}{2}\frac{Ee}{m}\left(\frac{x}{v}\right)^2 = \frac{Ee}{2mv^2}x^2$$

The only variables in the equation are x and y, so the equation shows that $y = kx^2$, which is the equation of a **parabola**. Therefore, the trajectory of a moving charged particle entering a uniform electric field initially at right angles is a parabola. On exiting the field the path resumes a straight line.

Worked example 2

A beam of electrons travelling at $5.0 \times 10^6 \text{ m s}^{-1}$ is fired horizontally from an electron gun into a uniform electric field of strength 2000 N C^{-1} at 90° to the field lines.

a. In this arrangement, determine

i. the vertical acceleration of an electron

ii. the time it takes for an electron to travel a horizontal distance of 5.0 cm in the field.

b. After an electron has travelled a horizontal distance of 5.0 cm within this field, determine

i. the vertical component of its velocity

ii. its vertical displacement

iii. its speed and direction.

a. i. Vertical acceleration is

$$a = \frac{F}{m} = \frac{EQ}{m} = \frac{2000 \times 1.6 \times 10^{-19}}{9.11 \times 10^{-31}}$$
$$= 3.513 \times 10^{14} \text{ m s}^{-2}$$
$$= 3.5 \times 10^{14} \text{ m s}^{-2} \text{ (to 2 s.f.)}$$

ii. Time to travel 5.0 cm horizontally is

$$t = \frac{\text{distance}}{\text{horizontal speed}} = \frac{0.05}{5.0 \times 10^6}$$
$$= 1.0 \times 10^{-8} \text{ s}$$

b. i. Using $v = u + at$ gives the vertical velocity:

$$v = 3.513 \times 10^{14} \times 1.0 \times 10^{-8}$$
$$= 3.513 \times 10^6 = 3.5 \times 10^6 \text{ m s}^{-1}$$

ii. Vertical displacement is

$$y = \frac{1}{2}at^2 = \frac{1}{2} \times 3.513 \times 10^{14} \times (1.0 \times 10^{-8})^2$$
$$= 1.757 \times 10^{-2} \text{ m} = 1.8 \text{ cm (to 2 s.f.)}$$

iii. The speed of an electron in the beam can be found by combining the horizontal and vertical components of velocity using Pythagoras's theorem:

$$\text{speed} = \sqrt{(5.0 \times 10^6)^2 + (3.513 \times 10^6)^2}$$
$$= 6.1 \times 10^6 \text{ m s}^{-1}$$

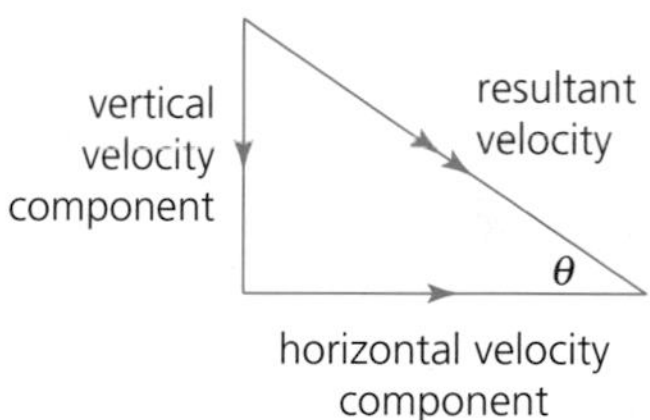

Figure 21

The angle θ between the direction of the electron beam and the horizontal can be found using a component triangle (see Figure 21). Thus

$$\theta = \tan^{-1}\left(\frac{\text{vertical velocity}}{\text{horizontal velocity}}\right)$$
$$= \tan^{-1}\left(\frac{3.513 \times 10^6}{5.0 \times 10^6}\right) = 35°$$

QUESTIONS

14. A beam of electrons travelling horizontally at a speed of $4.0 \times 10^6 \text{ m s}^{-1}$ enters a uniform electric field initially at right angles to the field lines. The electric field is produced by two metal plates that are 8.0 cm long, 5.0 cm apart and positioned in an evacuated tube. The voltage between the plates is 50 V, with the lower plate being positive with respect to the upper plate.

a. Determine

i. the strength of the electric field

ii. the time the electron spends in the field.

b. As the electron exits the field, determine

i. its vertical acceleration

ii. its vertical velocity

iii. its vertical displacement

iv. its velocity.

Stretch and challenge

15. A linear accelerator (linac) uses electric fields to accelerate charged particles to high energies. The length of the linac depends on its planned use, and can vary from 1 m to about 2 km. The linac shown in Figure 22 has electrons travelling through a series of 'drift tubes' and being accelerated by electric fields in the gaps between the tubes.

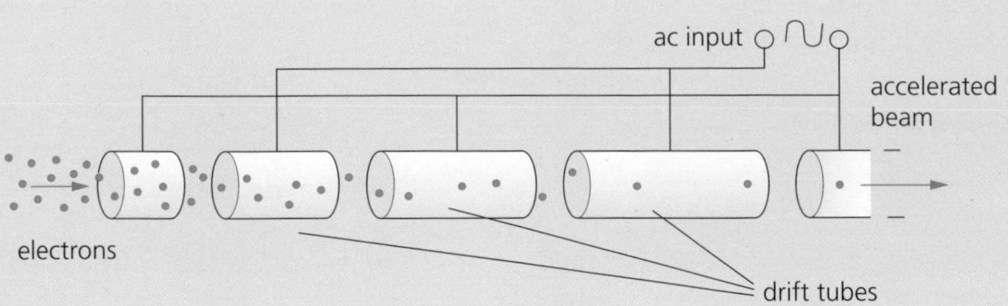

Figure 22 Drift tubes in a linear accelerator

Determine the speed reached by electrons accelerated by a linac with 200 drift tubes, given that the voltage across each gap is 25 kV. There is a complication here. The speed reached by the electrons is likely to be close to the speed of light, so the following relativistic equation will have to be used:

kinetic energy $E_k = (\gamma - 1)m_0c^2$, where $\gamma = \dfrac{1}{\sqrt{1 - \dfrac{v^2}{c^2}}}$

[Rest mass of an electron $m_0 = 9.11 \times 10^{-31}$ kg; speed of light $c = 3.00 \times 10^8\,\mathrm{m\,s^{-1}}$]

ASSIGNMENT 2: ANALYSING THE PATH OF AN ELECTRON BEAM

(MS 3.1, MS 3.2, MS 3.3, PS 2.2, PS 2.3, PS 3.1, PS 3.2, PS 4.1)

To observe the path followed by an electron beam that has been fired horizontally at a constant speed at right angles to a uniform electric field requires an evacuated deflection tube containing two horizontal metal plates across which is maintained a constant potential difference (Figure A1). The tube contains a graduated fluorescent screen, which makes the path of the electron beam visible and enables the measurement of the vertical deflection y for various horizontal values x.

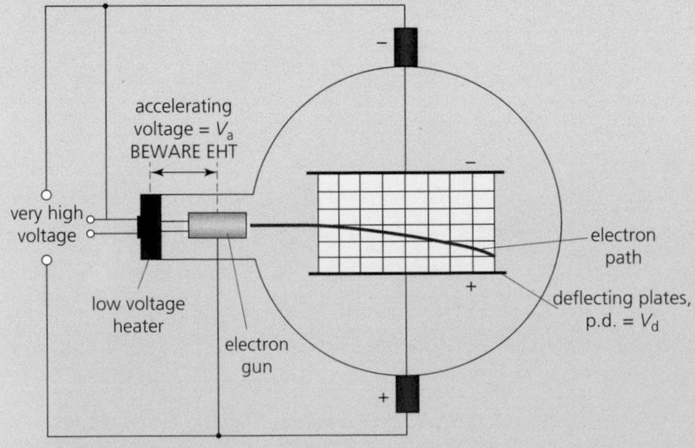

Figure A1 Electron deflection tube

High-voltage power supply units provide the voltages required for the electron gun and the electric field, and should be adjusted to produce a beam that produces as large a deflection as possible but still exits the field. The theory shows that the path of the beam should be parabolic, which means that the y and x values should have a relationship of the form $y = kx^2$. This relationship can be tested by plotting a graph of log y versus log x.

Questions

A1 Explain how a log y versus log x graph can be used to show the power-law relationship $y = kx^2$.

A2 A student makes measurements of x and y and obtains the data shown in Table A1.

x	y
2	0.2
3	0.3
4	0.4
5	0.5
6	0.8
7	1.1
8	1.3
9	1.8
10	2.1

Table A1

a. Which measurements are likely to be unreliable?

b. Use the student's data to plot a graph of y versus x^2. Explain whether the graph demonstrates that the electron beam follows a parabolic path.

Stretch and challenge

c. Which graph do you think is a better test of the power-law relationship: log y versus log x or y versus x^2? Explain your reasoning.

KEY IDEAS

- The electric field strength E at a point in the field is defined as the force per unit charge on a positive test charge placed at that point:

$$E = \frac{F}{Q}$$

Electric field strength is a vector quantity with unit $\mathrm{N\,C^{-1}}$ or alternatively $\mathrm{V\,m^{-1}}$.

- The arrow on an electric field line shows the direction of the force on a positive test charge placed in the field at that point.
- The electric field strength of a radial field at a distance r from a point charge Q is given by

$$E = \frac{Q}{4\pi\varepsilon_0 r^2}$$

- The electric field strength within a uniform field between two plates a distance d apart with a potential difference V between them is given by

$$E = \frac{V}{d}$$

- The work done in moving a charge Q between the plates of a uniform field is $Fd = QV$.
- The path of a beam of electrons entering a uniform field initially at right angles to the field lines is in the shape of a parabola.

5.3 ELECTRIC POTENTIAL

The concept of electric potential *difference* is familiar. To understand the concept of *absolute* **electric potential**, consider the electric field of an isolated positive charge, Q. A location distant from this charge where its electric field is too weak to have any influence is referred to as 'infinity'. If a small positive test charge q is to be brought from infinity towards charge Q (Figure 23), external work has to be done against the repulsive force between the two charges, and therefore the system gains electric potential energy.

The electric potential V at a point is defined as the work done per unit charge in bringing a small positive test charge from infinity to the point.

The electric potential in an electric field created by an isolated positive charge is therefore always positive. Electric potential is a scalar quantity and its unit is the volt. The zero of electric potential is chosen to be at infinity, as was the case with gravitational potential (see section 4.3 of Chapter 4).

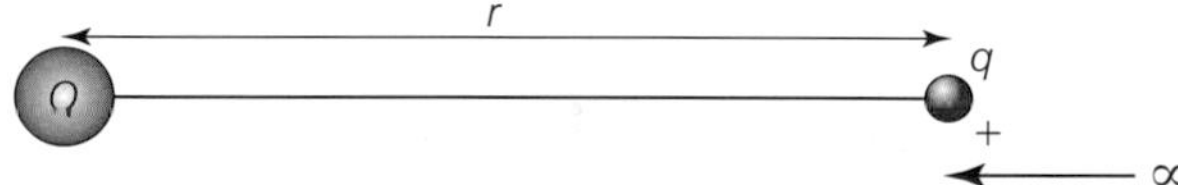

Figure 23 *Defining electric potential*

Now consider the electric field of an isolated negative charge, $-Q$. To bring a small positive test charge q would not require external work to be done. It is the electric field itself that does the work accelerating charge q towards $-Q$, resulting in a loss of electric potential energy. Consequently, the electric potential in an electric field created by an isolated negative charge is always negative.

An analogy can be drawn between the example of an electric field created by an isolated negative charge and the gravitational field created by a mass, like the Earth, for example. Consider a mass held a metre above the Earth's surface. If the mass is released, the Earth's gravitational field does work in accelerating the mass towards the ground, resulting in a loss of gravitational potential energy. Consequently, the gravitational potential in a gravitational field is always negative (see section 4.3 of Chapter 4). This is a characteristic of an attractive force field.

Relationship between field strength and potential

Consider a graph of electric field strength versus distance for a point charge Q (Figure 24).

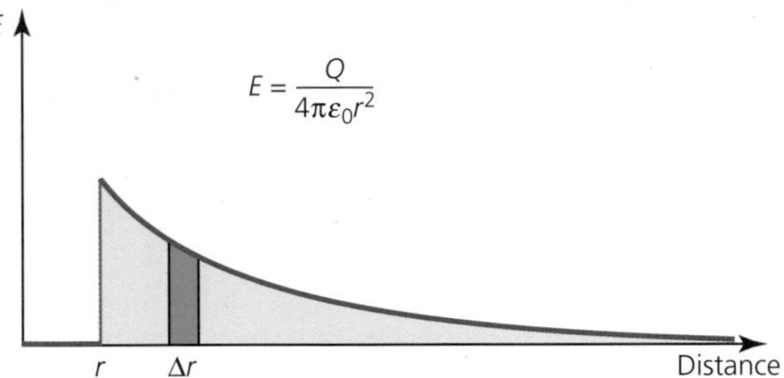

Figure 24 *Graph of electric field strength E against distance from a point charge Q*

The work done ΔW in moving a small test charge q a very small distance Δr towards charge Q is given by $\Delta W = F\Delta r$, where F is the force needed to move the charge. However, since $F = Eq$, the work done ΔW can be written as $\Delta W = Eq\Delta r$.

Dividing ΔW by the charge q gives the work done per unit charge:

$$\frac{\Delta W}{q} = E\Delta r$$

But the work done per unit charge is the potential, so the small change in potential ΔV over distance Δr can be written as

$$\Delta V = \frac{\Delta W}{q}$$

Hence

$$\Delta V = E\Delta r$$

which corresponds to the darker shaded area on the graph in Figure 24. Therefore, the total area between the curve and the distance axis on a graph of electric field strength versus distance from infinity to a distance r is equal to the potential at r from the charge Q. From the equation for the electric field strength

$$E = \frac{Q}{4\pi\varepsilon_0 r^2}$$

the mathematical process of integration can be used to determine the total area, resulting in this expression for the electric potential V at a distance r from a point charge Q:

$$V = \frac{Q}{4\pi\varepsilon_0 r}$$

The variations of electric potential with distance from a both a positive and a negative charge are shown in Figure 25. The curve for a negative charge is the same as that for gravitational potential (see Figure 11 of Chapter 4).

Having shown that $\Delta V = E\Delta r$, which rearranges to give

$$E = \frac{\Delta V}{\Delta r}$$

it can be seen that the gradient of the potential versus distance graph is equal to the magnitude of the electric field strength E.

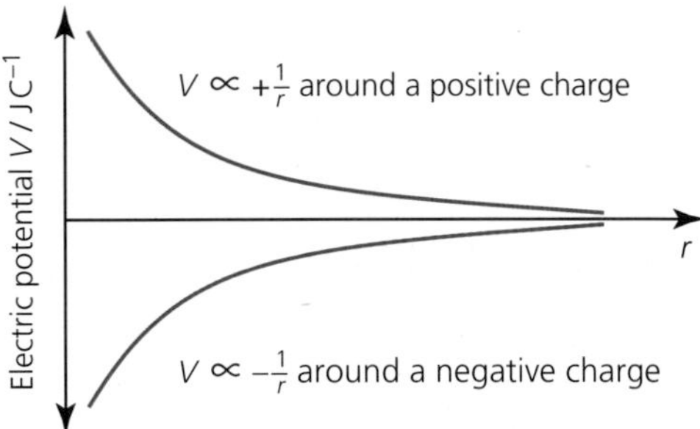

Figure 25 *Graph of electric potential V against distance from a point charge*

It is important to note that, when considering electric potentials, strictly the symbol V is used for potential and the symbol ΔV is used for potential difference. But when considering the electric field strength in a uniform field, $E = \frac{V}{d}$, the symbol V represents the potential difference between the plates.

Worked example

Determine the electric potential at point P at a distance of 20 cm from two insulated charged spheres, as shown in Figure 26, each sphere carrying a charge of 10 nC.

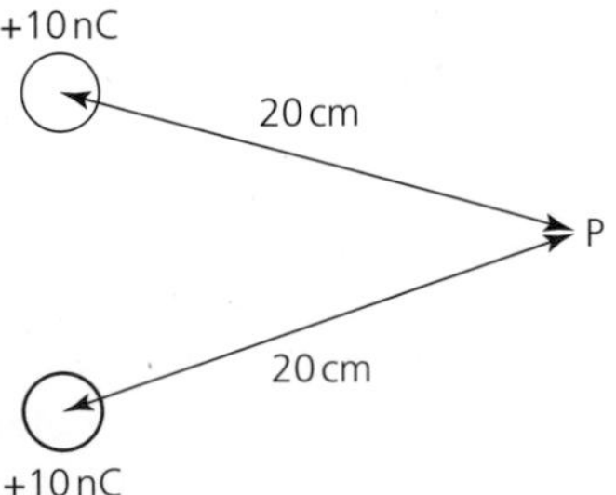

Figure 26

The potential at point P due to the upper charged sphere is

$$V = \frac{Q}{4\pi\varepsilon_0 r} = \frac{10 \times 10^{-9}}{4\pi \times 8.85 \times 10^{-12} \times 0.2}$$

$$= 4.496 \times 10^2 = 450\,\text{V}$$

The potential at point P due to the lower charged sphere is also 4.496×10^2 V .

Potential is a scalar quantity, so the overall potential at P can be found by adding the two potentials, which gives

potential at P $= 4.496 \times 10^2 + 4.496 \times 10^2 = 900\,\text{V}$

QUESTIONS

16. Determine the electric potential at point P due to the three 10 nC charges arranged as in Figure 27. [Permittivity of free space $= 8.85 \times 10^{-12}\,\text{F m}^{-1}$]

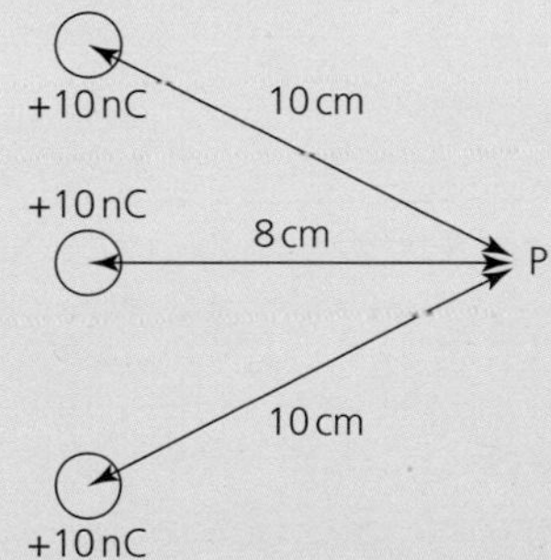

Figure 27

17. Two charges are positioned as in Figure 28. How far from the –5.0 nC charge, along the line of the two charges, is the electric potential zero?

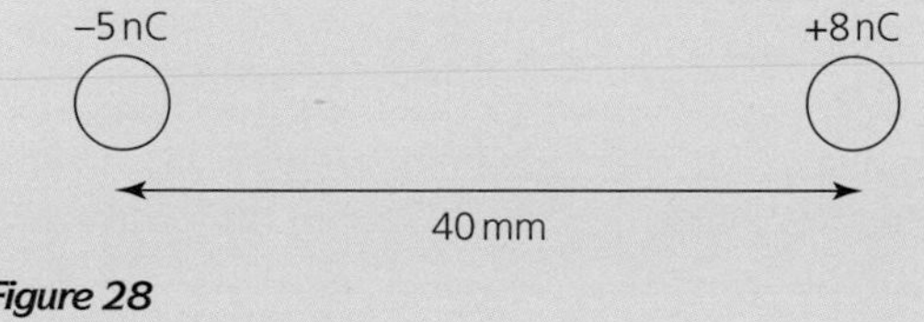

Figure 28

Equipotentials

Just like gravitational fields, electric fields can be represented both by field lines and by equipotential surfaces. On an equipotential surface, the potential is the same everywhere, so no work is done in moving a charge along an equipotential surface. A point charge $+Q$ creates an electric field that has spherical equipotential surfaces, which in two dimensions appear as concentric circles (Figure 29). The potential difference between adjacent equipotentials is constant.

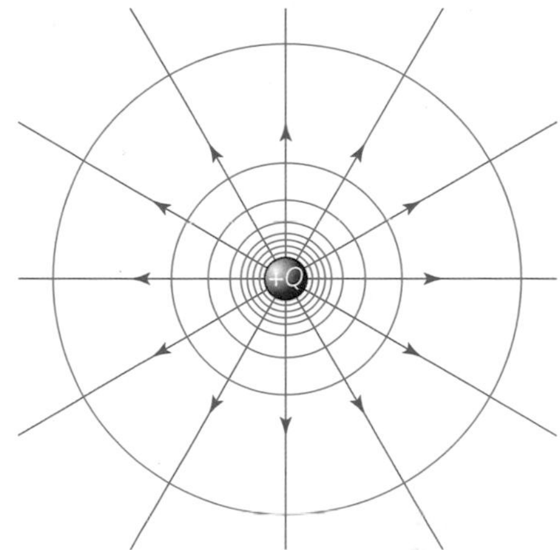

Figure 29 *Electric field lines and equipotentials for a point positive charge*

A field made up of a positive and a negative charge, represented both by electric field lines and by equipotentials, is shown in Figure 30. Notice that equipotentials meet field lines at 90°.

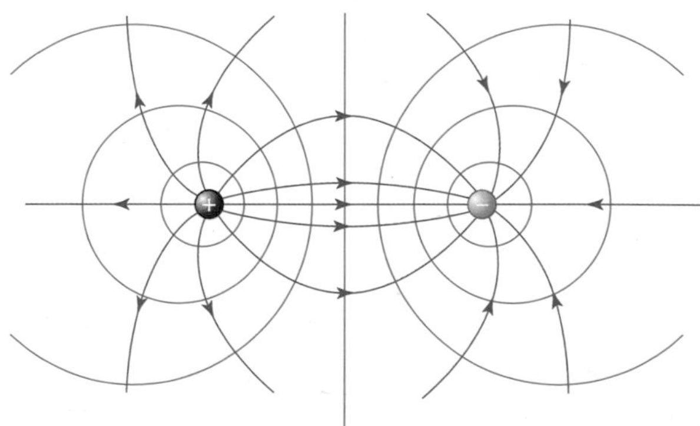

Figure 30 *Electric field lines and equipotentials for a positive point charge near to a point negative charge*

Equipotential surfaces within a uniform field are equally spaced planes parallel to the plates.

We have seen that the field outside an isolated charged conducting sphere is the same as if the charge were concentrated at the centre of the sphere. But what happens inside the sphere? The electric field and potential variations with distance from the centre of a charged spherical conductor are shown in Figure 31.

Charge on the surface of the conductor will always distribute so that the potential is the same over the whole surface. The electric field strength inside the sphere is therefore zero, since there is no difference in potential from one side of the sphere to the other. In general, there cannot be an electric field within any isolated conductor – if the field were not zero, charge would move to equalise the potential.

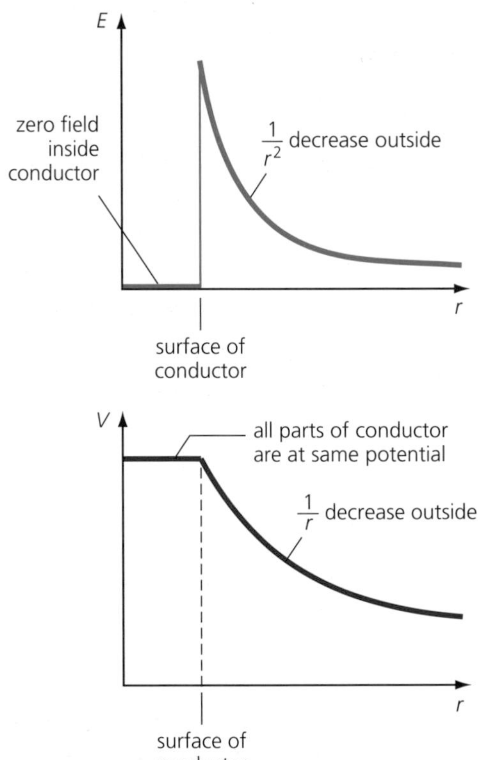

Figure 31 *The E–r and V–r graphs for a charged metal sphere*

In some electric circuits it is convenient to connect a point in the circuit to earth. This point is denoted by the symbol shown in Figure 32. Although in the theory of electric potential we chose infinity as the zero of potential, in practice it is useful to treat an earthed point as being at 0 V. This is acceptable because voltmeter measurements are of potential differences, not absolute potential. The Earth is a conductor, so its surface is an equipotential surface, which can reasonably be given the value zero.

Figure 32 *The earth connection symbol*

QUESTIONS

18. A positively charged metal sphere is placed a short distance from an uncharged metal sphere connected to earth (Figure 33). A negative charge is induced on the initially uncharged sphere, by electrons flowing from earth on to the sphere. The electric field created is represented in the diagram by equipotentials.

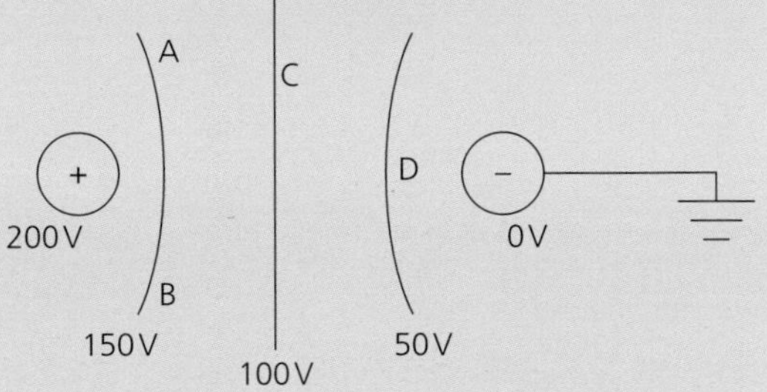

Figure 33 *Equipotentials between a positive and a negative sphere*

a. Determine the potential difference between

i. A and B

ii. B and C

iii. C and D

iv. A and D

b. Determine the work done in moving a charge of 1 nC from

i. A to B

ii. B to C

iii. C to D

iv. A to D

c. Find the net work done in moving a charge of 1 nC from A to C to B.

19. The electric field just outside an isolated charged conducting sphere of radius r is E. Which row of the table gives the correct values of the electric field strength at points that are $r/2$ and $2r$ from the centre of the sphere?

	At $r/2$	At $2r$
A	0	$E/4$
B	0	$E/2$
C	E	$E/4$
D	$2E$	$E/4$

20. Explain why a metal object such as the dome of a Van de Graaff generator is an equipotential surface.

KEY IDEAS

- The electric potential V at a point is defined as the work done (or energy needed) per unit charge in bringing a small positive test charge from infinity to the point. It is a scalar quantity and is measured in volts.
- The electric potential at a point in a radial field due to a point charge Q is given by

$$V = \frac{Q}{4\pi\varepsilon_0 r}$$

- The work done ΔW in moving a charge Q through a potential difference of ΔV is given by

$$\Delta W = Q\Delta V$$

- Electric fields can be represented by equipotential surfaces.
- No work is done when a charge moves along an equipotential surface.
- The gradient of a potential–distance graph is equal to the magnitude of the electric field strength:

$$E = \frac{\Delta V}{\Delta r}$$

- The area between the curve and the distance axis on a graph of electric field strength versus distance is equal to potential difference ΔV.

5.4 COMPARING GRAVITATIONAL FIELDS AND ELECTRIC FIELDS

It is clear that there are many similarities, and some differences, between gravitational fields and electric fields. Table 1 summarises the equations from Chapter 4 and this chapter, and the lists below summarise similarities and differences in these two areas of physics.

Gravitational	Electric
Force between masses, $F = \frac{Gm_1m_2}{r^2}$ (always attractive)	Force between charges, $F = \frac{Q_1Q_2}{4\pi\varepsilon_0 r^2}$
Field strength, $g = \frac{F}{m}$	Field strength, $E = \frac{F}{Q}$
Radial field strength, $g = \frac{GM}{r^2}$	Radial field strength, $E = \frac{Q}{4\pi\varepsilon_0 r^2}$
Radial field potential, $V = -\frac{GM}{r}$	Radial field potential, $V = \frac{Q}{4\pi\varepsilon_0 r}$
Work done in moving a mass, $\Delta W = m\Delta V$	Work done in moving a charge, $\Delta W = Q\Delta V$
Potential gradient, $g = -\frac{\Delta V}{\Delta r}$	Potential gradient (magnitude), $E = \frac{\Delta V}{\Delta r}$

Table 1 *Equations for gravitational and electric fields*

Similarities

- The gravitational force between two masses and the electrostatic force between two charges both have laws that vary inversely with the square of separation.
- The gravitational force between two masses is directly proportional to the product of the masses, and the electrostatic force between two charges is directly proportional to the product of the two charges.
- Gravitational fields and electric fields can both be represented by field lines, which indicate the direction of the force on a test point mass or point positive charge in the field, and whose density indicates the strength of the field.
- The expressions for gravitational field strength near a point mass and electric field strength near a point charge both vary inversely with the square of distance from the mass and charge, respectively.
- Gravitational potential at a point is the work done per unit mass in bringing a mass from infinity to the point. Electric potential at a point is the work done per unit charge in bringing a small positive test charge from infinity to the point.

- Both gravitational potential and electric potential are inversely proportional to distance from the centre of the source of the field.
- Gravitational fields and electric fields can both be represented by equipotential surfaces, which are at right angles to field lines.

Differences

- There are positive and negative charges, but only positive masses.
- The gravitational force is always attractive, whereas the electrostatic force may be attractive or repulsive, depending on the polarity of the charges present.
- The gravitational force is independent of the medium between the masses, whereas the size of the force between two charges can be affected by the material between them.

QUESTIONS

21. Explain, in terms of the definition of *potential*, why the electric potential–distance graph for a point charge has two possible forms, but the gravitational potential–distance graph for a point mass has only one form. Sketch the graphs.

PRACTICE QUESTIONS

1. a. Define the *electric potential* at a point in a field.

   b. Figure Q1 shows part of the region around a small positive charge.

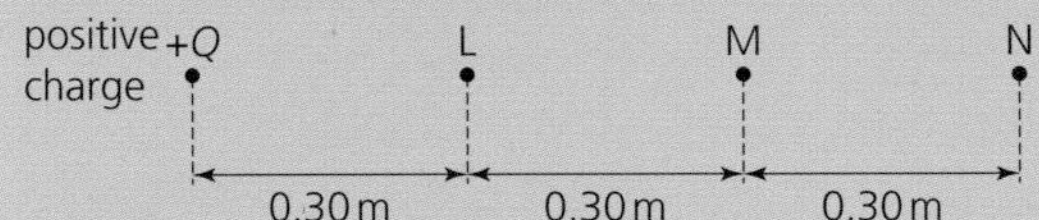

Figure Q1

   i. The electric potential at point L due to this charge is +3.0 V. Calculate the magnitude *Q* of the charge. Express your answer to an appropriate number of significant figures.

   ii. Show that the electric potential at point N, due to the charge, is +1.0 V.

   iii. Show that the electric field strength at point M, which is mid-way between L and N, is 2.5 V m^{-1}.

   c. R and S are two charged parallel plates, 0.60 m apart, as shown in Figure Q2. They are at potentials of +3.0 V and +1.0 V, respectively.

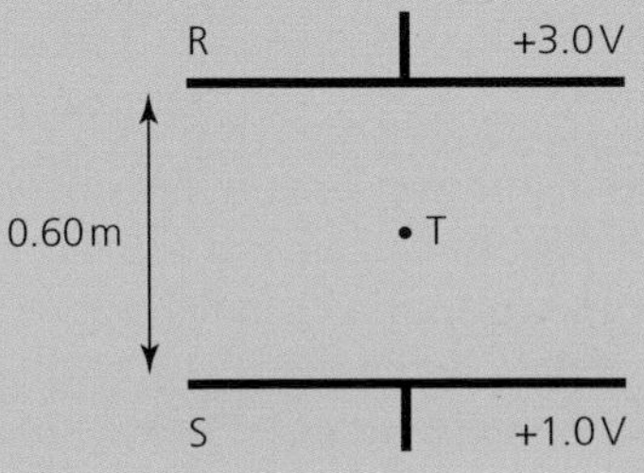

Figure Q2

   i. Copy Figure Q2 and sketch the electric field between R and S, showing its direction.

   ii. Point T is mid-way between R and S. Calculate the electric field strength at T.

   iii. Parts **b iii** and **c ii** both involve the electric field strength at a point mid-way between potentials of +1.0 V and +3.0 V. Explain why the magnitudes of these electric field strengths are different.

AQA Unit 4 section B January 2012 Q1

2. Figure Q3 shows a small polystyrene ball which is suspended between two vertical metal plates P_1 and P_2, 80 mm apart, that are initially uncharged. The ball carries a charge of −0.17 μC.

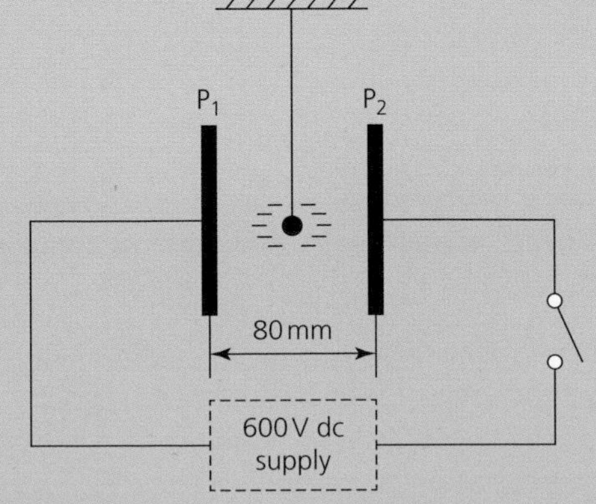

Figure Q3

a. i. A pd of 600 V is applied between P_1 and P_2 when the switch is closed. Calculate the magnitude of the electric field strength between the plates, assuming it is uniform.

ii. Show that the magnitude of the electrostatic force that acts on the ball under these conditions is 1.3 mN.

b. Because of the electrostatic force acting on it, the ball is displaced from its original position. It comes to rest when the suspended thread makes an angle θ with the vertical as shown in Figure Q4

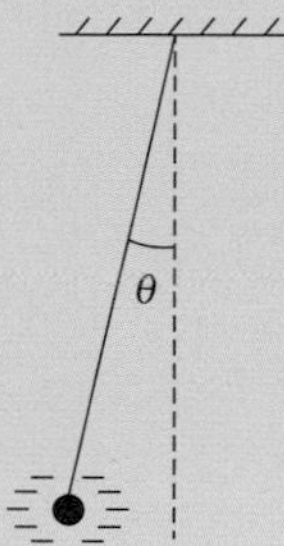

Figure Q4

i. Copy Figure Q4, and mark and label the forces that act on the ball when in this position.

ii. The mass of the ball is 4.8×10^{-4} kg. By considering the equilibrium of the ball, determine the value of θ.

AQA Unit 4 section B January 2011 Q4

3. A cathode ray tube uses a high-voltage anode to accelerate a beam of electrons across a vacuum tube. The potential difference between the cathode and anode is 2000 V (Figure Q5).

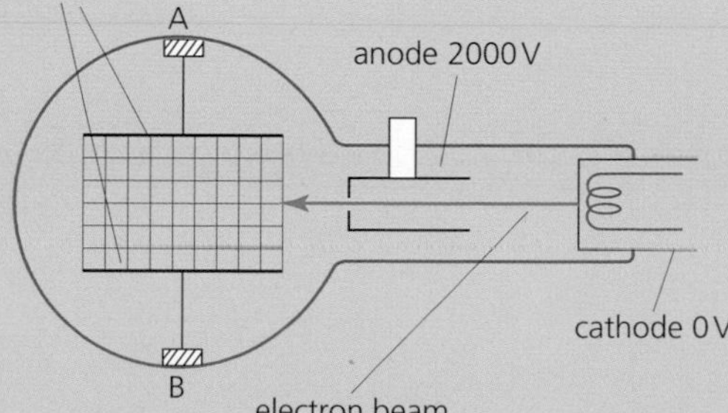

Figure Q5

a. Calculate the energy gained by an electron as it accelerates from the cathode to the anode. Give your answer in joules.

b. Show that the velocity of an electron as it reaches the anode is $2.7 \times 10^7 \text{ m s}^{-1}$.

c. The deflecting plates are 5.0 cm apart. A potential of +3.0 kV is applied to the top plate. The bottom plate is connected to 0 V.

i. Describe the path of the electron as it travels between the plates.

ii. Calculate the electric field strength between the plates.

iii. Calculate the acceleration of the electron as it moves between the deflection plates.

4. Figure Q6 shows the field lines and equipotential lines around an isolated point charge.

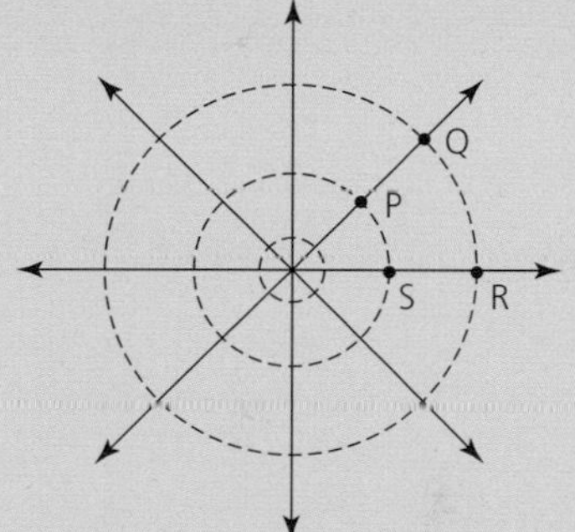

Figure Q6

Which one of the following statements, concerning the work done when a small charge is moved in the field, is **incorrect**?

A When it is moved from either P to Q or S to R, the work done is the same in each case.

B When it is moved from Q to R, no work is done.

C When it is moved around the path PQRS, the overall work done is zero.

D When it is moved around the path PQRS, the overall work done is equal to twice the work done in moving from P to Q.

AQA Unit 4 section A June 2014 Q16

5. Two fixed parallel metal plates X and Y are at constant potentials of +100 V and +70 V respectively (Figure Q7). An electron travelling from X to Y experiences a change of potential energy ΔE_p.

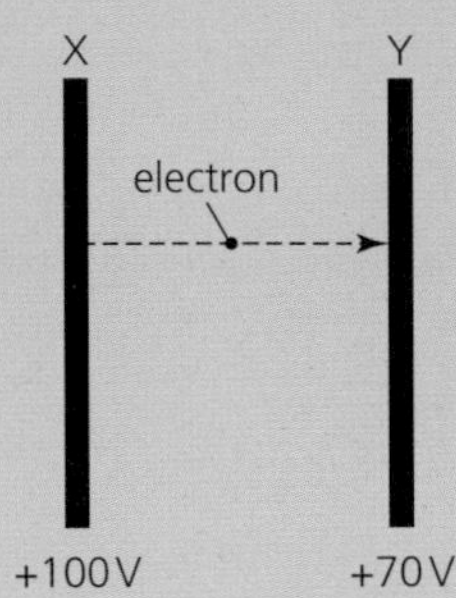

Figure Q7

Which line **A** to **D** in the table shows correctly the direction of the electrostatic force *F* on the electron and the value of ΔE_p?

	Direction of *F*	ΔE_p
A	Towards X	+30 eV
B	Towards Y	−30 eV
C	Away from X	+30 eV
D	Away from Y	−30 eV

AQA Unit 4 section A June 2014 Q17

6. Two fixed charges $+Q$ and $+3Q$ repel each other with a force F. If an additional charge of $-2Q$ is given to each of them, what is the force between them?

A $F/6$ attraction **B** $F/3$ attraction

C $F/6$ repulsion **D** $F/3$ repulsion

6 CAPACITANCE

PRIOR KNOWLEDGE

You will be familiar with electric current, potential difference and electrical resistance from (*Chapters 13 and 14 in Year 1 Student Book*), and will have experience of building electric circuits. You will have an understanding of uniform electric fields from the previous chapter.

LEARNING OBJECTIVES

In this chapter you will learn about a versatile electronic component called the capacitor. You will learn about the capacitor's properties, in particular, what is meant by its capacitance, its behaviour during its charging and discharging, and some of its many applications.

(Specification 3.7.4.1 to 3.7.4.4)

Toyota's TS040 Hybrid (Figure 1) has been designed to compete in the FIA World Endurance championships. The TS040 uses a 3.7 litre V8 petrol engine alongside a supercapacitor system mounted on both the front and rear axles to provide a temporary four-wheel drive facility.

Figure 1 *The Toyota TS040 Hybrid*

The supercapacitor, also known as ultracapacitor, is an electrochemical capacitor that can have huge capacitance values – of up to 1000 F. It is a charge storage device that can be used to provide back-up power, but is particularly useful when rapid bursts of energy are required. A selection of supercapacitors is shown in Figure 2. Supercapacitors store energy in an electric field, unlike batteries, which are a store of chemical energy. They can be charged much more rapidly and last much longer than a lithium-ion battery, but at present the amount of energy that they can store per unit mass is much less. Consequently, the supercapacitor has been unable to take over from the battery in electric vehicles. However, research continues to try to improve the supercapacitor, and possibly, with the use of nanomaterials such as graphene, they may play a more significant role in the cars of the future.

Figure 2 *A range of supercapacitors*

6.1 CAPACITANCE

A **capacitor** (Figure 3) is an electrical component that can be found in almost all electronic devices. There are many types of capacitor, with many different applications, but fundamentally their function is to store electric charge.

Figure 3 *There are many different types of capacitor component.*

A simple parallel-plate capacitor consists of two metal plates separated by a layer of electrical insulator called a **dielectric** (Figure 4). The plates may be flat sheets of metal or in the form of a foil. The electrical symbol for a capacitor is also shown in Figure 4.

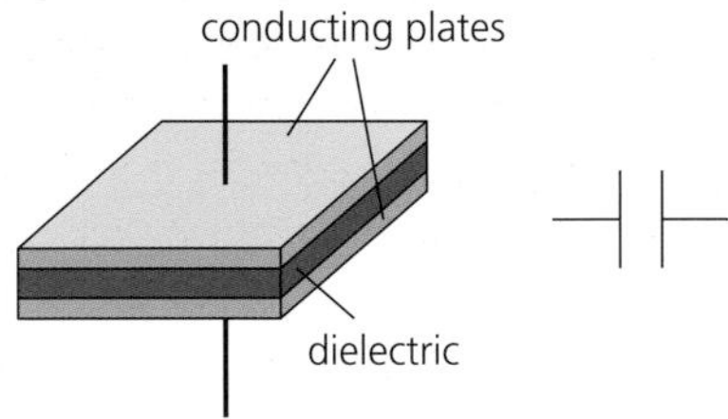

Figure 4 *Parallel-plate capacitor and the circuit symbol*

Since the capacitor plates are made of metal, they contain large numbers of electrons moving freely within a lattice of positive metal ions. The number of free electrons in the metal is balanced by the positive charge of the metal ions. However, when a capacitor is connected to a cell (Figure 5), electrons flow from the negative terminal of the cell onto one metal plate, giving that plate an excess of electrons and so making it negatively charged. An equal number of electrons flow off the other plate into the positive terminal of the cell, so making that plate positively charged, since it now has fewer free electrons than positive metal ions. It is important to understand that the dielectric, being an electrical insulator, prevents any electrons from actually passing *through* the capacitor itself. As more electrons accumulate on the negative plate, they oppose further electrons flowing on to it. Once the potential difference across the capacitor is equal to the emf ε of the cell (*see section 14.1 in Chapter 14 of Year 1 Student Book*), the flow of electrons stops. The capacitor is now fully charged, with a charge of $+Q$ on one plate and a charge of $-Q$ on the other plate.

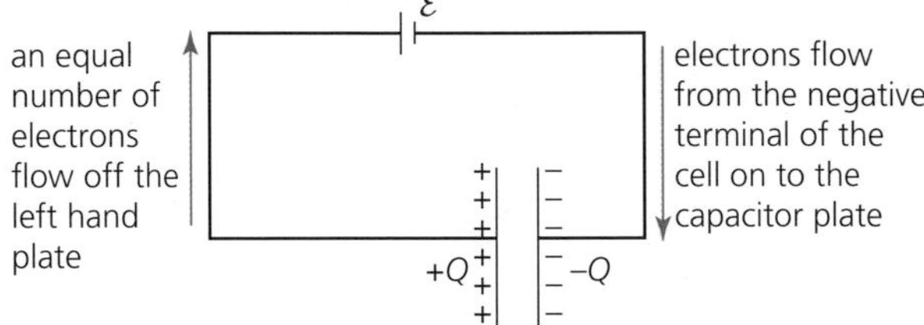

Figure 5 *A capacitor being charged*

The capacitor retains its charge when it is disconnected from the cell. If the terminals of the capacitor are subsequently connected to a bulb (Figure 6), the electrons flow from the negatively charged plate to the positively charged plate via the bulb, causing the bulb to light briefly. Once electrons amounting to charge $-Q$ have flowed through the lamp and onto the other plate, no more current flows and the capacitor has fully discharged. Since, during the discharge, a total charge of $-Q$ flowed from the negative plate round the circuit to the positive plate, the capacitor is described as having initial charge of magnitude Q.

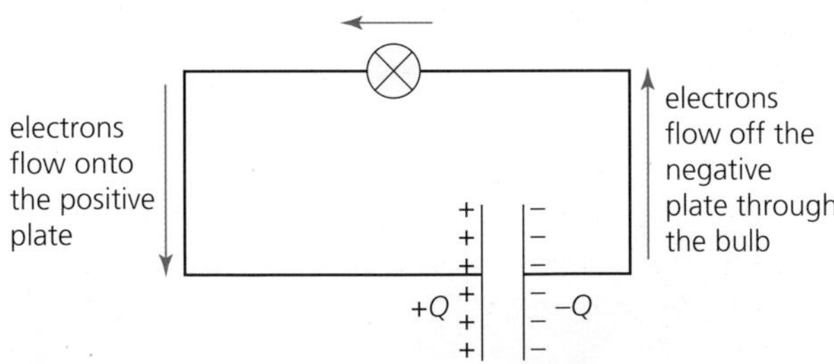

Figure 6 *A capacitor discharging and lighting a bulb*

It is possible to store more charge on the capacitor by increasing the pd across the capacitor. A particular capacitor will store double the charge if two cells in series are connected to its terminals rather than just one. However, depending on the design of the capacitor, there is a limit to the pd that can be applied. If the pd applied across the plates is too large, the insulating material of the dielectric will break down electrically and conduction will occur across the gap. Capacitors are usually labelled with their maximum permitted pd (Figure 7).

Figure 7 Capacitor showing its capacitance value and its maximum pd

The **capacitance** C of a capacitor is defined as the charge stored per unit potential difference across the plates:

$$C = \frac{Q}{V}$$

The unit of capacitance is the farad (F), and 1 F is equal to $1\,CV^{-1}$, where C is the coulomb. The farad is a large unit, and most capacitors used in electronic circuits have capacitance values very much smaller, so prefixes such as micro µ (10^{-6}), nano n (10^{-9}) and pico p (10^{-12}) are usually required (see Figure 7).

Figure 8 shows the graph of charge stored on a capacitor versus the pd applied across its plates. Since $Q = CV$, the gradient of the graph is equal to the capacitance of the capacitor. Since the graph is a straight line, this demonstrates that the capacitance of a particular capacitor is constant.

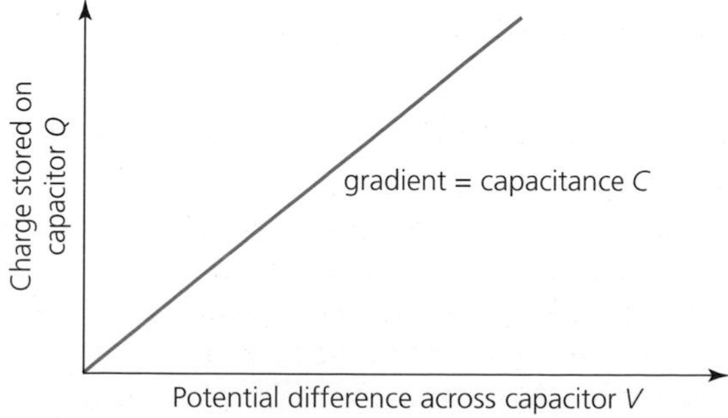

Figure 8 Graph of charge stored versus pd across a capacitor

Worked example

A capacitor is labelled 100 nF 10 V. Calculate the maximum charge that can be stored on it.

The maximum pd V that can be applied across the capacitor is 10 V.

$$\text{Capacitance } C = \frac{Q}{V} = 100\,\text{nF} = 100 \times 10^{-9}\,\text{F}$$

So the maximum charge stored Q will be

$$Q = CV = 100 \times 10^{-9} \times 10 = 1 \times 10^{-6}\,\text{C}$$

QUESTIONS

1. Calculate the maximum charge that can be stored by a 1000 µF capacitor that has a maximum pd of 35 V.
2. Calculate the voltage that would have to be applied to a 100 µF capacitor in order to store 500 µC of charge.

A capacitor with a large capacitance can store a large amount of charge with a relatively low pd across its plates. In contrast, a capacitor with a small capacitance would not store much charge even with a large pd across its plates. The capacitance of a parallel-plate capacitor depends on the area and on the separation of the plates, as well as on the type of dielectric that is sandwiched between the plates. If the capacitance of a particular capacitor is not known, it can be measured using a capacitance meter (Figure 9).

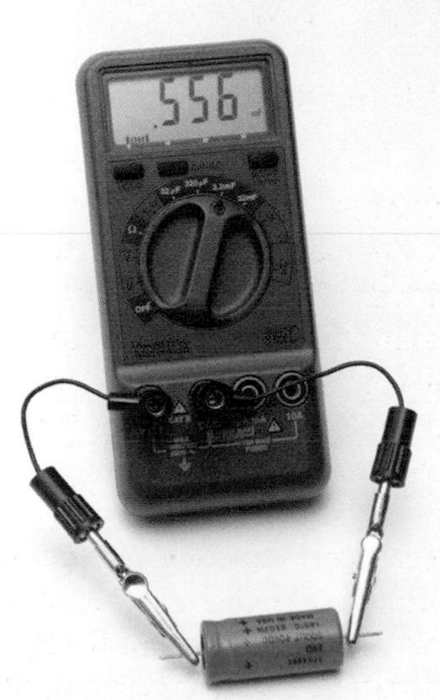

Figure 9 Capacitance meter

ASSIGNMENT 1: INVESTIGATING THE FACTORS AFFECTING THE CAPACITANCE OF A PARALLEL-PLATE CAPACITOR

(PS 1.2, PS 2.1, PS 2.4, PS 4.1)

The aim of this assignment is to consider how to investigate the factors that affect the capacitance of an air-gap capacitor made up of two flat metal plates positioned parallel to each other and supported vertically (Figure A1). The changes that can be made to the arrangement in Figure A1 include changing the plate separation *d* and the area of overlap *A*.

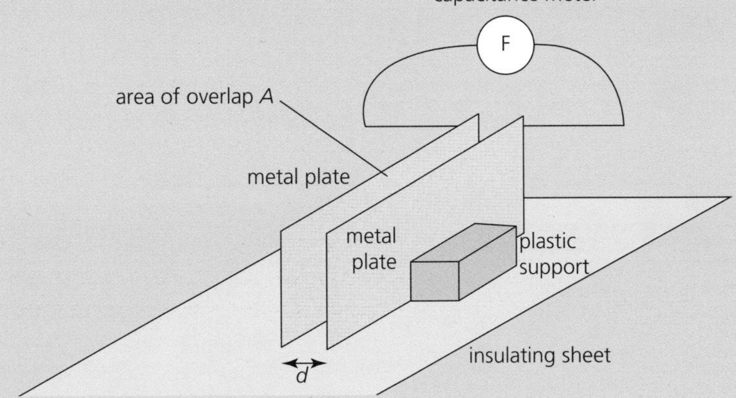

Figure A1 *Flat-plate capacitor arrangement*

The distance *d* can be measured using the internal jaws of a set of digital callipers (Figure A2). The separation of the internal jaws is adjusted as required, and there is a screw that can be tightened to lock the jaws, should this be necessary.

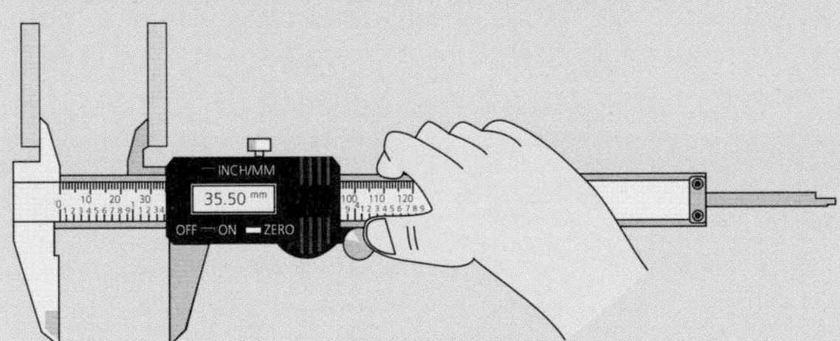

Figure A2 *Digital calliper measurement using the internal jaws*

The capacitance for various arrangements of the air-gap flat-plate capacitor can be measured using a digital capacitance meter (see Figure 9), which, since the capacitance of this arrangement is small, needs to be able to measure to 0.1 pF. It is not necessary to charge the plates with an external electromotive force (emf).

Questions

A1 Consider how an experiment to determine the effect on capacitance of varying the plate separation *d* could proceed. First, identify the dependent, the independent and the control variables. You should explain how you would ensure that the plate separation was the same for the full overlap area of the plates. State the graph that you would plot, specifying the quantities on the *x*-axis and the *y*-axis.

A2 Consider how an experiment to determine the effect on capacitance of varying the area of overlap *A* of the plates could proceed. First, identify the dependent, the independent and the control variables. You should explain with the aid of a diagram how you would determine the area of overlap.

A3 Consider two metal plates positioned as in Figure A1, with the maximum area of overlap. Determine the area of overlap with its uncertainty if the plate dimensions are measured using a 30 cm rule and found to be 12.6 cm × 9.1 cm.

A4 An alternative method for measuring the capacitance of the parallel-plate capacitor is to apply a high voltage to the plates and measure the charge on the plates using a coulombmeter, and then calculate capacitance from $C = \frac{Q}{V}$. Suggest an advantage that the capacitance meter method has compared with this coulombmeter method.

Factors affecting the capacitance of a parallel-plate capacitor

Experiments show that the capacitance C of a parallel-plate capacitor is directly proportional to the area A of the overlap of the plates and inversely proportional to the plate separation d (Figure 10). Therefore

$$C \propto \frac{A}{d}$$

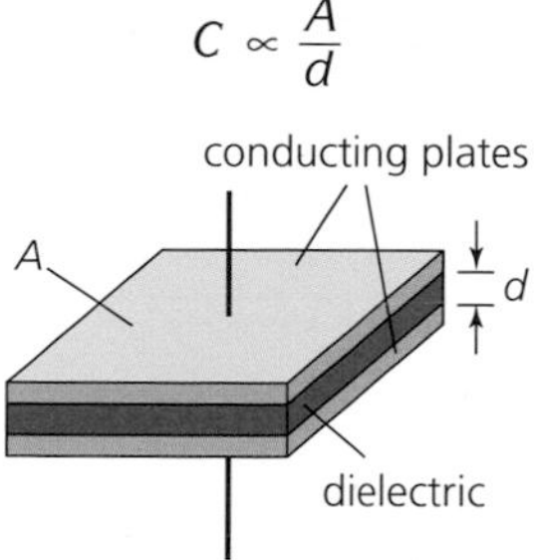

Figure 10

Capacitors of different capacitance can therefore be produced by changing the overlap area of the plates and the plate separation. A variable air capacitor, used for example in a radio tuner, is made up of interleaved parallel metal plates. The amount of plate surface that overlaps and the plate separation can be changed by rotating the rotor into the stator (Figure 11).

Figure 11 *Variable air capacitor*

QUESTIONS

3. By what factor would the capacitance of a parallel-plate capacitor change if the length and width of each plate were doubled and the plate separation was halved?

The capacitance of any arrangement of parallel plates also depends on the type of dielectric. Consider first a charged parallel-plate capacitor with air in the gap between the plates. The charge on the plates creates a uniform electric field in the region between the plates, and the capacitance is equal to $\frac{Q}{V}$. To increase the capacitance, a dielectric material is placed in the gap between the plates. The dielectric must be a good electrical insulator to prevent the flow of electrons between the plates – suitable materials include polyester, polycarbonate, mica and polythene. The materials used as dielectrics in capacitors contain **polar molecules**. A polar molecule has a partial positive charge in one part of the molecule and a partial negative charge in another part of the molecule, although overall the molecule is electrically neutral. The polar molecules in the dielectric material have random orientations before the capacitor is charged, but, on experiencing the electric field of a charged capacitor, they rotate and align themselves with the field (Figure 12), with the positive end of the molecule facing the negative plate and the negative end of the molecule facing the positive plate.

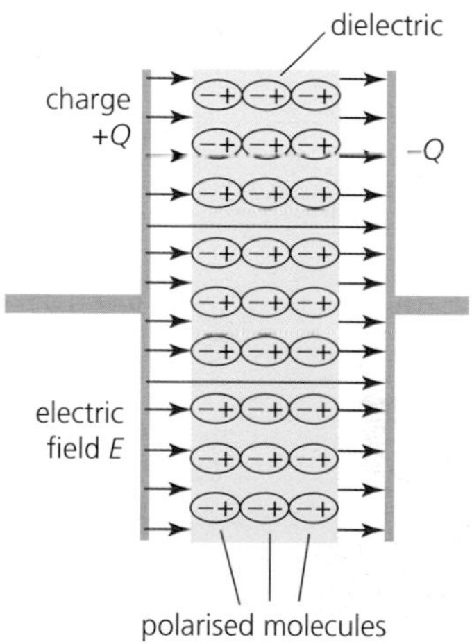

Figure 12 *Polarised dielectric*

This **polarisation** of the dielectric creates an internal electric field, which partially cancels the electric field created by the charge on the plates, so reducing the potential difference across the capacitor. To return the potential difference to its original value requires the addition of more charge onto the plates. The overall effect of the dielectric is therefore to increase the amount of charge a capacitor can store for a given potential difference, hence increasing its capacitance.

The **relative permittivity**, ε_r, of a dielectric is the factor by which the electric field between two charges is decreased by the presence of the dielectric, relative to a vacuum. In the case of a capacitor, the relative permittivity is also equal to the factor by which the capacitance is increased when using that material as a dielectric compared with a similar capacitor that has a vacuum between its plates. Relative permittivity is also known as the **dielectric constant**. It is a number with no unit. Values of the relative permittivity of some of the materials used as capacitor dielectrics are shown in Table 1.

Material	Relative permittivity
Polyester resin	3.5
Polycarbonate	3
Mica	7
Polythene	2.4

Table 1 *Values of relative permittivity for common dielectric materials*

The equation for the capacitance C of a parallel-plate capacitor with plate area A and plate separation d is

$$C = \frac{A\varepsilon_0\varepsilon_r}{d}$$

where ε_0 is a constant called the 'permittivity of free space', equal to $8.85 \times 10^{-12}\,\mathrm{F\,m^{-1}}$. The relative permittivity of air is 1.0006, so a capacitor with air in the gap between its plates can be treated as having a vacuum between its plates.

QUESTIONS

4. A flat-plate capacitor like the one in Figure A1, with plates approximately 20 cm × 20 cm separated by air, has a very small capacitance, typically of the order of picofarads (pF). Figure 13 shows the internal structure of a typical capacitor component, composed of layers of foil and plastic rolled up like a Swiss roll. It is just a few centimetres long. Explain why this type of capacitor could have a much larger capacitance than the flat-plate capacitor.

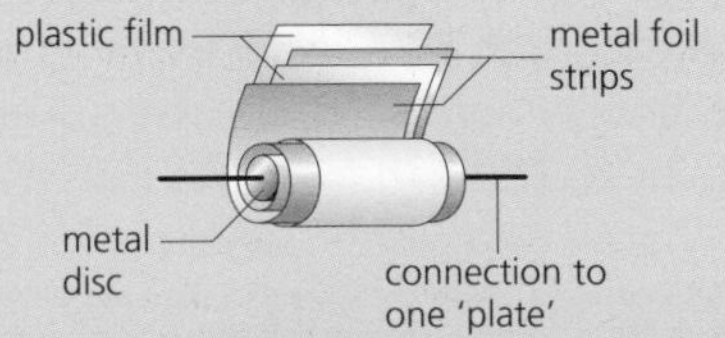

Figure 13 *Capacitor made from layers of foil and plastic*

Electrolytic capacitors

Capacitors produced using the dielectrics listed in Table 1 come in various designs, but have capacitances typically less than 100 μF. However, **electrolytic capacitors** are capable of storing much more charge, and have capacitances of up to 0.1 F. They consist of two aluminium plates either side of a sheet of paper soaked in aluminium borate (Figure 14). When the capacitor is charged up, a chemical reaction occurs, which deposits a layer of aluminium oxide on the positive plate. This very thin oxide layer acts as the dielectric.

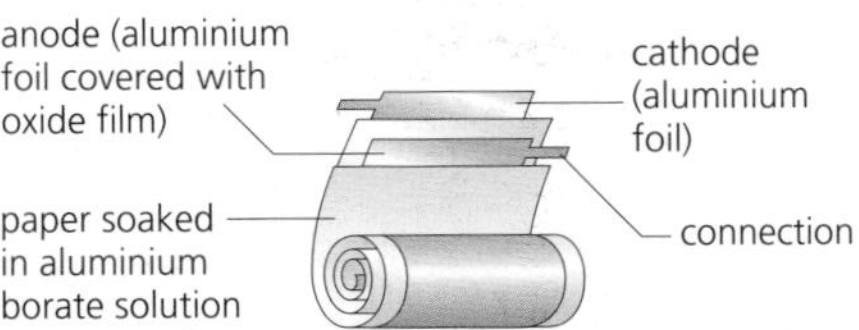

Figure 14 *Construction of an electrolytic capacitor*

Unlike other capacitors, electrolytic capacitors have a definite polarity and must be connected the correct way round when being charged. To help avoid making a mistake with the polarity, electrolytic capacitors carry a label identifying the negative terminal (Figure 15), along with the capacitance and maximum voltage labels. Electrolytic capacitors are used in audio amplifiers from hi-fi to mobile phones and in power supply units.

Figure 15 *Negative terminal labelling on an electrolytic capacitor*

QUESTIONS

5. Calculate the capacitance in pF of an air-filled parallel-plate capacitor with plate dimensions 15 cm × 20 cm and plate separation of 10 mm.
6. A capacitor consists of two flat metal plates separated by a gap of 5.0 mm. The area of each plate is 0.04 m^2. Determine its capacitance for each of the dielectrics listed in Table 1, giving your answers in picofarads (pF).
7. a. Describe what happens to a polar molecule of dielectric when its capacitor is charged.
   b. What effect does the internal field created by the dielectric in a charged capacitor have on the overall electric field of the capacitor?
8. Determine the maximum amount of charge that can be safely stored by a electrolytic capacitor labelled 4700 μF 16 V.

KEY IDEAS

- A capacitor stores electric charge.
- Capacitance C is defined as the charge Q stored per unit potential difference V across the capacitor plates:

$$C = \frac{Q}{V}$$

Capacitance is measured in farad, F, where 1 F = 1 CV^{-1}.

- The capacitance C of a parallel-plate capacitor with plate area A and plate separation d is

$$C = \frac{A\varepsilon_0\varepsilon_r}{d}$$

where ε_0 is the permittivity of free space and ε_r is the relative permittivity, or dielectric constant, of the dielectric between the plates.

- A dielectric is an insulating material with polar molecules that orient themselves in the electric field between the capacitor plates. This results in the capacitor with a particular applied pd V across it storing more charge.

6.2 ENERGY STORED BY A CAPACITOR

Since a charged capacitor can light a bulb or deliver an electric current to other components in a circuit, it is clearly a store of electrical energy. Capacitors can therefore be used as back-up power supplies for electronic systems such as computers.

When a battery is first connected to a capacitor, electrons start to flow onto one capacitor plate and off the other. As charge accumulates on the capacitor, the battery has to work harder to get more electrons onto the negative plate because of the repulsion from the electrons already on the plate. If the potential difference at some instant during the charging is V_1, and an additional small amount of charge ΔQ is added to the capacitor, then the work done ΔW by the battery in adding this additional charge is $\Delta W = V_1\Delta Q$ (see section 5.2 of Chapter 5). On a graph of potential difference versus charge for a capacitor (Figure 16), work ΔW corresponds to the small darker shaded area.

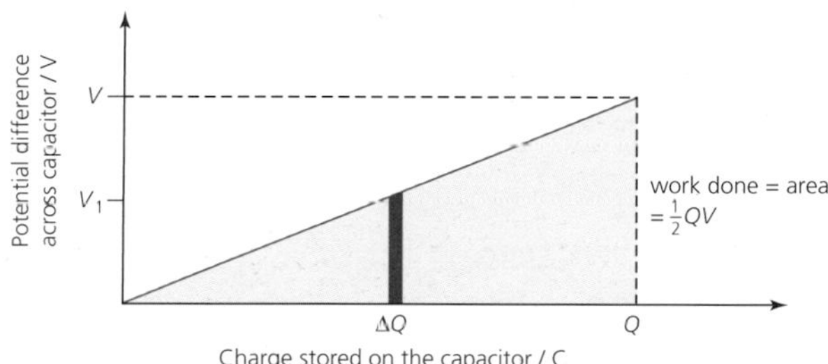

Figure 16 *Graph of pd versus charge for a capacitor*

Therefore, to fully charge a capacitor from zero charge to charge Q requires the battery to do a total amount of work W equal to the lighter shaded area between the line and the charge axis of the graph in Figure 16. This area is a triangle of base Q and height V, where V is the final pd across the charged capacitor. The work done, W, by the battery in charging the capacitor is therefore given by

$$W = \frac{1}{2}QV$$

The work done in charging the capacitor is equal to the energy stored in the capacitor, and therefore

$$\text{energy stored } E = \frac{1}{2}QV$$

Since $C = \frac{Q}{V}$, this can also be expressed as

$$\text{energy stored } E = \frac{1}{2}CV^2 = \frac{1}{2}\frac{Q^2}{C}$$

The energy stored in a capacitor is often given the symbol E, so care has to be taken not to confuse this with E used for electric field strength.

Capacitors can be particularly useful when a high power output is required for a short time interval. A camera flash unit contains a capacitor that is charged by a battery to its full potential difference. The capacitor then discharges through a xenon flash tube in a very short period of time, converting the electrical energy stored in the capacitor to light energy emitted from the flash tube. The power output from the capacitor is high but lasts less than a millisecond.

Worked example

The capacitor used to store charge in a camera flash unit has a capacitance of 470 mF and can be charged to a potential of 30 V.

a. How much energy is stored by the capacitor when it is fully charged?

b. If the capacitor discharges through the flash in a time of 0.20 ms, calculate the average power.

a. The energy stored is $E = \frac{1}{2}QV = \frac{1}{2}CV^2$

$= 0.5 \times 470 \times 10^{-3} \times 30^2 = 211.5\text{ J}$

b. Power is rate of energy transfer $= \dfrac{211.5}{0.20 \times 10^{-3}}$

$= 1.06\text{ MW}$

QUESTIONS

9. a. Calculate the theoretical energy stored in a charged 1000 μF capacitor when there is a potential difference of 30 V across it.

   b. A real capacitor is labelled 1000 μF 25 V. Determine the maximum energy that the capacitor can safely store.

10. A 1000 μF capacitor is charged by being connected to a battery of emf 6.0 V.

    a. Determine

       i. the total charge that flows from the battery to the capacitor

       ii. the energy stored in the capacitor when fully charged

       iii. the energy supplied by the battery.

    b. Account for the difference between your answers to **a ii** and **a iii**.

11. A bank of capacitors has a total capacitance of 50 μF and is charged up to 20 kV. If the bank of capacitors is discharged in 5.0 ms, what would be the average power output?

    A 2 MW **B** 200 MW **C** 200 kW **D** 2 kW

12. Figure 17 shows a graph of charge stored versus pd across the plates of a capacitor. Determine the energy stored in the capacitor when the pd across the plates is 20 V.

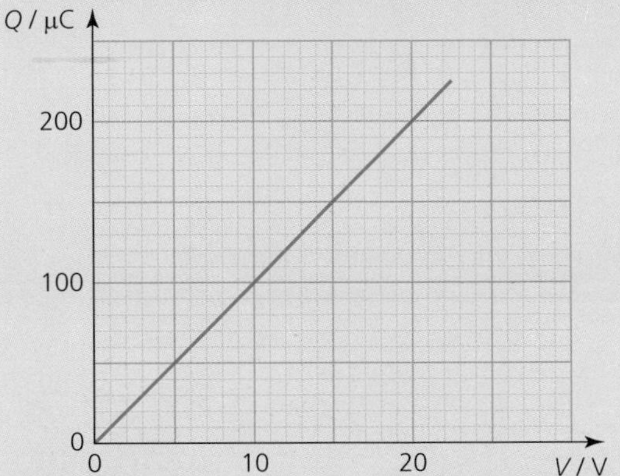

Figure 17

13. A portable defibrillator (Figure 18) is an electronic device that uses a supercapacitor to deliver a therapeutic dose of electrical energy to the heart.

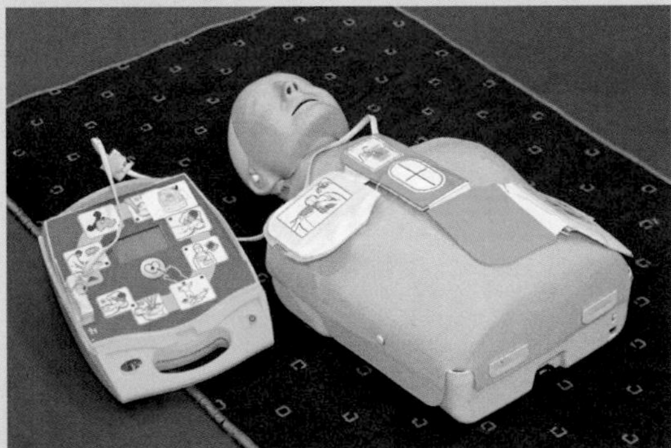

Figure 18 *A portable defibrillator works by using a supercapacitor, which can deliver its stored energy very quickly and give an electric shock.*

    a. How much energy is delivered in one dose if the supercapacitor has a capacitance of 100 F and is charged to 2.7 V?

    b. Why should supercapacitor components be stored and handled with care?

When a thunder cloud forms, positive ions collect near the top of the cloud and electrons collect near the bottom (Figure 19). The electrons that have accumulated at the bottom of the cloud exert a repulsive force on free electrons in the surface of the Earth and push them deeper into the ground, leaving a net positive charge on the surface of the Earth in the area beneath the cloud. The negative lower section of the cloud and the surface of the Earth effectively becomes a charged capacitor. Air is an insulator and in dry conditions requires an electric field of 3000 kV m^{-1} to make it conduct. However, in the damp conditions leading to the formation of a thunder cloud, an electric field of 300 kV m^{-1} will make air conduct, leading to a lightning strike between the cloud and the ground.

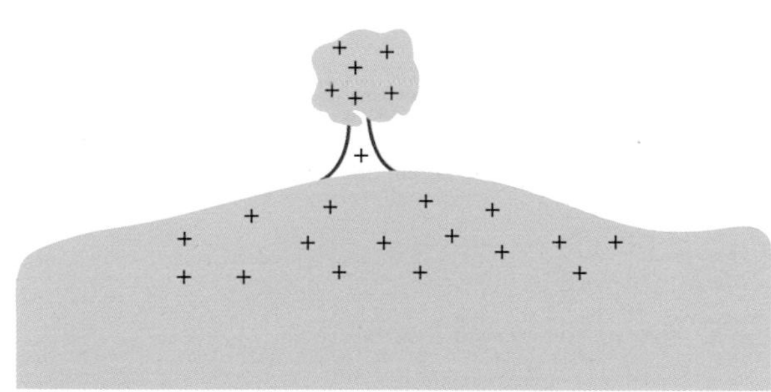

Figure 19 *Charge distribution in a thunder cloud*

QUESTIONS

14. a. Determine the energy stored in the 'capacitor' formed from the underside of a thunder cloud and the ground beneath, given that the distance from the Earth's surface to the underside of the cloud is 2000 m and the charge on the underside of the cloud is about 40 C. Assume that an electric field of 300 kV m^{-1} will make air conduct in damp conditions.

b. If the cloud discharged fully during a lightning strike lasting 30 μs, what would be the power output of the strike?

KEY IDEAS

- A charged capacitor is a store of electrical energy.
- The energy stored is equal to the area under the charge–pd graph.
- The energy stored E in a capacitor is given by

$$E = \frac{1}{2}QV = \frac{1}{2}CV^2 = \frac{1}{2}\frac{Q^2}{C}$$

6.3 CAPACITOR CHARGING

For many capacitor applications, for example, when a time delay is required, the rate that a capacitor becomes charged and then subsequently discharges needs to be carefully controlled. This can be achieved by connecting a resistor in series with the capacitor. The greater the resistance of the resistor, the more slowly the charging or discharging takes place.

Consider the capacitor charging circuit of Figure 20 in which a capacitor of capacitance C is connected in series with a switch, a cell and a resistor. Just as the switch is closed, the current is initially at its maximum value, which we will call I_0, because at this instant, before any charge builds up on the capacitor, the pd across the resistor R is equal to the emf, ε, of the cell. Therefore, assuming that the cell's internal resistance is negligible (*see section 14.1 in Chapter 14 in Year 1 Student Book*), we have

$$I_0 = \frac{\text{emf of cell}}{\text{circuit resistance}} = \frac{\varepsilon}{R}$$

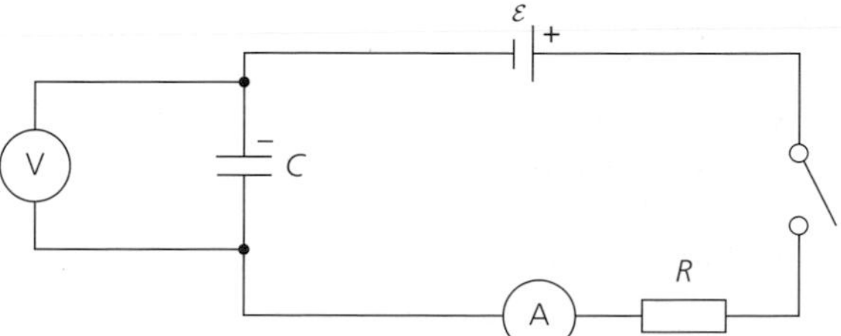

Figure 20 *Capacitor charging circuit*

The current decreases with time, approaching zero when the capacitor is fully charged. If the charging current I is measured as time passes, the graph of charging current versus time (Figure 21) is an **exponential decay** curve – the current is decreasing exponentially. This means that the charging current reduces by the same factor in equal time intervals – the graph has a 'constant ratio property'. Specifically, the current halves in equal time intervals. The time for a quantity that is decreasing exponentially to halve is called the **half-life**, $T_{1/2}$.

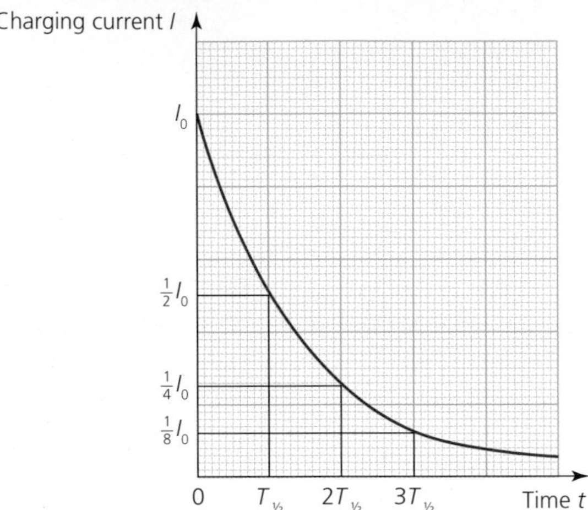

Figure 21 *Capacitor charging current versus time: an exponential decay curve*

The gradient of the curve in Figure 21 at any point is the rate of change of current at that moment in time, which, as can be seen from the graph, decreases as time passes.

The equation of the curve is an exponential and is given by

$$I = I_0\, e^{-t/RC}$$

Since the half-life, $T_{1/2}$, of the charging circuit is the time for the charging current to halve, and the initial charging current is I_0, then $T_{1/2}$ is the time for the current to fall to $\frac{1}{2}I_0$. Therefore

$$\frac{1}{2}I_0 = I_0\, e^{-T_{1/2}/RC}$$

Taking natural logs of both sides and rearranging gives

$$\log_e 2 = \frac{T_{1/2}}{RC}$$

and therefore

$$T_{1/2} = 0.69\,RC$$

While the half-life, $T_{1/2}$, is a useful indicator of how quickly, current, pd and charge decrease in a capacitor discharge circuit, another quantity, the **time constant**, is particularly useful in calculating time delays required in electronic circuits.

The time constant is the time for the charging current to fall to $\frac{1}{e}$ of its original value.

Consider the charging current at time $t = RC$. Then

$$I = I_0\, e^{-t/RC} = I_0\, e^{-RC/RC} = I_0\, e^{-1} = \frac{I_0}{e}$$

Therefore, time constant = RC. This can be illustrated in a graph of current versus time (Figure 22). Two useful facts are:

- $\frac{1}{e}$ of the original current value is approximately 37% of I_0.
- A reasonable estimate for the time for a capacitor to fully charge is 5 time constants = $5RC$.

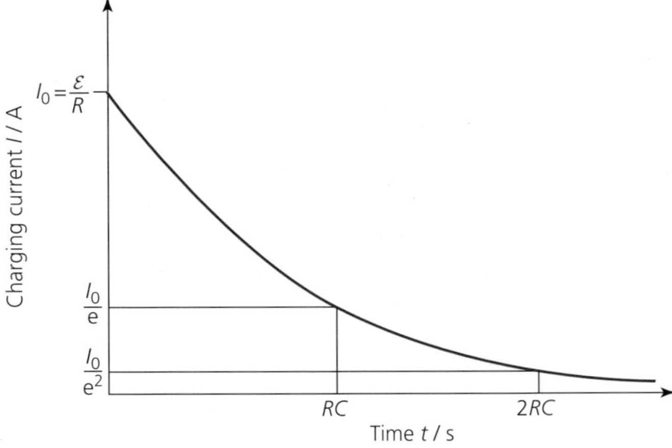

Figure 22 *Capacitor charging current versus time, showing the time in intervals of the time constant* RC

Charging current versus time data can also be analysed using a log–linear plot of $\log_e I$ versus time. Taking logarithms of the equation

$$I = I_0\, e^{-t/RC}$$

gives

$$\log_e I = \log_e I_0 + \log_e e^{-t/RC}$$

which can be simplified to

$$\log_e I = \log_e I_0 - \frac{t}{RC}$$

Rearranging to

$$\log_e I = -\frac{t}{RC} + \log_e I_0$$

helps in making a comparison with the equation of a straight line, $y = mx + c$. A graph of $\log_e I$ versus t (Figure 23) is a straight line with gradient of $-\frac{1}{RC}$ and y intercept of $\log_e I_0$. Since the time constant of the circuit is equal to RC, then

$$\text{time constant} = -\frac{1}{\text{gradient}}$$

Note that $\log_e I$ can also be written $\ln I$.

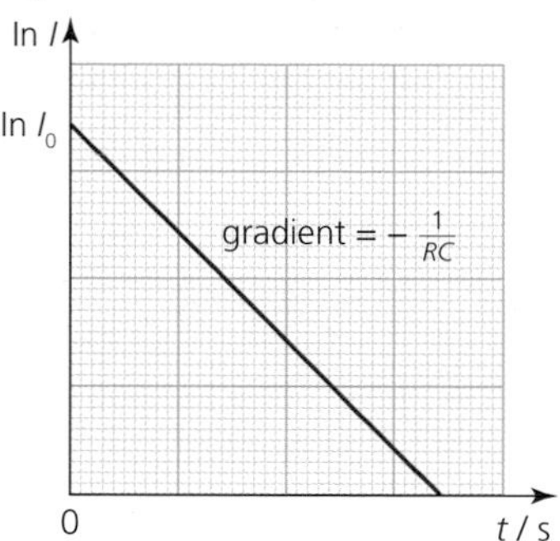

Figure 23 Graph of $\log_e$ I (ln I) versus time

Worked example 1

A 1000 µF capacitor is connected in series with a switch, a 22 kΩ resistor and a battery of emf 6.0 V. Calculate:

a. the maximum current
b. the current flowing 10 s after the switch was closed
c. the time for the initial current to halve
d. the time for the current to fall to 50 µA
e. the time constant.
f. Make a reasonable estimate of the time for the capacitor to fully charge.

a. The initial current is the maximum current:

$$I_0 = \frac{\text{emf of cell}}{\text{circuit resistance}} = \frac{\varepsilon}{R} = \frac{6}{22 \times 10^3}$$
$$= 2.727 \times 10^{-4} = 2.7 \times 10^{-4}\text{ A}$$

b. After 10 s the current is

$$I = I_0\, e^{-t/RC}$$
$$= 2.727 \times 10^{-4} \times e^{-10/(22 \times 10^3 \times 1000 \times 10^{-6})}$$
$$= 2.727 \times 10^{-4} \times e^{-10/22}$$
$$= 1.7 \times 10^{-4}\text{ A}$$

c. Half-life is $T_{1/2} = 0.69RC$
$$= 0.69 \times 22 \times 10^3 \times 1000 \times 10^{-6} = 15\text{ s}$$

d. Charging current is given by $I = I_0\, e^{-t/RC}$. Substituting current values gives

$0.5 \times 10^{-4} = 2.7 \times 10^{-4}\, e^{-t/RC}$. Therefore $\frac{0.5}{2.7} = e^{-t/RC}$, and taking natural logarithms of both sides

gives $-1.686 = -\frac{t}{RC}$. Finally,

$$t = RC \times 1.686$$
$$= 22 \times 10^3 \times 1000 \times 10^{-6} \times 1.686 = 37\text{ s}$$

e. Time constant $RC = 22 \times 10^3 \times 1000 \times 10^{-6} = 22$ s.

f. Estimate of time for capacitor to become fully charged is $5RC = 5 \times 22 = 110$ s.

QUESTIONS

15. A 1000 µF capacitor is charged by connecting it in series with a 9.0 V battery and an 18 kΩ resistor.
    a. Calculate the maximum current.
    b. Calculate the time for the charging current to halve.
    c. Calculate the time constant of the circuit.
    d. Estimate the time for the capacitor to be fully charged.

Change of pd while charging

During the charging process, the pd V across the capacitor of capacitance C increases as time passes, and, once the pd across the capacitor has reached the emf ε of the cell, the current effectively becomes zero. For the circuit in Figure 20, the maximum pd V_0 across the capacitor is equal to ε.

At any instant, the pd across resistor R is equal to $\varepsilon - V$. Therefore, current I at any instant is given by

$$I = \frac{\text{pd across } R}{R} = \frac{\varepsilon - V}{R}$$

and, at any instant, the potential difference V across the capacitor is given by $V = \varepsilon - IR$. Since current I is decreasing, the equation predicts that pd V should increase as time passes, eventually reaching a maximum value equal to the emf ε of the cell.

Figure 24 shows the predicted graph of pd across the charging capacitor versus time. The gradient of the graph decreases as time passes, showing that the pd across the capacitor increases at a reduced rate.

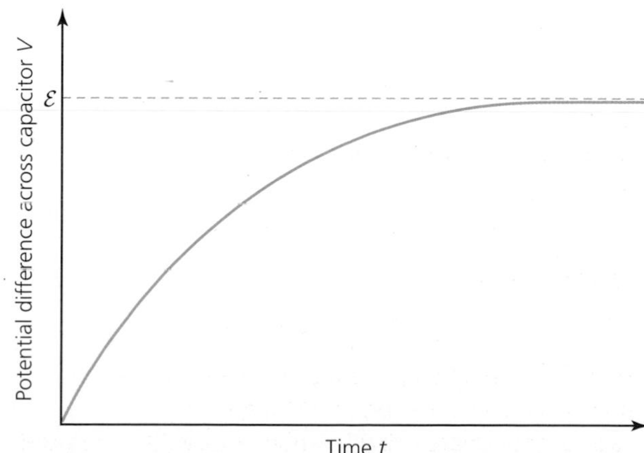

Figure 24 The pd across a charging capacitor versus time

REQUIRED PRACTICAL: APPARATUS AND TECHNIQUES

Part 1: Investigation of the variation of capacitor charging current with time

The aim of this practical is to show that a capacitor charging current decreases exponentially as time passes. Measurements of charging current at specific time intervals allow it to be shown that a capacitor charging current decreases exponentially and also allow the half-life and time constant of the circuit to be determined.

The practical gives you the opportunity to:

- correctly construct circuits from circuit diagrams using cells and a range of circuit components including those where polarity is important
- use appropriate digital instruments to obtain a range of measurements (to include time, current).

Apparatus

The circuit in Figure P1 can be used to generate current data during the charging of a 1000 μF electrolytic capacitor. To avoid damage to the electrolytic capacitor, the negative terminal of the capacitor **must** be connected to the negative terminal of the cell. The negative terminal of an electrolytic capacitor is marked on it (see Figure 15).

A multimeter set as a microammeter capable of reading to 0.1 μA would be suitable for measuring the charging current. A digital stopwatch should be used.

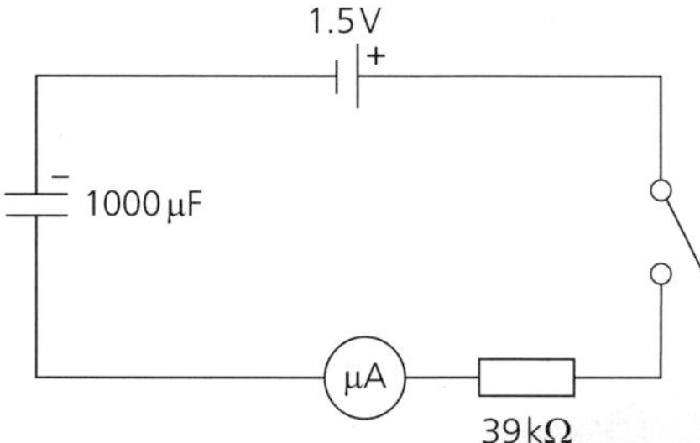

Figure P1 *Capacitor charging circuit*

Technique

Current measurements should be taken at least every 10 s after closing the switch, for at least 60 s.

Analysis 1: Using a graph of charging current versus time

The y variable of an exponential decay curve halves for a constant increase in the x variable. This is called a constant ratio property, and the constant increase in the x variable is called the half-life. To show that the capacitor charging current decreases exponentially with time, a graph of current versus time should be plotted and then the times for the current to fall by half, for example, from 30 μA to 15 μA, from 20 μA to 10 μA and from 10 μA to 5 μA, determined. If these three time measurements are sufficiently close in value, then the constant ratio property is confirmed, and it can be concluded that the charging current decreases exponentially. An average of the three time measurements gives a value for the half-life of the circuit.

Analysis 2: Using a log–linear plot of $\log_e I$ versus time

A log–linear plot involves plotting the log of the dependent variable on the y-axis, and the independent variable, usually time t, on the x-axis. A scientific calculator is needed to obtain natural log values of charging current to enable a graph of $\log_e I$ versus time t to be plotted.

Although logarithms are powers and therefore do not have units, it is good practice to label the log axis in such a way that shows the units of the quantity to which logarithms have been applied. So the y-axis of the graph should be labelled $\log_e (I/\mu A)$, since the current will be measured in microamps. This graph will be a straight line if the charging current decreases exponentially with time. The gradient of the line can be used to determine the time constant of the circuit, because from the exponential decay equation

$$\text{time constant} = -\frac{1}{\text{gradient}}$$

QUESTIONS

The circuit in Figure P1 was used to obtain time and current measurements after closing the switch to charge the capacitor. The results shown in Table P1 were obtained.

Time / s	Charging current / µA
0	38.5
10.00	29.2
20.00	22.8
30.00	17.6
40.00	13.5
50.00	10.3
60.00	8.2
70.00	6.4
80.00	4.9
90.00	3.8
100.00	2.9

Table P1

P1 Plot a graph of charging current on the *y*-axis versus time on the *x*-axis. Demonstrate that the charging current decreases exponentially with time and obtain a value for the half-life of the circuit.

P2 Since the analysis requires a graph of current on the *y*-axis versus time on the ***x***-axis, why has the data been presented with time in the first column of the table and current in the second column?

P3 Plot a graph of $\log_e (I/\mu A)$ on the y-axis versus time on the *x*-axis. Demonstrate that the charging current decreases exponentially with time and obtain a value for the time constant of the circuit.

P4 **a.** Describe how you would expect the data to change if a resistor of larger resistance were used in the charging circuit.

b. Explain your answer to part **a** using your knowledge of the theoretical equation for the exponential decay of the current.

Since the charging current $I = I_0\, e^{-t/RC}$ and the pd across the charging capacitor is $V = \varepsilon - IR$, the equation of the pd versus time graph shown in Figure 24 is

$$V = \varepsilon - RI_0\, e^{-t/RC}$$

In the capacitor charging circuit in Figure 20, RI_0 is equal to the pd across resistor R at the instant the switch is closed and is equal to the emf ε of the cell. The equation for the pd across the capacitor therefore becomes

$$V = \varepsilon(1 - e^{-t/RC})$$

This equation is usually written with V_0, the maximum pd across the capacitor, since $V_0 = \varepsilon$, and therefore

$$V = V_0(1 - e^{-t/RC})$$

Change of charge while charging

Since the charge on the capacitor $Q = CV$, we can write

$$Q = CV = CV_0(1 - e^{-t/RC})$$

The maximum charge on the capacitor is denoted by $Q_0 = CV_0$, and therefore the equation for the charge on a capacitor during charging is

$$Q = Q_0(1 - e^{-t/RC})$$

A graph of the charge stored on the capacitor versus time during charging is shown in Figure 25.

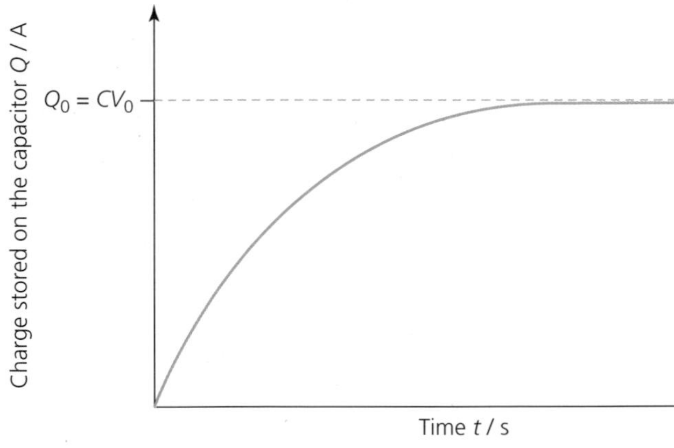

Figure 25 *Charge versus time for a charging capacitor*

The time constant RC is the time for either the charge or the pd to reach $\left(1 - \frac{1}{e}\right)$ of its maximum value, that is, 63% of Q_0 and V_0, respectively.

Charge stored and charging current are related, since current is equal to the rate of flow of charge, $I = \frac{\Delta Q}{\Delta t}$. This means that the gradient of a tangent at any point on a graph of charge stored versus time gives the current. It also means that, for a current versus time graph, the area between the curve and the time axis is equal to the charge stored. Consider a short time interval Δt during the charging. During this short time interval, current I barely changes. Therefore, $I \times \Delta t$, which corresponds to the shaded area of the I–t graph (Figure 26), is equal to ΔQ, which is the charge that flows onto the capacitor during time Δt. Therefore, the full area between the curve and the time axis is equal to the total charge that flows onto the capacitor, and is equal to the total charge stored.

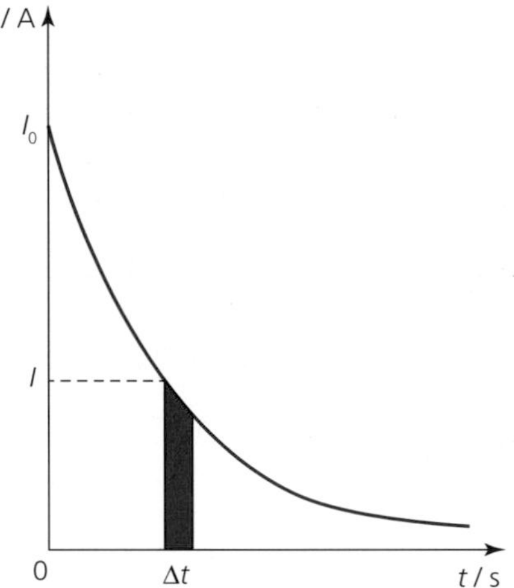

Figure 26 *Considering the area below the charging current versus time graph*

Worked example 2

As in Worked example 1, a 1000 μF capacitor is connected in series with a switch, a 22 kΩ resistor and a battery of emf 6.0 V. Calculate:

a. the charge on the capacitor after 10 s
b. the pd across the capacitor after 10 s
c. the time for the pd to reach 63% of its maximum value (= the time constant).

a. Charge is
$Q = Q_0(1 - e^{-t/RC}) = CV_0(1 - e^{-t/RC})$
$= 1000 \times 10^{-6} \times 6 \times (1 - e^{-10/(22\times10^3 \times 1000 \times 10^{-6})})$
$= 1000 \times 10^{-6} \times 6 \times (1 - e^{-10/22})$
which gives $Q = 2.192 \times 10^{-3} = 2.2 \times 10^{-3}$ C.

b. The pd is $V = \frac{Q}{C} = \frac{2.192 \times 10^{-3}}{1000 \times 10^{-6}} = 2.192 = 2.2\,\text{V}$

c. The time for the pd to reach 63% of its maximum value is the time constant for a charging circuit, which is equal to $RC = 22 \times 10^3 \times 1000 \times 10^{-6} = 22\,\text{s}$.

QUESTIONS

16. A 1000 μF capacitor is charged by connecting it in series with a 4.5 V battery and a 47 kΩ resistor. Calculate

a. the maximum pd across the capacitor

b. the charge on the capacitor 10 s after the circuit is completed

c. the time for the charge stored to reach 63% of its maximum value (= the time constant).

17. A graph of capacitor charging current versus time is shown in Figure 27. Use the graph to estimate the total charge that has flowed onto the capacitor in the first 60 s.

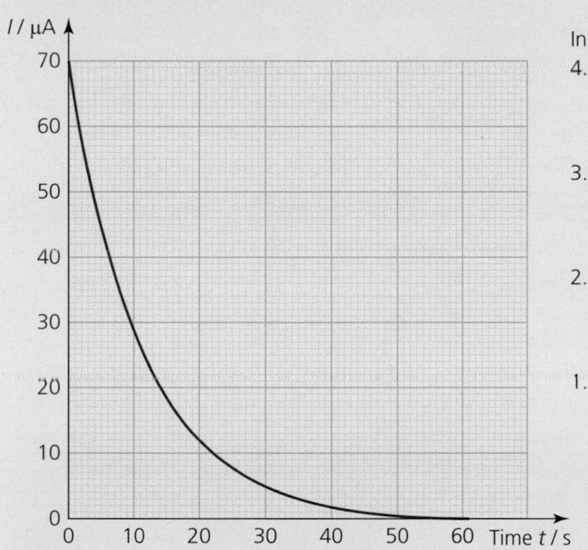

Figure 27 *An I–t graph for a charging capacitor*

18. A capacitor of unknown size is charged using the circuit in Figure 28. On closing the switch, the current starts to flow. To prevent the current from decreasing exponentially, the variable resistor is continuously adjusted so that the current remains constant at 20 μA. The charging continues for 70 s, at which point the current suddenly drops to zero and the voltmeter reads 3.0 V. Determine the capacitance of the unknown capacitor.

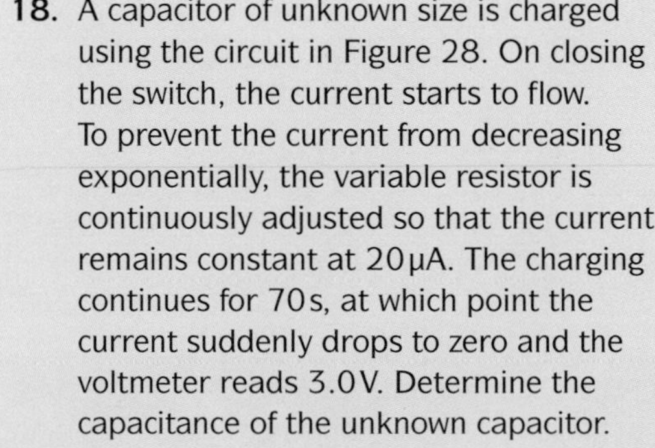

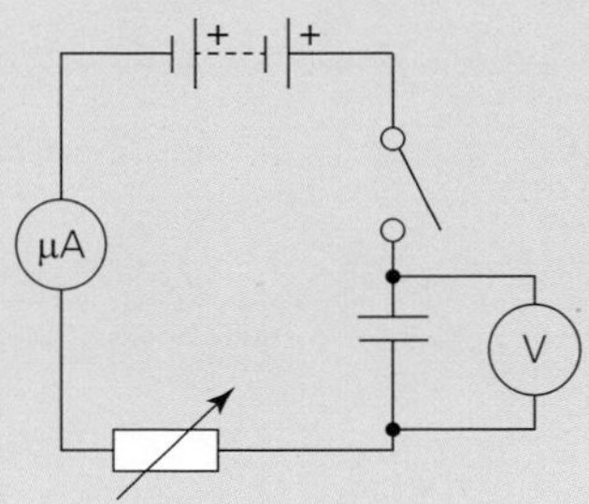

Figure 28 *Constant charging current circuit*

19. Figure 29 shows a graph of $\log_e I$ versus time t, where I is the charging current. Use the graph to determine the time constant of the circuit.

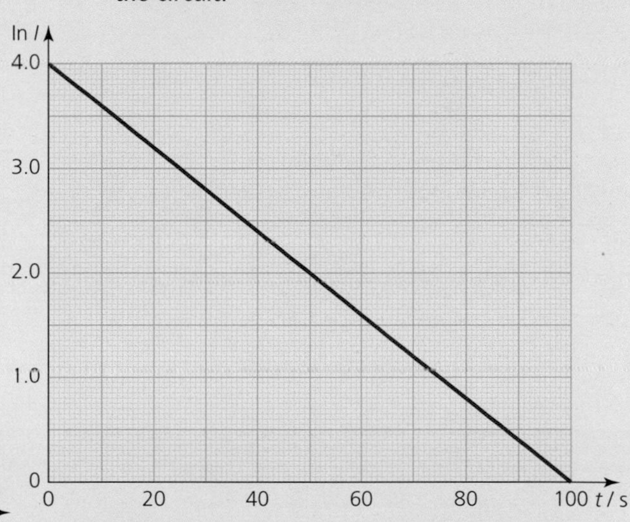

Figure 29 *A graph of ln I versus t for a capacitor charging circuit*

20. Figure 30 is a graph of charge stored versus time for a charging capacitor. A tangent has been drawn corresponding to time 20 s. Estimate the charging current flowing at 20 s.

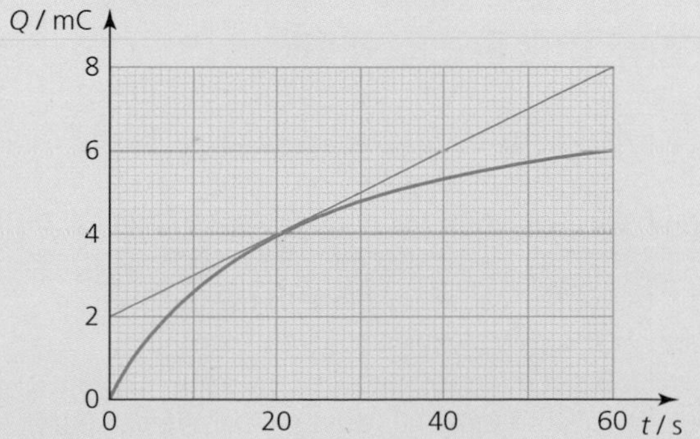

Figure 30 *Graph of charge versus time for a charging capacitor*

KEY IDEAS

- As a capacitor is charged, the charging current I decreases exponentially according to the equation

$$I = I_0\, e^{-t/RC}$$

- The area between the curve and the time axis of an I–t graph is equal to the charge stored.
- The half-life $T_{1/2}$ is the time for the charging current to halve: $T_{1/2} = 0.69RC$.
- RC is the time constant of the circuit, which is the time for the charging current to decrease to $\frac{1}{e}$ of its original value.
- The gradient of a graph of $\log_e I$ versus time is equal to $-\frac{1}{\text{time constant}} = -\frac{1}{RC}$.
- The potential difference V across the capacitor increases during charging and is given by

$$V = V_0(1 - e^{-t/RC})$$

which means that the circuit time constant, RC, is the time for V to reach $\left(1-\frac{1}{e}\right)$ of its final value.
- The charge Q on a capacitor increases during charging and is given by

$$Q = Q_0(1 - e^{-t/RC})$$

which means that the circuit time constant, RC, is the time for Q to reach $\left(1-\frac{1}{e}\right)$ of its final value.
- The gradient of the tangent at a point on a Q–t graph for a capacitor charging is equal to the current at that point in time.

6.4 CAPACITOR DISCHARGING

Consider the charging/discharging circuit of Figure 31. Initially, the capacitor is uncharged. When switch S1 is closed, electrons flow from the negative terminal of the cell onto the lower capacitor plate, and electrons flow off the upper capacitor plate to the positive terminal of the cell. In this circuit, the charging happens almost instantly because the only resistance in the circuit is the internal resistance of the cell plus the resistance of the wires and the contacts, which will probably be roughly 1Ω.

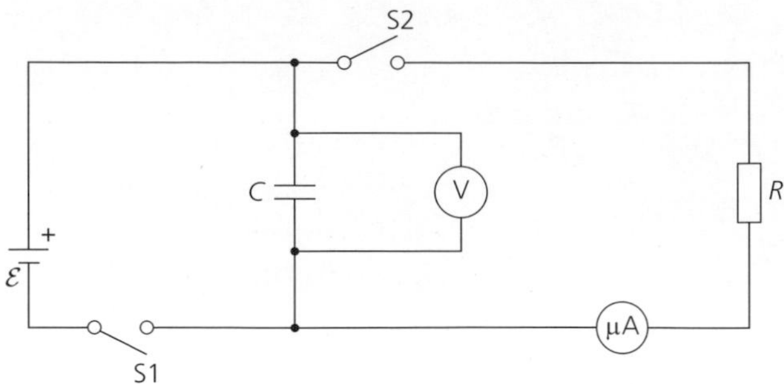

Figure 31 *Charging/discharging circuit*

The capacitor is now storing charge, Q_0, and the maximum pd across the capacitor, V_0, is equal to the emf, ε, of the cell. Switch S1 is now opened. At the instant when switch S2 is closed, the pd across resistor R is equal to V_0 and the current is at its maximum, I_0, such that $I_0 = \frac{V_0}{R}$. Electrons flow from the negative plate of the capacitor through the resistor and ammeter to the positive plate of the capacitor. The capacitor is now in the process of discharging.

The discharging current, like the charging current, decreases exponentially with time according to the equation

$$I = I_0\, e^{-t/RC}$$

This relation, and hence the graph of current versus time, is the same as that for the capacitor charging current, shown in Figure 26. The total area between the curve and the time axis is equal to the total charge that flows off the capacitor and must equal the original charge that was stored on the capacitor, which is $Q_0 = CV_0$.

Change in pd during discharging

Throughout the discharge, the pd across the resistor R in Figure 31 is equal to the pd, V, across the capacitor, and therefore $V = IR$, showing that, at any instant, the discharging current I is directly proportional to the pd V across the capacitor. The potential difference across the capacitor must therefore also decrease exponentially with time according to the equation

$$V = V_0\, e^{-t/RC}$$

The half-life $T_{1/2}$ is the time for *both* the current and the pd to halve during the capacitor's discharge, and is illustrated on the V–t graph in Figure 32. Similarly, the time constant = RC is the time for both the current and the pd to decrease to $\frac{1}{e}$ (37%) of their original value.

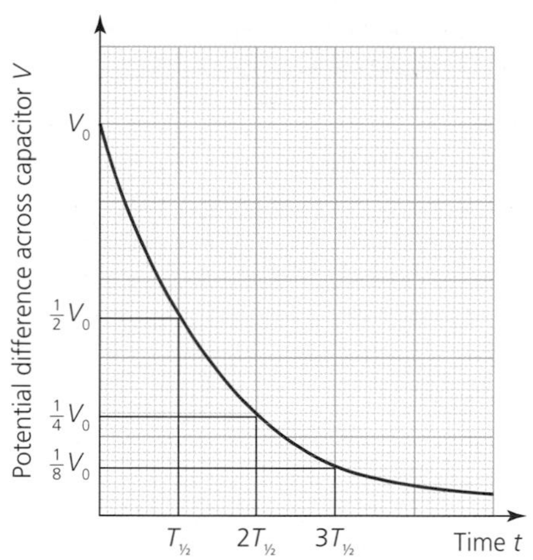

Figure 32 *The pd across a discharging capacitor versus time: an exponential decay curve*

Taking natural logarithms of $V = V_0\ e^{-t/RC}$ gives

$$\log_e V = \log_e V_0 + \log_e (e^{-t/RC})$$

which can be simplified to

$$\log_e V = \log_e V_0 - \frac{t}{RC}$$

Rearranging to

$$\log_e V = -\frac{t}{RC} + \log_e V_0$$

helps in making a comparison with the equation of a straight line, $y = mx + c$. A graph of $\log_e$ V versus time t (Figure 33) is a straight line with gradient of $-\frac{1}{RC}$ and y intercept of $\log_e V_0$. Since the time constant of the circuit is equal to RC, then

$$\text{time constant} = -\frac{1}{\text{gradient}}$$

Note that $\log_e V$ can also be written as $\ln V$.

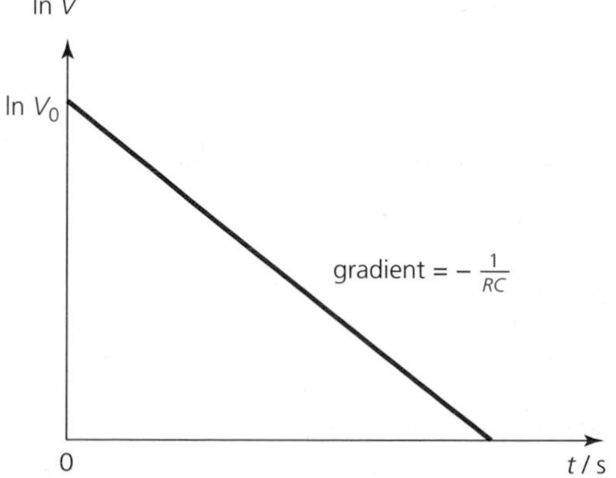

Figure 33 *Graph of $\log_e V$ versus t*

Change in charge during discharging

Consider the relationship between the discharging current I and the charge Q remaining on the capacitor. During the capacitor's discharge, the potential difference V across the capacitor is equal to the pd across the resistor, IR, and also equal to $\frac{Q}{C}$.

Therefore

$$I = \frac{Q}{CR}$$

showing that, at any instant during the discharge, the charge on the capacitor is directly proportional to the discharging current.

Combining the discharging current equation

$$I = I_0\ e^{-t/RC}$$

with

$$I = \frac{Q}{CR}$$

gives

$$\frac{Q}{CR} = \frac{Q_0}{CR} e^{-t/RC}$$

Therefore the equation for the charge on the capacitor during discharge is given by

$$Q = Q_0\ e^{-t/RC}$$

showing that the charge Q on the capacitor also decreases exponentially with time. The time constant RC is equal to the time for the charge to fall to $\frac{1}{e}$ (37%) of Q_0, and the half-life is the time for the charge on the capacitor to halve.

A Q versus t curve (Figure 34) therefore has the same shape as the I–t and the V–t graphs. The gradient of a tangent drawn at a point on this graph is equal to the rate of flow of charge at that instant in time, which is equal to the current I.

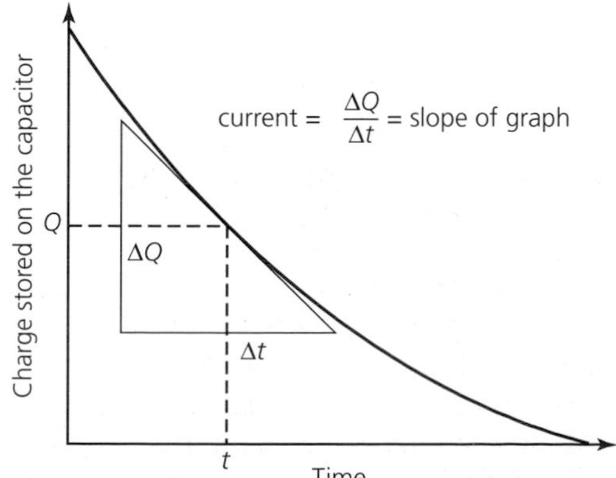

Figure 34 *Charge versus time for a discharging capacitor*

REQUIRED PRACTICAL: APPARATUS AND TECHNIQUES

Part 2: Investigation of variation of capacitor discharging current and pd with time

The aim of this practical is to obtain simultaneous current and pd measurements as time passes during the discharging of an electrolytic capacitor in order to produce graphs of I versus t, V versus t and $\log_e V$ versus t.

The practical gives you the opportunity to

- correctly construct circuits from circuit diagrams using cells and a range of circuit components, including where polarity is important
- use a data logger with a variety of sensors to collect data, and use software to process data.

Apparatus

The circuit in Figure P2 can be used. A 4700 μF electrolytic capacitor is charged by connecting it across a 6 V battery. The capacitor is then discharged through a 200 Ω resistor. The discharging current and pd are measured with ammeter and voltmeter sensors connected to a data logger. The data logger and the computer may be connected without cables, using Wi-Fi.

To protect the electrolytic capacitor from damage, it is necessary to ensure that the negative terminal of the capacitor is connected to the negative terminal of the battery.

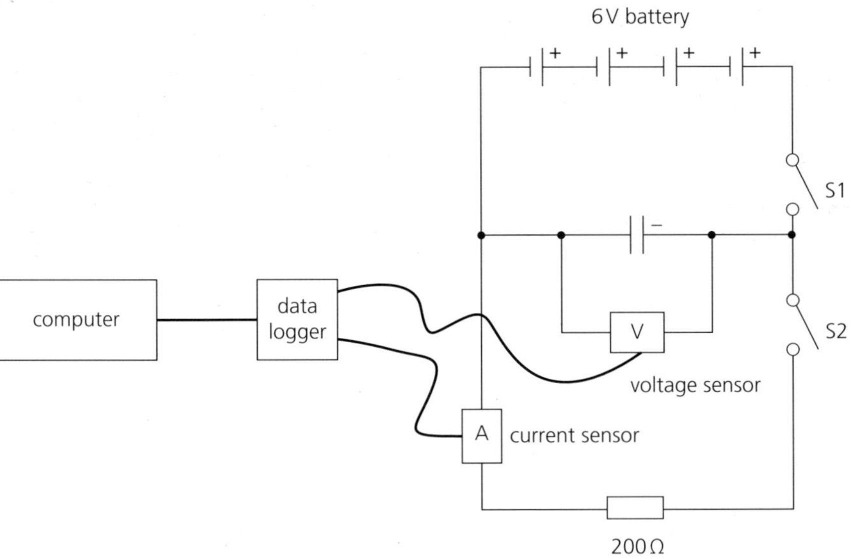

Figure P2 *Capacitor charging/discharging circuit*

Technique

The capacitor is charged, almost instantly, by closing switch S1. Switch S1 is then opened in order to isolate the charged capacitor. The data logger is then switched on and switch S2 is closed to enable the capacitor to discharge through the 200 Ω resistor.

The data logging software allows display of the discharging current versus time graph and the pd across the capacitor versus time graph, and should be able to determine the area between the current curve and the time axis. It can also be used to generate a graph of $\log_e V$ versus time. The exponential decrease of pd with time is confirmed if this graph is a straight line. The software should be able to generate a gradient, which can be used to determine the time constant of the circuit.

QUESTIONS

P5 A student has undertaken the experiment and obtained a value for the area between the curve and the time axis of an *I*–*t* graph of 0.0326 C, and a gradient of the $\log_e V$ versus time graph of $-0.893\,s^{-1}$. The emf of the battery was measured at 5.97 V. Use the data to determine

a. the circuit time constant

b. the size of the capacitor

c. the size of the resistor.

P6 The nominal capacitance and resistance values used in the circuit of Figure P2 have tolerances of 20% and 5%, respectively, as specified by the manufacturer. Determine whether the values of capacitance and resistance values calculated in question **P5** lie within these tolerances.

P7 Suggest a suitable sampling rate for the data logger in the experiment using the apparatus shown in Figure **P2**.

P8 A student who does not have access to data logging equipment uses a voltmeter and stopwatch to measure the potential difference across a discharging capacitor every 10 s and obtains the data in Table P2.

Time / s	0	10.0	20.0	30.0	40.0	50.0	60.0
Pd / V	12.0	8.67	6.31	4.56	3.27	2.39	1.71

Table P2

a. Use the data to plot a log–linear graph to determine the time constant of the circuit.

b. Suggest the disadvantages of this method compared with using sensors and data logging equipment and software.

In a capacitor discharge circuit, the current, the pd and the charge Q on the capacitor all decrease exponentially with time with the same time constant RC. We can summarise this as follows:

The gradients of the $\ln Q$ versus time graph, the $\ln I$ versus time graph and the $\ln V$ versus time graph for capacitor discharge are all equal: gradient $= -\frac{1}{\text{time constant}}$.

QUESTIONS

21. A graph of $\log_e V$ versus time for a capacitor discharge circuit is shown in Figure 35. Determine the time constant for the circuit.

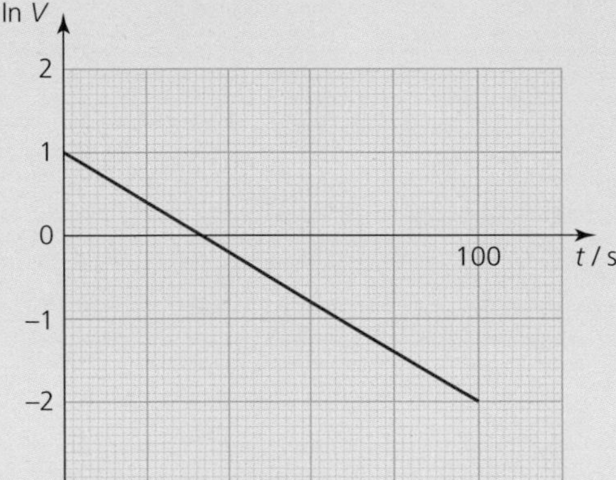

Figure 35 *A ln V versus t graph for capacitor discharge*

22. A 500 μF capacitor is discharging through a 60 kΩ resistor. At the instant that the current flowing is 60 μA, what is the amount of charge still stored on the capacitor plates?

Stretch and challenge

23. In a capacitor discharge circuit, the discharging current I and the charge Q on the capacitor are related by the equation

$$I = \frac{Q}{CR}$$

Since current is the rate of flow of charge,

$$I = -\frac{dQ}{dt}$$

The minus sign is present because the charge on the capacitor is decreasing. Therefore

$$-\frac{dQ}{dt} = \frac{Q}{CR}$$

which can be written as

$$-\frac{dt}{CR} = \frac{dQ}{Q}$$

Integrate both sides of this equation between time = 0, when charge = Q_0, and time = t, when charge = Q, and take inverse natural logs to show that $Q = Q_0 e^{-t/RC}$.

KEY IDEAS

- As a capacitor is discharged, the discharging current I decreases exponentially according to the equation

$$I = I_0 e^{-t/RC}$$

- The area between the curve and the time axis of an I–t graph is equal to the charge that has flowed off the capacitor.
- The potential difference V across the capacitor decreases exponentially during discharge and is given by

$$V = V_0 e^{-t/RC}$$

- The charge Q stored on a capacitor decreases exponentially during discharge and is given by

$$Q = Q_0 e^{-t/RC}$$

- The gradient of the tangent at a point on a Q–t graph for a capacitor discharging is equal to the current at that point in time.
- The half life $T_{1/2} = 0.69RC$ is the time for the discharging current, the pd across the capacitor and the charge stored on the capacitor to halve.
- The time constant of the circuit, RC, is the time for the discharging current, the pd across the capacitor and the charge stored on the capacitor to decrease to $\frac{1}{e}$ of their original values.
- The graphs of $\log_e$(current), $\log_e$(pd) and $\log_e$(charge) versus time for a discharging capacitor all have the same gradient, equal to $-\frac{1}{\text{time constant}}$.

PRACTICE QUESTIONS

1. a. Define the *capacitance* of a capacitor.
   b. The circuit shown in Figure Q1 contains a battery, a resistor, a capacitor and a switch.
   The switch in the circuit is closed at time $t = 0$. The graph in Figure Q2 shows how the charge Q stored by the capacitor varies with t.

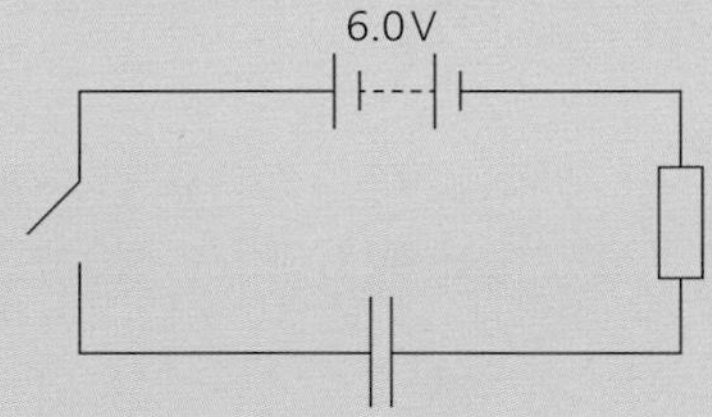

Figure Q1

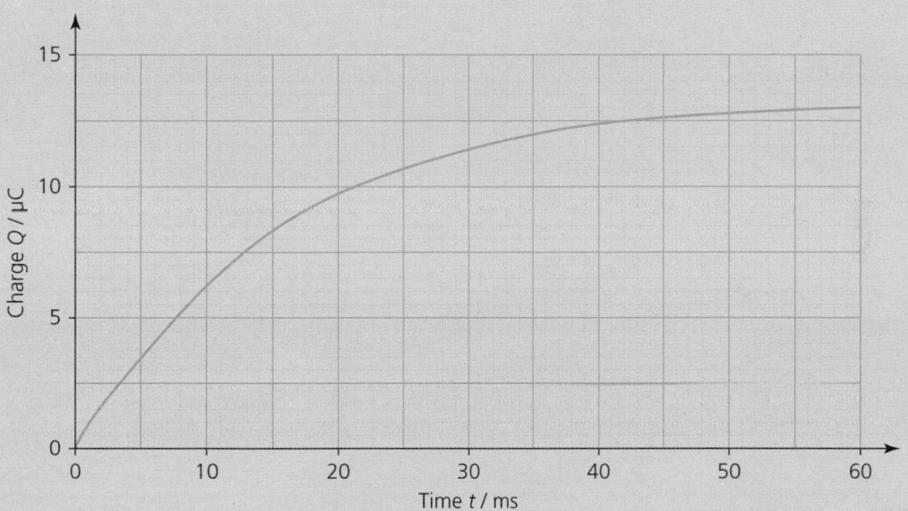

Figure Q2

   i. When the capacitor is fully charged, the charge stored is 13.2 μC. The electromotive force (emf) of the battery is 6.0 V. Determine the capacitance of the capacitor.
   ii. The time constant for this circuit is the time taken for the charge stored to increase from 0 to 63% of its final value. Use the graph to find the time constant in milliseconds.
   iii. Hence calculate the resistance of the resistor.
   iv. What physical quantity is represented by the gradient of the graph?

   c. i. Calculate the maximum value of the current, in mA, in this circuit during the charging process.
   ii. Sketch a graph to show how the current (in mA) varies with time (in ms up to 60 ms) as the capacitor is charged. Mark the maximum value of the current on your graph.

AQA January 2012 Unit 4 Section B Q2

2. Figure Q3 shows how the charge stored by a capacitor varies with time when it is discharged through a fixed resistor.

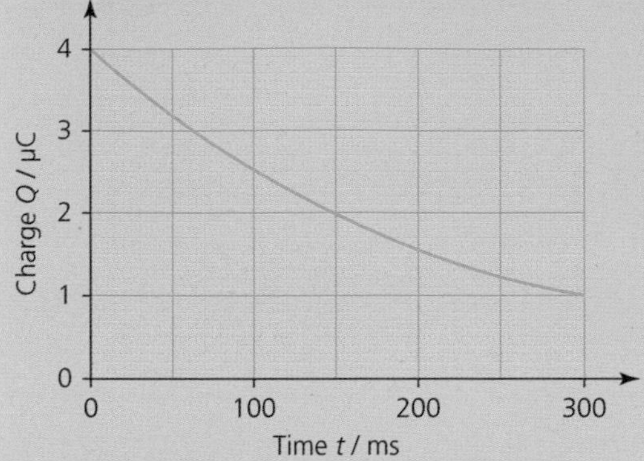

Figure Q3

   a. Determine the time constant, in ms, of the discharge circuit.
   b. Explain why the rate of discharge will be greater if the fixed resistor has a smaller resistance.

AQA January 2013 Unit 4 Section B Q3

3. Taking a photograph indoors requires the use of an electronic flash, which emits a short burst of light as the photograph is taken. The flash has a step-up circuit so that a 1.5 V cell can charge a 200 μF capacitor up to 200 V. The capacitor then completely discharges through a xenon gas discharge tube in a time of 1.0 ms.
   a. Calculate
      i. the charge stored
      ii. the energy stored on the capacitor just before it discharges.
   b. Calculate the average output power as the capacitor discharges through the xenon gas discharge tube.

4. Switch S in the circuit (Figure Q4) is held in position 1, so that the capacitor C becomes fully charged to a pd V and stores energy E.

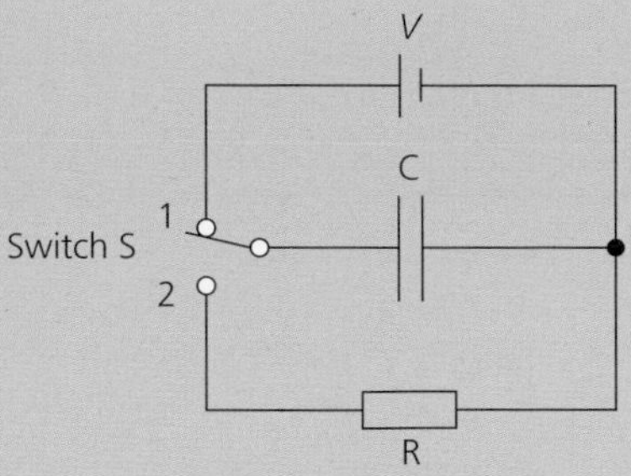

Figure Q4

The switch is then moved quickly to position 2, allowing C to discharge through the fixed resistor R. It takes 36 ms for the pd to fall to $\frac{V}{2}$. What period of time must elapse, after the switch has moved to position 2, before the energy stored by C has fallen to $\frac{E}{16}$?

A 51 ms **B** 72 ms **C** 432 ms **D** 576 ms

AQA June 2014 Unit 4 Section A Q20

5. Which one of the following statements about a parallel-plate capacitor is **incorrect**?
   A The capacitance of the capacitor is the amount of charge stored by the capacitor when the pd across the plates is 1 V.
   B A uniform electric field exists between the plates of the capacitor.
   C The charge stored on the capacitor is inversely proportional to the pd across the plates.
   D The energy stored when the capacitor is fully charged is proportional to the square of the pd across the plates.

AQA January 2013 Unit 4 Section A Q19

6. A 2200 μF capacitor and a 22 μF capacitor are charged to the same potential difference. If the charge stored in the 2200 μF capacitor is Q_1 and the charge stored in the 22 μF capacitor is Q_2, what is the ratio $\frac{Q_1}{Q_2}$?

A $\frac{1}{100}$ **B** 1 **C** 10 **D** 100

7 MAGNETIC FIELDS

PRIOR KNOWLEDGE

You will have experience of magnetic fields produced by permanent magnets. You should know that a magnetic field is produced when a current flows through a wire, with applications in electromagnets and motors. You will be familiar with the concept of a field of force and the use of field lines to provide a visual representation of a field's strength and direction.

LEARNING OBJECTIVES

In this chapter you will learn about magnetic flux and flux density, about the forces that arise when a current-carrying conductor or when moving charged particles are in a magnetic field and about the application of these effects in the cyclotron accelerator.

(Specification 3.7.5.1 to 3.7.5.3)

NASA's *Solar Dynamics Observatory* satellite (*SDO*) is on a mission to find out more about how the Sun's magnetic field is generated, why it flips every 11 years and how its stored magnetic energy is released into the heliosphere. Data and images (Figure 1) from *SDO* will help scientists learn how to predict more accurately the variations in the Sun's behaviour that affect life on Earth.

The Sun's interior is an inferno of hot ionised gases that rise and fall, creating electric currents, which generate the Sun's magnetic field. The smaller-scale movements of these ionised gases distort the magnetic field, which can emerge from the Sun's interior and appear at the surface, forming characteristic loops (Figure 1) and sunspots. When these magnetically active regions become over-stressed, a large amount of hot gas is released from the Sun's atmosphere into space in what is known as a coronal mass ejection (CME). While CMEs can damage satellites and disrupt radio communications, they also cause the spectacular 'Northern Lights' (Figure 2) and their counterpart in the southern hemisphere (the 'Southern Lights').

Figure 2 (background) *The Northern Lights*

Figure 1 *Image of the Sun from SDO*

7.1 MAGNETIC FIELDS AND THEIR EFFECTS

Electric currents circulating in the Earth's molten core are thought to be the origin of the Earth's magnetic field, which extends several thousands of kilometres beyond the Earth's surface into space. The strength of the Earth's magnetic field at its surface is 25 to 65 μT (microteslas), which is about 200 times weaker than a typical bar magnet. Magnetite, also known as lodestone, is a naturally occurring magnetic material. Pieces of lodestone, suspended so they were free to rotate, were used as the first magnetic compasses, as they aligned themselves with the Earth's magnetic field and could be used as an aid to navigation. A suspended bar magnet, or compass needle, aligns with the Earth's magnetic field. The end that points towards the Earth's North Pole is the N pole of the magnet, an abbreviation of 'north-seeking pole'.

In the mid-19th century, Michael Faraday experimented extensively with electricity and magnetism. In an experiment to demonstrate the lines of magnetic force around a bar magnet, he coated sheets of paper with wax, placed them on top of a bar magnet, then sprinkled iron filings onto the paper. The iron filings became aligned with the magnetic field and concentrated around the poles of the magnet (Figure 3). Faraday then warmed the paper so that the wax melted, and then allowed it to cool and set, fixing the iron filings in place, giving him a permanent record of the magnetic field for further study.

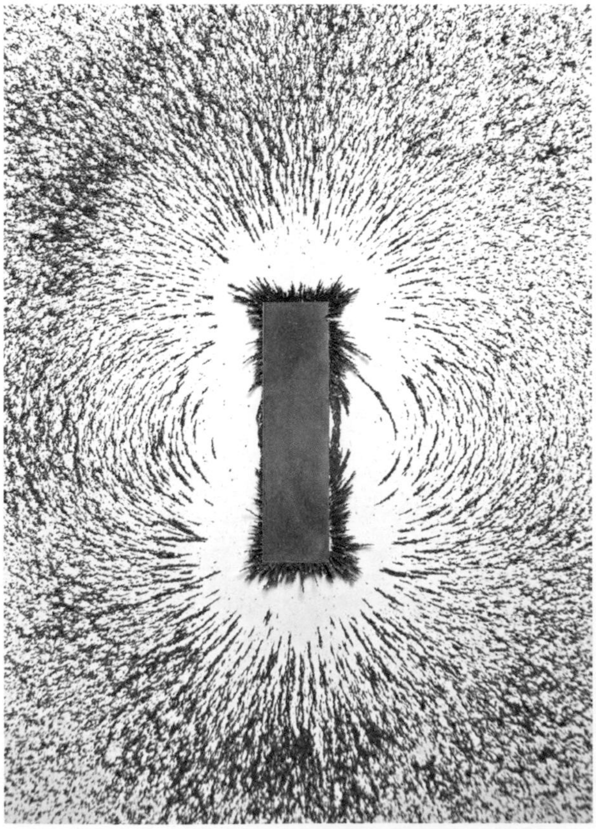

Figure 3 *The magnetic field of a bar magnet shown by iron filings*

The direction assigned to a magnetic field at a point is the direction of the force on the N pole of a plotting compass (Figure 4) at that point. The magnetic field lines can be plotted out in this way. They always leave a N pole and are directed into a S pole. The field is strongest where the field lines are most dense (see section 7.4).

The magnetic field created by a current flowing in a solenoid is similar in shape to the field of a permanent bar magnet, the N pole being at the end of the solenoid where the field lines are shown leaving the solenoid. It is possible to predict which end of a solenoid acts as a N pole by considering the direction of conventional current flow in the coil. If the coil is 'viewed' from one end, and the current is flowing anticlockwise, then that end acts as a N pole (Figure 5). The field is at its strongest inside the coil, well away from the ends. The field direction inside the coil is from south to north. A powerful magnetic field produced by a coil carrying a large electric current is used to magnetise materials such as iron, nickel, cobalt and neodymium in order to make permanent magnets.

Figure 4 *Magnetic field direction shown by plotting compasses*

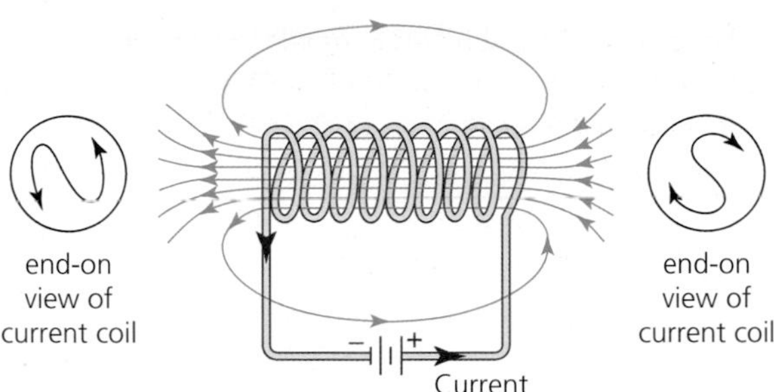

Figure 5 *A magnetic field produced by a current flowing in a solenoid*

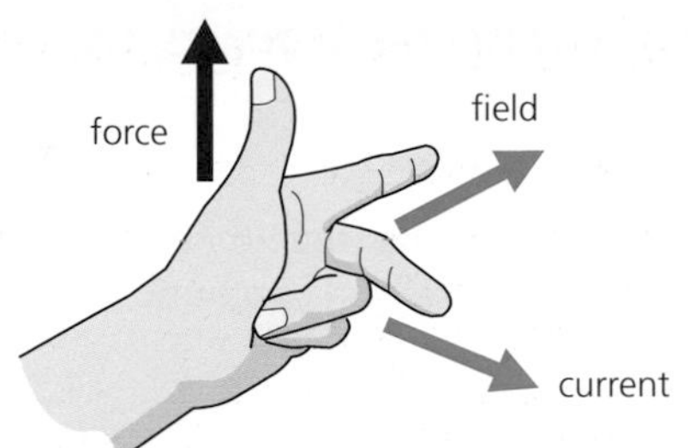

Figure 7 *Fleming's left hand rule*

The effect of a magnetic field on a current

We know that magnetic fields can be used to change the direction of a beam of moving charged particles, for example, in a synchrotron particle accelerator (see the introduction to Chapter 5). This is considered further in section 7.3. If a wire is carrying an electric current, electrons are moving along the wire. If the wire is placed in a magnetic field, the force exerted by the field on the moving electrons within the wire can move the wire itself, an effect called the motor effect. Figure 6 shows a brass rod resting on two metal rails situated between powerful magnet poles. When the switch is closed, so that current flows along the brass bar, the magnetic field pushes the bar so that it rolls to the left, out of the field.

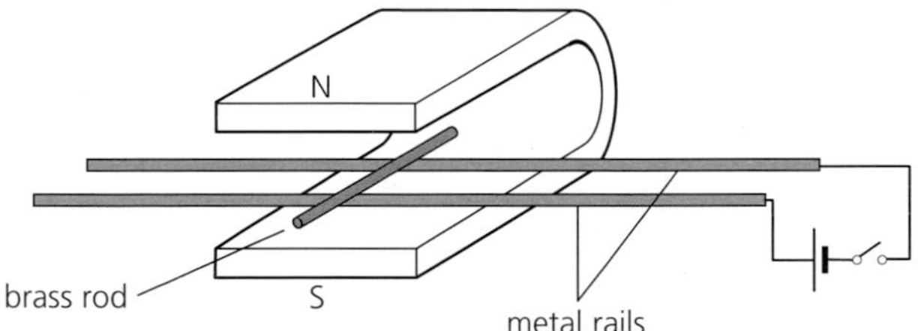

Figure 6 *Demonstrating the force on a current-carrying conductor in a magnetic field*

The direction of the force on the conductor can be predicted using **Fleming's left hand rule** (Figure 7).The first finger points in the direction of the magnetic field: out of a N pole into a S pole. The second finger must point in the direction of *conventional* current. The thumb then indicates the direction of the force on the conductor.

QUESTIONS

1. The solenoid in Figure 8 contains an iron bar, which strengthens the magnetic field created.

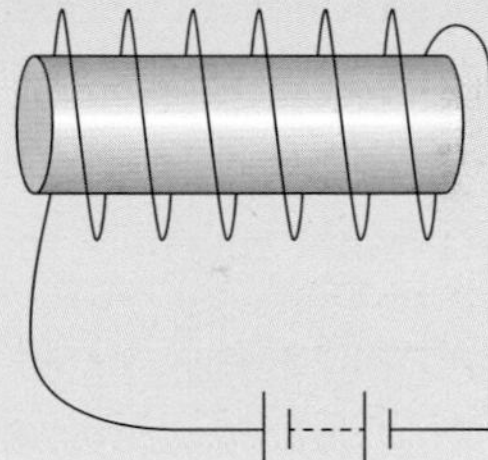

Figure 8

a. Which end of the solenoid acts as a N pole?

b. In what direction is the field in the iron bar?

2. A wire carrying a current is positioned between the poles of a U-shaped magnet (Figure 9). The direction of the current is shown by the arrow labelled I. What is the direction of the resulting magnetic force on the wire?

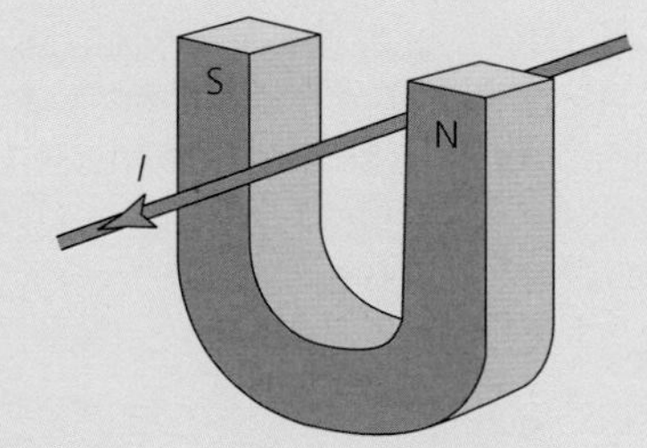

Figure 9

REQUIRED PRACTICAL: APPARATUS AND TECHNIQUES

Investigation of how the force on a wire varies with current, length of wire and magnetic flux density using a top-pan balance

The aim of this practical is

1 to demonstrate that the magnetic force on a wire carrying electric current, positioned perpendicular to the field lines in a magnetic field, is directly proportional to the size of the current
2 to determine how the magnetic force varies with length of a current-carrying wire positioned perpendicular to the field lines in a magnetic field.

The practical gives you the opportunity to show that you can:

› correctly construct circuits from circuit diagrams using dc power supplies and a range of circuit components
› use appropriate digital instruments, including electrical multimeters, to obtain a range of measurements (to include current, mass).

3 to determine how the magnetic force varies with magnetic flux density

Part 1: Investigation of how the force on a current-carrying wire in a magnetic field depends on the size of the current

Apparatus

A low-voltage variable dc power supply unit is required to deliver a current of up to 6 A through a thick piece of rigid bare copper wire (Figure P1). The wire, which is clamped and supported by two stands, is placed between four slab magnets supported on a yoke and positioned to create a uniform magnetic field. The magnets are positioned so that the exposed pole faces are N poles on one side of the yoke and S poles on the other side. The wire must be positioned parallel to the slab magnets, and so at right angles to the magnetic field direction. The yoke sits on a top-pan digital balance capable of reading to 0.01 g.

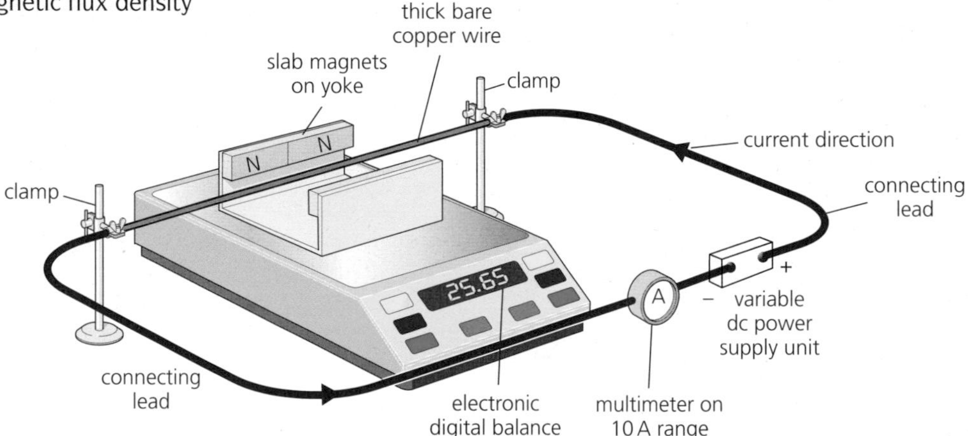

Figure P1 *Apparatus for measuring the magnetic force on a current-carrying wire of constant length in a magnetic field*

Care must be taken, since large currents are required for this experiment, so the bare copper wire will get hot. The variable power supply is connected in series with the wire and a multimeter set to read up to 10 A and capable of reading to 0.01 A.

Technique

The technique for measuring the magnetic force on the wire makes use of Newton's third law of motion. When a current is flowing in the bare copper wire, the magnetic field exerts an upward force on the wire. The wire, which is rigid and securely clamped, cannot move, but since the field exerts an upward force on the wire, the wire exerts an equal but downward force on the magnetic field (Newton's third law), causing the balance to change its mass reading. The magnetic force F on the wire can be calculated by converting the change in the mass reading, Δm, in grams, to kilograms and multiplying by 9.81 N kg^{-1}. Alternatively, if the balance has a tare facility, the balance display can be zeroed before switching on the power supply and therefore the actual mass reading is then used to determine F.

By adjusting the power supply, a series of values for F for different currents can be obtained.

Analysis

A graph of force F versus current I should be plotted to enable the relationship between F and I to be established.

QUESTIONS

P1 Identify the independent, dependent and control variables in this experiment.

P2 Suggest what procedure should be included to enable the precision of the measurements to be assessed.

P3 A student undertaking the above experiment is advised to take a minimum of eight data sets for currents in the range of 1 to 6 A. Draw a results table with titled columns that would be suitable for the experiment, and include suitably chosen current values that would satisfy the advice given to the student.

P4 The student zeroes the balance reading and then completes the experiment. The first data set is as follows: for a current of 1.00 A, the mass reading on the balance is 2.53 g. When the measurements are repeated, a value of 2.57 g is obtained for a current of 1.00 A. Determine the value of the magnetic force F, with an uncertainty, for a current of 1.00 A.

P5 State the key features of a graph of force F versus current I that would confirm that these quantities are directly proportional.

P6 The student plots the data generated from the experiment using error bars, then calculates the best gradient from the line that best fits the points, obtaining a value of 0.0251. This shows that the force per unit current on the wire was 0.0251 N A^{-1}. The student draws the steepest line on the graph that still fits within the error bars and the shallowest line that also still fits within the error bars, and obtains gradients of 0.0256 and 0.0244, respectively. State a percentage uncertainty for the gradient.

Part 2: Investigation of how the force on a current-carrying wire in a magnetic field depends on the length of the wire in the field

Apparatus

A low-voltage variable dc power supply unit, connected in series with a digital multimeter, is required to deliver an electric current to a U-shaped section of rigid bare copper wire (Figure P2). The wire is shaped so that it can be clamped with a known length l of the wire in the magnetic field (Figure P3), which is created by four slab magnets supported on a yoke that is sitting on a top-pan digital balance, as in Figure P1. At least seven U-shaped pieces of wire with various lengths are required.

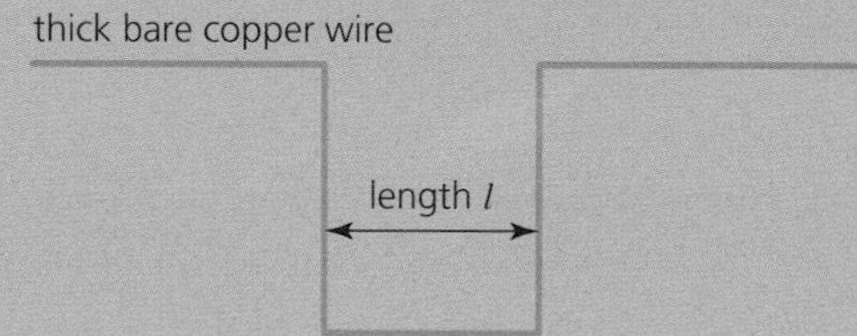

Figure P2 *U-shaped wire*

The wire is supported by stands so that length l is positioned parallel to the slab magnets and therefore perpendicular to the magnetic field direction.

Care must be taken, since large currents are required for this experiment, so the bare copper wire will get hot.

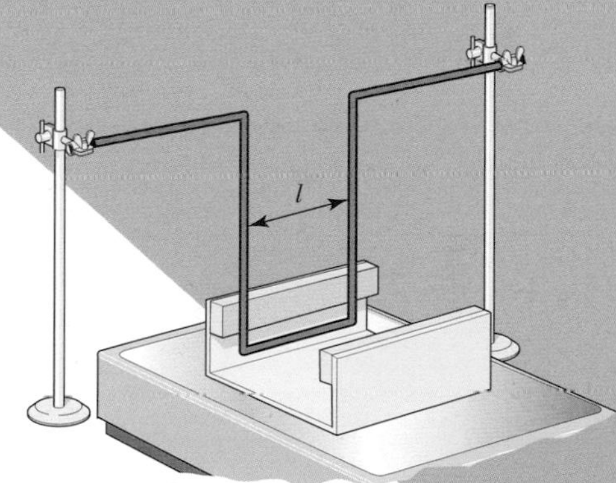

Figure P3 *Only length l of the copper wire is in the magnetic field.*

Technique

The length l should be measured between the vertical sections of the wire using a 30 cm rule. As in part 1, the technique for measuring the magnetic force F on the wire makes use of Newton's third law of motion. Assuming the balance has been zeroed, the force on the wire can be calculated from:

$$F = \frac{\text{mass reading}}{1000} \times 9.81$$

QUESTIONS

P7 Identify the independent, dependent and control variables in this part of the experiment.

P8 In parts 1 and 2, it is important that the wire between the magnets is positioned at 90° to the magnetic field direction. Suggest how this could be achieved.

Part 3: Investigation of how the force on a current carrying wire in a magnetic field depends on magnetic flux density.

The apparatus and techniques used in part 2 are suitable provided a range of slab magnets of varying strengths are available along with a magnetic field sensor or Hall probe with which the magnetic flux density can be measured.

QUESTIONS

P9 A student undertakes part 2 of the experiment and obtains a graph of *F* versus length *l* for a constant current of 5.00 A, shown in Figure P4. The graph is a straight line, with all the points close to the best-fitting line, but it does not pass through the origin.

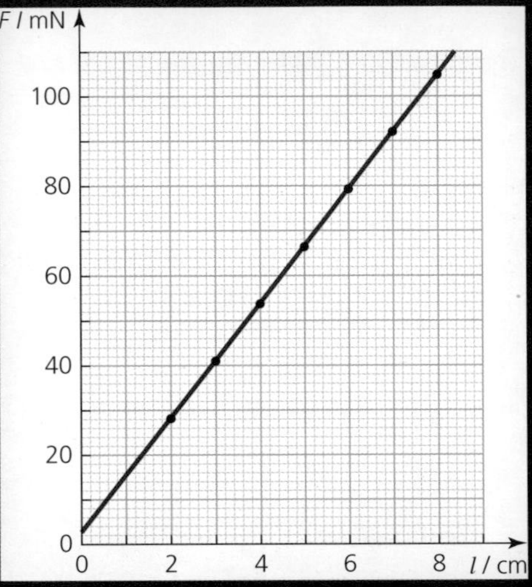

Figure P4

a. i. If an error has caused the straight line to miss the origin, what type of error is most likely?

ii. Suggest a possible source of this error.

b. i. Determine the magnetic force per unit current per unit length on the wire, given that the gradient of the graph is $1.26\,\text{N m}^{-1}$.

ii. Explain why the error considered in **a** has no impact on your answer to **b i**.

P10 Identify the independent, dependent and control variables for this part of the experiment.

P11 A student only has access to sufficient magnets to create magnetic fields of two different values of flux density. Discuss the limitations that having only two different flux densities available puts on this investigation.

7.2 MAGNETIC FLUX DENSITY

The quantity representing the strength of a magnetic field is called the **magnetic flux density** and is assigned the symbol B. The experimental observation that the force F on a wire, carrying a current and placed in a magnetic field perpendicular to the field lines, is directly proportional to both the current I and the length l of wire in the field is used to define magnetic flux density.

Since $F \propto I$ and $F \propto l$ then

$$F = constant \times Il$$

The constant of proportionality represents the strength of the magnetic field, since a field is considered to be stronger if it exerts a bigger force for a specific current and length. Therefore, the force on a current-carrying wire at 90° to a magnetic field becomes

$$F = BIl$$

Rearranging gives

$$B = \frac{F}{Il}$$

leading to the definition of magnetic flux density as follows:

Magnetic flux density B of a magnetic field is defined as the force per unit current per unit length on a wire placed at 90° to the direction of the magnetic field.

The unit of magnetic flux density is the **tesla**, symbol T, named after Nikola Tesla, a Serbian engineer. A flux density of 1 T causes a force of 1 N to be exerted on every 1 m length of a wire carrying a current of 1 A in a direction perpendicular to the field.

QUESTIONS

3. Figure 10 shows a clamped copper wire positioned perpendicular to a magnetic field created by slab magnets supported on a yoke. Before the wire is connected to the battery, the reading on the electronic balance is 284.591 g. When the circuit is complete, the balance reading increases to 298.100 g and the current flowing is 3.75 A. Determine the magnetic flux density of the field, given that the length of wire in the field is 11.5 cm.

4. Figure 11 shows an aluminium rod carrying an electric current of 5.2 A and positioned in a magnetic field of flux density 330 mT. Determine
   a. the direction (relative to the page) of the force on the rod
   b. the size of the force on the rod.

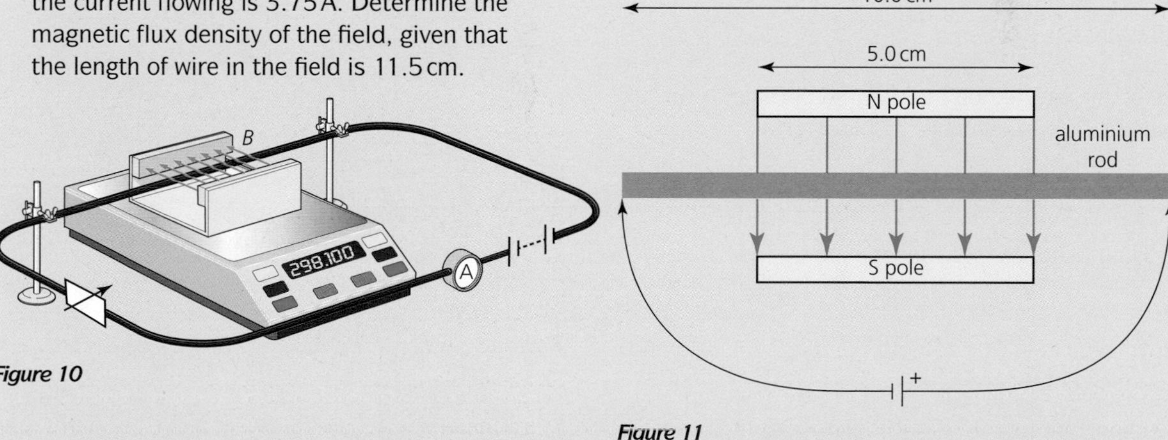

Figure 10

Figure 11

KEY IDEAS

- Fleming's left hand rule can be used to predict the direction of the force on a current-carrying wire in a magnetic field.
- Magnetic flux density, B, is defined as the force per unit current per unit length on a wire placed at 90° to the direction of the magnetic field.
- The unit of magnetic flux density is the tesla (T), where a flux density of 1 T causes a force of 1 N to be exerted on every 1 m length of a wire carrying a current of 1 A in a direction perpendicular to the field.
- The magnetic force F on a wire of length l carrying current I in a field of flux density B perpendicular to the wire is given by

$$F = BIl$$

7.3 MOVING CHARGES IN A MAGNETIC FIELD

A beam of charged particles moving through a vacuum is essentially an electric current. Consider one particle of charge Q moving at speed v at right angles to and through a magnetic field of flux density B and length l. The time t for the particle to travel the distance l in the field is given by $t = \frac{l}{v}$ and therefore $l = vt$. The equivalent current is $I = \frac{Q}{t}$.

The force F exerted on the charged particle by the magnetic field is given by

$$F = BIl = B \times \frac{Q}{t} \times vt$$

which simplifies to give the force on the particle as

$$F = BQv$$

This force acts at 90° to both the magnetic field and the velocity of the particle.

It is important to note that the magnetic field does not exert a force on a stationary charged particle, nor on a charged particle that is moving parallel to the field.

Consider a beam of *positively* charged particles travelling in a vacuum tube and entering a region where a magnetic field acts at 90° to the velocity of the particles (Figure 12). The particles enter the field horizontally at the bottom left of the diagram, moving from left to right. Fleming's left hand rule confirms that the force initially acts upwards, causing the path of the beam to bend. However, the force continues to act at 90° to the velocity, causing further bending. Since the force is always at 90° to the path of the particles, their speed remains constant and therefore, since $F = BQv$, the magnetic force remains constant. An object, in this case a charged particle, experiencing a constant force at 90° to its velocity follows a circular path (see Chapter 1).

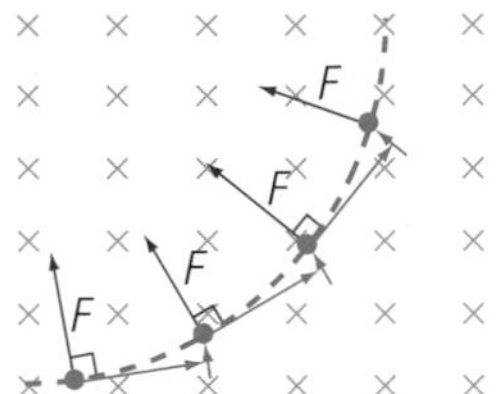

Figure 12 *A charge particle following a circular path in a magnetic field. The crosses indicate that the magnetic field is into the page.*

In Figure 13, an electron gun (see section 5.2 of Chapter 5) injects electrons into a flask containing hydrogen at low pressure. The presence of the hydrogen makes the beam visible. A magnetic field is created by two circular coils on either side of the flask. The field, which acts into the page, makes the electrons follow a circular path. When applying Fleming's left hand rule to determine the direction of the magnetic force on an electron beam, it is necessary to remember that the direction of conventional current is opposite to the direction of motion of the electrons.

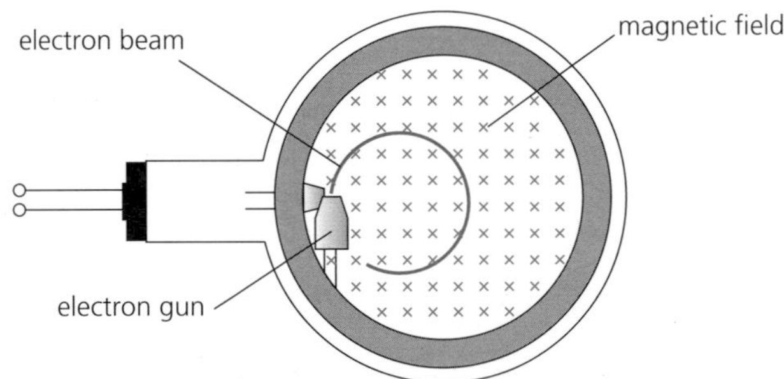

Figure 13 *Electrons follow a circular path in a magnetic field.*

Since the electrons are following a circular path, the equation for centripetal force (see section 1.2 of Chapter 1) can be equated to the force due to the magnetic field:

$$BQv = \frac{mv^2}{r}$$

which rearranges to give

$$r = \frac{mv}{BQ}$$

The radius of curvature of the path therefore depends on the speed of the particle, the magnetic flux density of the field and the inverse of the specific charge of the particle. Since each type of particle has a constant value for its specific charge, magnetic fields can be used to help identify the many particles created in high-energy particle collisions. The above equation also shows that the radius of curvature of the path of a particular particle in a magnetic field depends on its momentum, mv.

The ATLAS detector (Figure 14) is one of the four detectors at CERN's Large Hadron Collider. It uses a magnetic field to deflect charged particles into curved paths to enable their identification.

Worked example 1

Electrons ejected from an electron gun at a speed of $1.4 \times 10^7 \text{ms}^{-1}$ enter a magnetic field of flux density 2.4 mT acting at 90° to the beam, in an apparatus similar to that in Figure 13. Determine the radius of curvature of the path followed by the electron beam, given that the specific charge (charge/mass ratio, e/m) of the electron is $1.76 \times 10^{11} \text{C kg}^{-1}$.

Equating magnetic force and centripetal force gives

$$BQv = \frac{mv^2}{r}$$

which rearranges to give the radius of curvature

$$r = \frac{mv}{BQ} = \frac{1.4 \times 10^7}{2.4 \times 10^{-3} \times 1.76 \times 10^{11}} = 0.033\text{m}$$

QUESTIONS

5. Calculate the speed of an antiproton created in a high-energy collision if the radius of curvature of its track is 0.11 m and the detector's magnetic field strength is 3.0 T.
6. A charged particle moving at right angles to a constant magnetic field follows a circular path. Show that the radius of curvature of the particle's path is directly proportional to its momentum.
7. Particles of mass m and charge Q travel in a circular path of radius r and speed v in a magnetic field of flux density B. How many of the quantities m, Q, v and B, if increased one at a time, would increase the radius of the path?

   A one **B** two **C** three **D** four
8. A singly charged particle and a doubly charged particle enter the same uniform magnetic field, which is perpendicular to their direction of travel. The singly charged particle has a speed 10 times that of the doubly charged particle. What is the value of the following ratio?

$$\frac{\text{magnetic force on the singly charged particle}}{\text{magnetic force on the doubly charged particle}}$$

   A 5 **B** 10 **C** 15 **D** 20

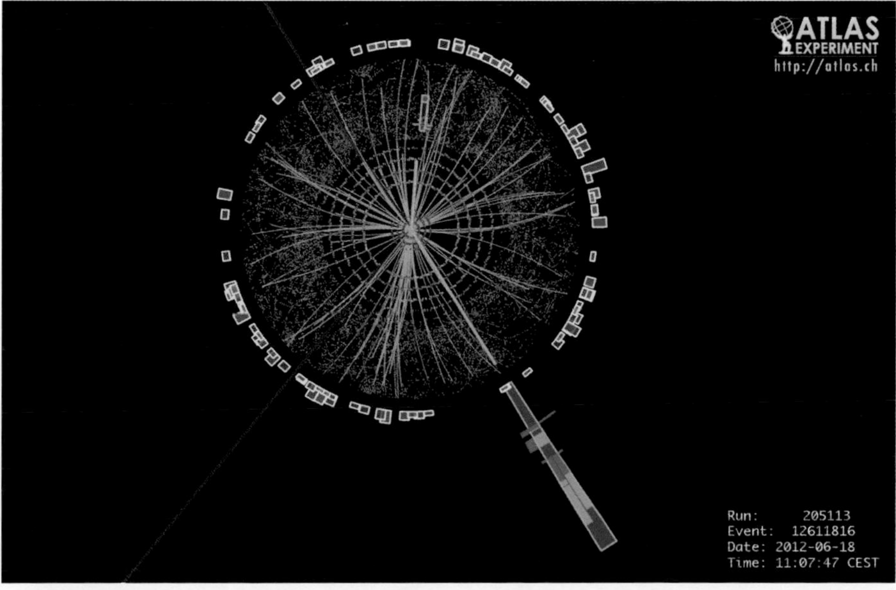

Figure 14 *ATLAS collision image*

ASSIGNMENT 1: IDENTIFYING IONS USING A MASS SPECTROMETER

The aim of this assignment is to appreciate how the magnetic deflection of charged particles is used in a mass spectrometer, which is an instrument used to identify different isotopes.

Chemists sometimes need accurate information on the different elements or isotopes present in a sample of material. If ions from the sample are injected into a magnetic field, at a chosen speed, the magnetic flux density required to deflect them into a circular path of specific radius depends on the charge/mass ratio of the ions, hence enabling their identification (Figure A1).

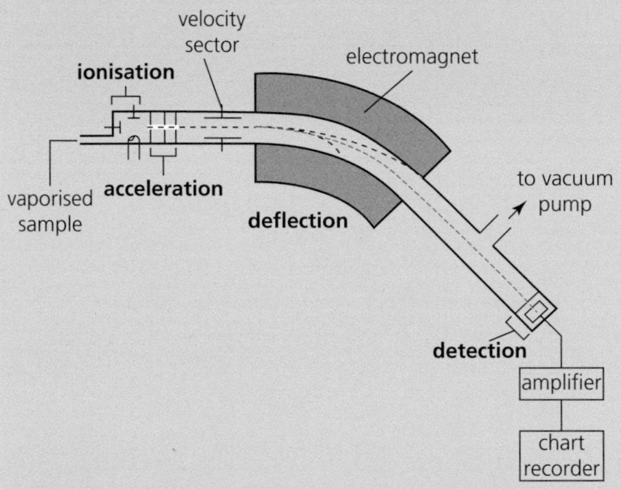

Figure A1 *Mass spectrometer*

A small sample of material is first vaporised so that it is in the form of a gas. The gas atoms are then ionised by knocking one or more electrons off the atoms. The resulting positive ions are then accelerated by an electric field, and those with a particular speed are picked out by a velocity selector. The ions then pass through the uniform magnetic field of the mass spectrometer, which deflects them into a semicircular path, with lighter particles being deflected more than heavier particles. The amount of deflection also depends on the charge on the ion: the greater the charge, the greater the deflection. The strength of the magnetic field can be adjusted in order to direct the beam onto the ion detector, which detects the ions electrically.

The mass spectrometer is able to separate ions of different mass /change ratio and determine their relative abundance. The results are displayed on a mass spectrum (Figure A2), which shows the distribution of ions based on their mass/charge ratio which is represented by $\frac{m}{z}$ with m representing the mass of an ion in atomic mass units and z representing its charge in units of electronic charge (e).

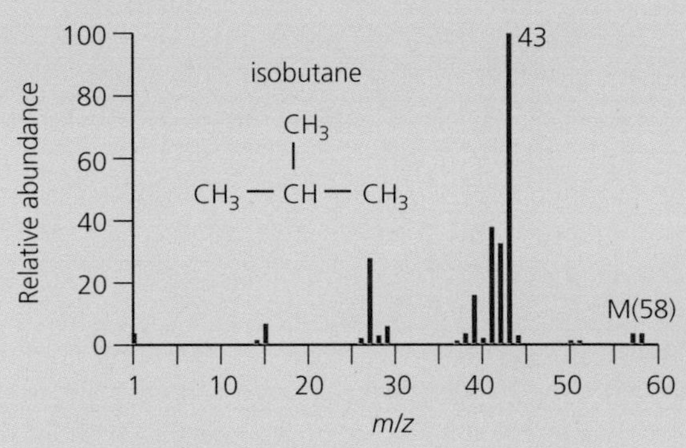

Figure A2 *Mass spectrum*

Questions

A1 Show that, in a mass spectrometer, the value of the magnetic flux density required to make a particle follow a curved path of a specific radius at a specific speed is directly proportional to the mass/charge ratio of the particle.

A2 Determine the magnetic flux density required to deflect a singly charged helium ion travelling at $1.2 \times 10^5\,\text{m s}^{-1}$ around a curve of radius 5.0 cm inside a mass spectrometer. [Mass of He atom = 6.64×10^{-27} kg

A3 Explain why it could be difficult for a mass spectrometer to distinguish between a $^{20}_{10}Ne^{2+}$ ion and a $^{10}_{5}B^{+}$ ion.

The cyclotron

In addition to their use in detectors for the identification of particles produced in high-energy collisions, magnetic fields have an important role in the particle accelerators that produce the beams of particles that actually undergo the high-energy collisions. The early particle accelerators, such as the Van de Graaff accelerator (developed in 1929) were linear and did not require the use of a magnetic field. But in the early 1930s Ernest Lawrence, an American nuclear physicist, developed the **cyclotron**, the first circular particle accelerator, in which charged particles, produced at a central source, are accelerated as they follow an outward spiralling path. Lawrence's first cyclotron had a diameter of only 10 cm (*see Assignment 2 in Chapter 4 in Year 1 Student Book*). He went on to build larger cyclotrons with diameters of up to 5 m.

Figure 15 shows a cyclotron accelerating protons. The cyclotron consists of two hollow D-shaped sections referred to as dees. An alternating electric field is maintained across the gap between the two dees by an alternating applied potential difference. A constant uniform magnetic field (B) acts through the whole of the accelerator. The charged particles to be accelerated are injected into the dees from their source at the centre of the cyclotron. The magnetic force, $F = BQv$, provides the centripetal force to make the particles follow a circular path, so that they repeatedly pass through the gap between the dees. Each time a particle passes through the gap, it is accelerated by the electric field and moves faster, so spirals further outwards (because $r \propto v$). When the radius of the path of charged particles reaches the radius of the dees, the particles exit the cyclotron and strike the target material.

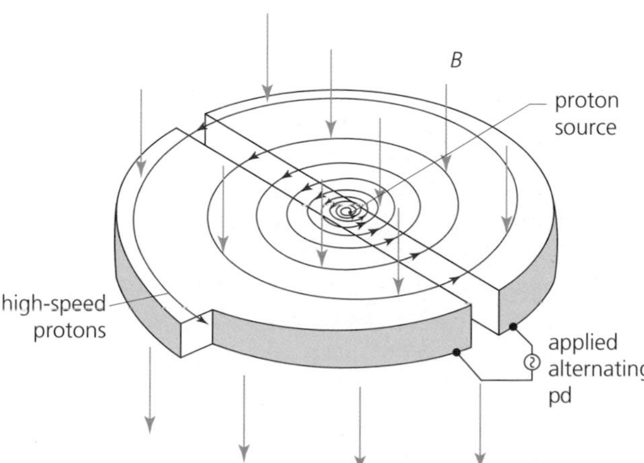

Figure 15 *The acceleration of protons in a cyclotron*

Consider a positively charged particle approaching the gap between the dees. The direction of the electric field between the dees needs to be in the direction that the positive charged particle is moving in order to accelerate the particle. Now suppose the particle has travelled in a semicircle and reaches the gap between the dees at the other side of the spiral. Since the particle is now travelling in the opposite direction, the direction of the electric field between the dees must be reversed if it is to further accelerate the particle. This is achieved by reversing the polarity of the pd applied to the dees. Consequently, to continue accelerating the particles, an alternating pd must be applied to the dees at just the right frequency to coincide with the particles arriving at the gap between the dees.

Crucial to Lawrence's invention of the cyclotron was his realisation that, even though the particles are getting faster, they take the same time to travel each semicircular path inside a dee.

If t represents the time it takes for a particle travelling at speed v to travel one semicircle of radius r, then

$$t = \frac{\pi r}{v}$$

However, since the centripetal force required for circular motion is provided by the magnetic field:

$$BQv = \frac{mv^2}{r}$$

which rearranges to give

$$\frac{v}{r} = \frac{BQ}{m}$$

Substituting into the equation for t gives

$$t = \frac{\pi r}{v} = \frac{m\pi}{BQ}$$

Since B, Q and m are constant, the time t for a particle to complete a semicircle is also constant and does not depend on the radius of the path. The time T for a particle to complete a full circle is therefore given by

$$T = 2t = \frac{2m\pi}{BQ}$$

which means that the frequency f of the alternating pd required for the acceleration of the particles is given by

$$f = \frac{BQ}{2m\pi}$$

Although the cyclotron is more cost- and space-effective than a linear accelerator, it does have its limitations.

The spiral path of the particles being accelerated needs to be synchronised with the frequency of the alternating applied pd. This is easily achieved provided the accelerated particles do not reach relativistic speeds. However, from the theory of special relativity, the mass of a particle increases as it gets closer to the speed of light. As a result, the approach of the particle at each dee becomes out of phase with the alternating electric field and the particle no longer accelerates. A classical cyclotron with a constant magnetic field and a constant-frequency alternating pd can therefore only accelerate particles to a few per cent of the speed of light.

Worked example 2

By 1936, Lawrence had built a cyclotron of diameter 0.94 m, which he used to accelerate deuterons (deuterium nuclei, consisting of a proton and a neutron) to an energy of 8.0 MeV. The deuterons were directed out of the cyclotron when the radius of their spiral path reached the radius of the dees. Determine

a. the speed of the deuterons as they exit the cyclotron
b. the magnetic flux density of the magnetic field required for the operation of this cyclotron.

[Deuteron mass = 3.34×10^{-27} kg; deuteron charge = 1.6×10^{-19} C

a. The exit velocity of the deuterons can be found by equating their final energy in joules to the kinetic energy equation. *From section 3.1 in Chapter 3 in Year 1 Student Book*,

$$1\,\text{eV} = 1.60 \times 10^{-19}\,\text{J}$$

Therefore the kinetic energy is

$$E_k = \frac{1}{2}mv^2 = 8.0 \times 10^6 \times 1.60 \times 10^{-19}$$
$$= 1.28 \times 10^{-12}\,\text{J}$$

which gives

$$v = \sqrt{\frac{2E_k}{m}} = \sqrt{\frac{2 \times 1.28 \times 10^{-12}}{3.34 \times 10^{-27}}} = 2.77 \times 10^7\,\text{m s}^{-1}$$

b. The centripetal force required for circular motion is provided by the magnetic field:

$$BQv = \frac{mv^2}{r}$$

Rearranging gives

$$B = \frac{mv}{rQ} = \frac{3.34 \times 10^{-27} \times 2.77 \times 10^7}{0.47 \times 1.6 \times 10^{-19}} = 1.2\,\text{T}$$

QUESTIONS

9. A proton is accelerated in a cyclotron of diameter 40 cm and magnetic field of flux density 1.1 T.

   Determine

   a. the energy, in MeV, that could be reached by the proton

   b. the frequency of the alternating potential difference applied to the dees.

10. A proton is accelerated to an energy 5.0 MeV in a cyclotron with an alternating pd of 1.0 kV and a frequency of 8 MHz applied to the dees.

    Calculate

    a. the number of times a proton would have to pass between the dees to reach an energy of 5.0 MeV

    b. the time that a proton spends spiralling in the cyclotron before exiting.

    [*Hint*: The proton passes between the dees twice in each cycle of the applied pd.]

11. A singly charged particle of mass 6.4×10^{-27} kg has been accelerated to a speed of $1.6 \times 10^7\,\text{m s}^{-1}$ in a cyclotron with a magnetic field of 2.8 T.

    Calculate

    a. the magnetic force on the particle

    b. the radius of the particle's circular path as it reached the speed of $1.6 \times 10^7\,\text{m s}^{-1}$.

By the 1940s, technology had progressed and it had become possible to produce an alternating voltage at a frequency that could be varied to keep track of the change in mass of the accelerated particles as they reach relativistic speeds. This technological advance led to the development of the 'synchrocyclotron' (Figure 16), which could accelerate particles to hundreds of MeV. The drive to achieve even higher particle energies led to the design of the modern **synchrotron** in which both the frequency of the accelerating voltage and the strength of the magnetic field could be varied and synchronised with the increasing mass of the particles, so that, instead of spiralling outwards, the particles' path could have a constant radius. Particle energies of 7 TeV (7×10^{12} eV) have been achieved with the synchrotron accelerator at CERN.

Figure 16 The first synchrocyclotron, built by Lawrence and his team at the Lawrence Berkeley Laboratory in California. It was completed in 1946.

QUESTIONS

Stretch and challenge

12. The flux density of the magnetic field of a synchrotron particle accelerator is increased in line with the increase in the inertial mass of the particles as they approach the speed of light. Use Einstein's equation for total energy $E = mc^2$ to determine the increase in the inertial mass of a proton that has been accelerated in a synchrotron so that its kinetic energy has reached 1.0 GeV. [*Hint*: Total energy is equal to kinetic energy plus rest energy.] [Proton rest energy = 938 MeV; rest mass = 1.67×10^{-27} kg]

KEY IDEAS

- The magnetic force F on a particle of charge Q moving with speed v at 90° to a magnetic field of flux density B is given by $F = BQv$.
- A charged particle moving at right angles to a uniform magnetic field follows a circular path at constant speed.
- For a particle of mass m and charge Q moving at speed v in a circular path of radius r at 90° to a magnetic field of flux density B,

$$BQv = \frac{mv^2}{r}$$

- In a cyclotron, the time for a particle to travel through a dee following a semicircular path is constant and does not depend on the radius of the path.

7.4 MAGNETIC FLUX LINKAGE

The strength of a magnetic field is measured by its flux density, in tesla. The **magnetic flux**, Φ, passing through an area A is defined as the product of the magnetic flux density B (normal to the area) and the area:

$$\Phi = BA$$

where B is normal to area A.

The unit of magnetic flux is the **weber** (Wb), where 1 Wb = 1 T m^2.

Visually, it is helpful to picture magnetic flux as being related to (but not equal to) the number of magnetic fields lines and the flux density as the number of field lines passing through unit area (Figure 17), so that where the magnetic field is at its strongest the flux lines are more tightly packed.

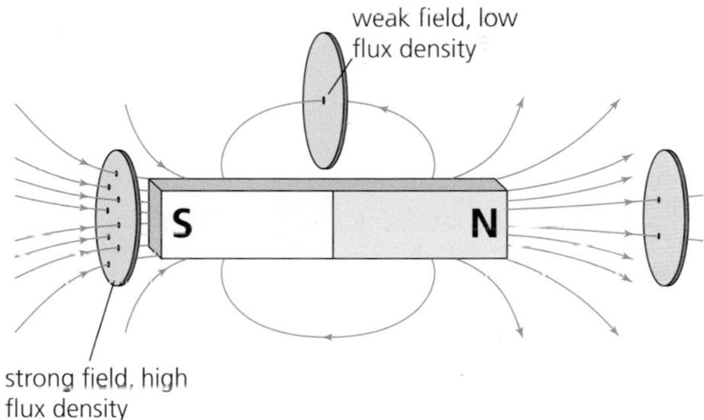

Figure 17 Magnetic flux

QUESTIONS

13. Calculate the magnetic flux passing through a small circular coil of diameter 1.5 cm positioned between the poles of two slab magnets (Figure 18), given that the magnetic flux density between the magnet poles is 300 mT.

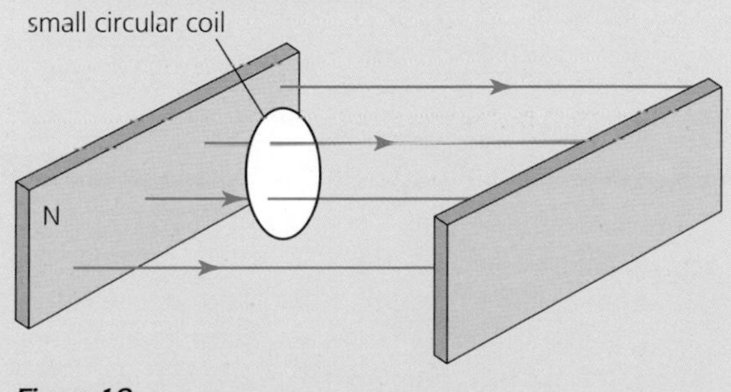

Figure 18

The definition of magnetic flux $\Phi = BA$ applies specifically to a situation where the magnetic flux density B is normal to area A (as in Figures 17 and 18). However, in a situation where the magnetic flux density is not normal to the area of the coil (as in Figure 19a), it is often necessary to determine the component of the magnetic flux density at right angles to the coil area. If the angle between the direction of the magnetic field and the normal to the plane of the coil is θ (see Figure 19b), the component of B at right angles to the coil is $B\cos\theta$ and therefore the magnetic flux through the coil is

$$\Phi = BA\cos\theta$$

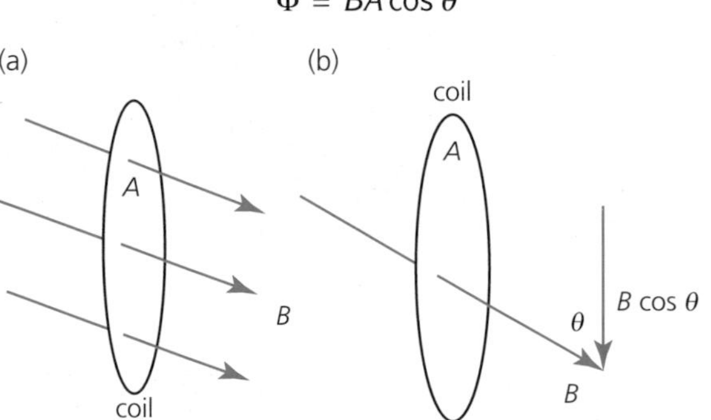

Figure 19 *Finding the component of B normal to an area A*

When we consider electromagnetic induction in Chapter 8, we will need to use the concept of magnetic **flux linkage**. This is defined as the product of the magnetic flux Φ passing through a conducting coil and the number of turns N of wire on the coil. Magnetic flux linkage does not have its own symbol, but is simply written as $N\Phi$. The unit is the 'Wb turn'.

QUESTIONS

14. Calculate the magnetic flux passing through a circular coil of diameter 2.0 cm if the magnetic flux density is 150 mT and the angle between the normal to the plane of the coil and the direction of the magnetic field is 40°.

15. Calculate the magnetic flux linkage through a 20-turn circular coil given that the magnetic flux density is 150 mT, the angle between the normal to the plane of the coil and the direction of the magnetic field is 16° and the radius of the coil is 5.0 mm.

A rotating coil

Consider a rectangular coil positioned in a magnetic field (Figure 20). If the coil is rotated to different orientations relative to the direction of the magnetic field, the magnetic flux linkage through the coil changes. For example, when the plane of the coil is parallel to the field lines (Figure 20a), the flux linkage is zero, but when the plane of the coil is at right angles to the field lines (Figure 20b), the magnetic flux linkage is a maximum.

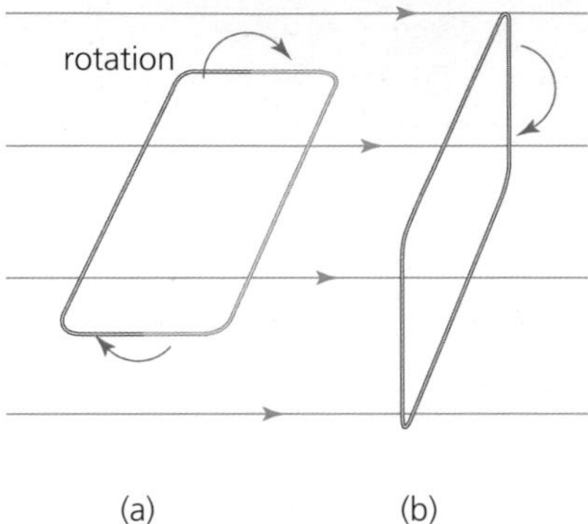

Figure 20 *Rotating a rectangular coil with respect to the field direction*

If the normal to the plane of the coil is at an angle θ to the direction of the magnetic field (Figure 21), the magnetic flux linkage through the coil is $N \times BA\cos\theta$. This is usually written as

$$N\Phi = BAN\cos\theta$$

If the coil is continuously rotated (with constant angular speed), the magnetic flux linkage varies as a cosine curve (Figure 22).

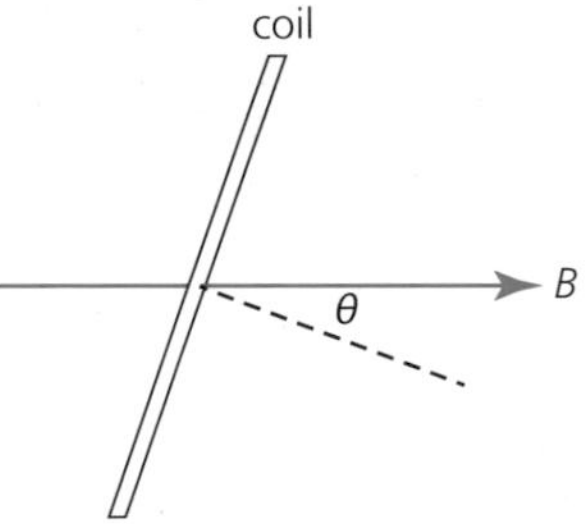

Figure 21 *Side view of a rectangular coil in a magnetic field*

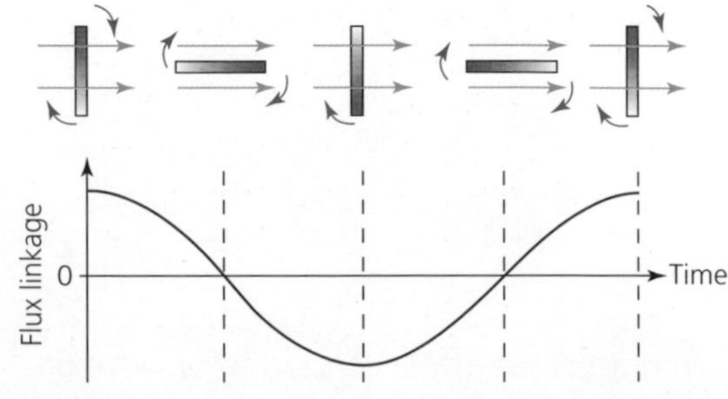

Figure 22 *Variation in flux linkage as a coil rotates*

QUESTIONS

16. Determine
   a. the magnetic flux and
   b. the magnetic flux linkage through a rectangular coil of dimensions 5.0 cm × 3.0 cm, given that the coil has 100 turns, the magnetic flux density is 120 mT and the angle θ between the normal to the plane of the coil and the magnetic field direction is 30°.
17. A small circular coil, made up of 5000 turns and with a radius of 5.0 mm, is positioned in a magnetic field of flux density 1.1 T, with the plane of the coil perpendicular to the field direction (Figure 23).

Determine
   a. the flux linkage through the coil
   b. the flux linkage if the coil is rotated so that the normal to the plane of the coil is at an angle of 30° to the magnetic field direction.

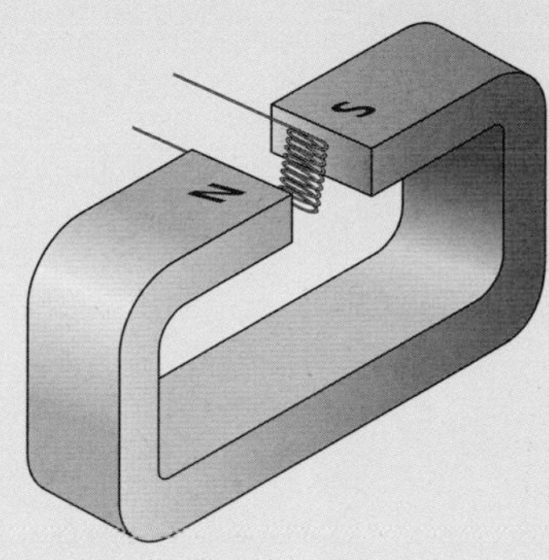

Figure 23

KEY IDEAS

- Magnetic flux, Φ, passing through an area A is defined as $\Phi = BA$, where the magnetic field of flux density B is normal to area A.

  The unit of magnetic flux is the weber (Wb), where 1 Wb = 1 T m^2.
- Magnetic flux linkage through a conducting coil is equal to $N\Phi$, where Φ is the flux through the coil and N is the number of turns on the coil. The unit of magnetic flux linkage is the Wb turn.
- The flux linkage through a coil of N turns and area A rotating in a magnetic field of flux density B is

$$N\Phi = BAN\cos\theta$$

  where θ is the angle between the normal to the plane of the coil and the direction of the magnetic field.

PRACTICE QUESTIONS

1. a. The equation $F = BIl$ gives the magnetic force that acts on a conductor in a magnetic field. Identify the physical quantities that are represented by the four symbols in the equation, and state the corresponding unit for each.
   b. Figure Q1 shows a horizontal copper bar of cross-section 20 mm × 30 mm and length 140 mm carrying a current of 60 A.

Figure Q1

      i. Calculate the weight of the copper bar. [Density of copper = 8900 kg m^{-3}]
      ii. Determine the magnetic flux density and direction of a magnetic field that could support the bar so that it is suspended above the bench.

2. a. i. State **two** situations in which a charged particle will experience no magnetic force when placed in a magnetic field.
   ii. A charged particle moves in a circular path when travelling perpendicular to a uniform magnetic field. By considering the force acting on the charged particle, show that the radius of the path is proportional to the momentum of the particle.
   b. In a cyclotron designed to produce high-energy protons, the protons pass repeatedly between two hollow D-shaped containers called 'dees'. The protons are acted on by a uniform magnetic field over the whole area of the dees. Each proton therefore moves in a semicircular path at a constant speed when inside a dee. Every time a proton crosses the gap between the dees it is accelerated by an alternating electric field applied between the dees. Figure Q2 shows a plan view of this arrangement.

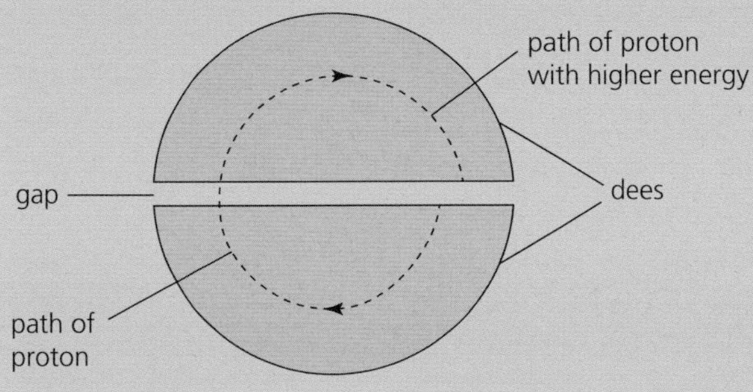

Figure Q2

   i. State the direction in which the magnetic field should be applied in order for the protons to travel along the semicircular paths inside each of the dees as shown in Figure Q2.
   ii. In a particular cyclotron the flux density of the uniform magnetic field is 0.48 T. Calculate the speed of a proton when the radius of its path inside the dee is 190 mm.
   iii. Calculate the time taken for this proton to travel at a constant speed in a semicircular path of radius 190 mm inside the dee.
   iv. As the protons gain energy, the radius of the path they follow increases steadily, as shown in Figure Q2. Show that your answer to part **b iii** does not depend on the radius of the proton's path.
   c. The protons leave the cyclotron when the radius of their path is equal to the outer radius of the dees. Calculate the maximum kinetic energy, in MeV, of the protons accelerated by the cyclotron if the outer radius of the dees is 470 mm.

AQA June 2013 Unit 4 Section B Q3

3. a. The equation $F = BQv$ may be used to calculate magnetic forces.
   i. State the condition under which this equation applies.
   ii. Identify the physical quantities that are represented by the four symbols in the equation.
   b. Figure Q3 shows the path followed by a stream of identical positively charged ions, of the same kinetic energy, as they pass through the region between two charged plates. Initially the ions are travelling horizontally and they are then deflected downwards by the electric field between the plates.

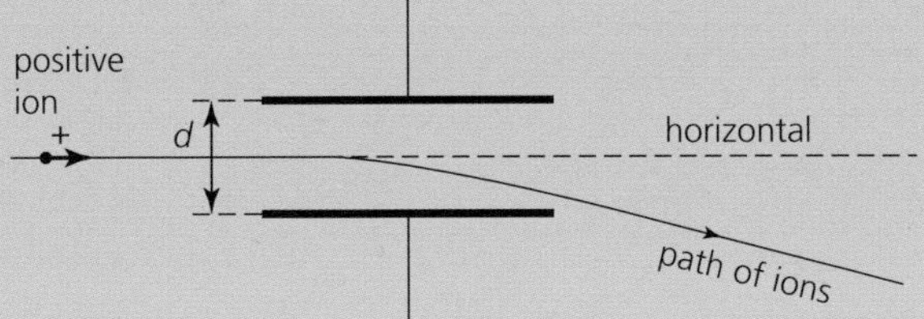

Figure Q3

Whilst the electric field is still applied, the path of the ions may be restored to the horizontal, so they have no overall deflection, by applying a magnetic field over the same region as the electric field. The magnetic field must be of a suitable strength and has to be applied in a particular direction.

i. State the direction in which the magnetic field should be applied.

ii. Explain why the ions have no overall deflection when a magnetic field of the required strength has been applied.

iii. A stream of ions passes between the plates at a velocity of $1.7 \times 10^5 \text{m s}^{-1}$. The separation d of the plates is 65 mm and the pd across them is 48 V. Calculate the value of B required so that there is no overall deflection of the ions, stating an appropriate unit.

c. Explain what would happen to ions with a velocity higher than $1.7 \times 10^5 \text{m s}^{-1}$ when they pass between the plates at a time when the conditions in part **b iii** have been established.

AQA June 2011 Unit 4 Section B Q4

4. The path followed by an electron of momentum p, carrying charge $-e$, which enters a magnetic field at right angles, is a circular arc of radius r. What would be the radius of the circular arc followed by an alpha particle of momentum $2p$, carrying charge $+2e$, which entered the same field at right angles?

A $\frac{r}{2}$ **B** r **C** $2r$ **D** $4r$

AQA June 2014 Unit 4 Section A Q21

5. Figure Q4 shows a rigidly clamped straight horizontal current-carrying wire held mid-way between the poles of a magnet on a top-pan balance. The wire is perpendicular to the magnetic field direction.

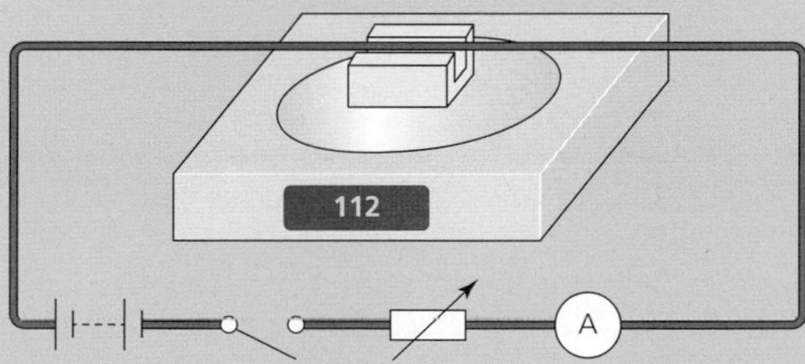

Figure Q4

The balance, which was zeroed before the switch was closed, reads 112 g after the switch is closed. If the current is reversed and doubled, what will be the new reading on the balance?

A −224 g **B** −112 g **C** zero **D** 224 g

AQA June 2010 Unit 4 Section A Q19

6. A jet of air carrying positively charged particles is directed horizontally between the poles of a strong magnet, as shown in Figure Q5.

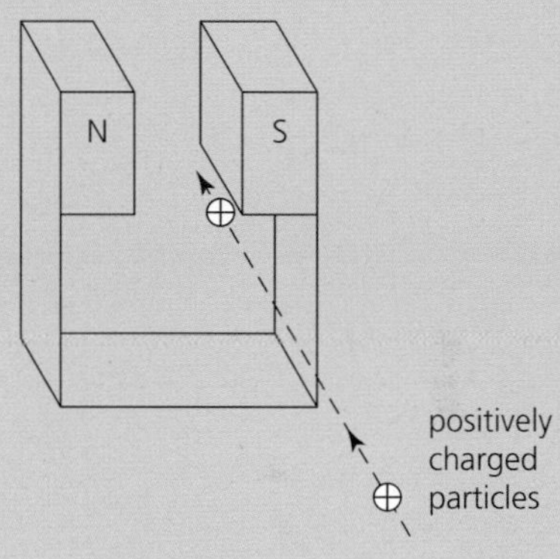

Figure Q5

In which direction are the charged particles deflected?

A upwards

B downwards

C towards the N pole of the magnet

D towards the S pole of the magnet

AQA June 2010 Unit 4 Section A Q21

8 ELECTROMAGNETIC INDUCTION AND ALTERNATING CURRENT

PRIOR KNOWLEDGE

You will already be familiar from Chapter 7 with the concept of a magnetic field, magnetic flux and magnetic flux linkage. You will be experienced in building electric circuits and have experience of using a digital multimeter as both an ammeter and a voltmeter.

LEARNING OBJECTIVES

In this chapter you will gain an understanding of Faraday's crucial work on producing electricity from magnetism, and how this led to our generation and distribution of electricity today. You will learn about the mathematics of alternating current (ac) and of transformers, and become skilful in using an oscilloscope.

(Specification 3.7.5.4 to 3.7.5.6)

Inductive charging, or wireless charging, uses the phenomenon of electromagnetic induction to transfer electrical energy between two objects. The mains-powered charging pad (Figure 1) contains an induction coil. This coil creates an alternating magnetic field that induces an electric current in a coil inside the mobile phone or notebook, which then recharges the batteries.

Figure 1 A wireless charging pad

A type of wireless charging pad is now being developed for use with electric vehicles. One of the problems with electric cars is that the driver has to remember to plug the vehicle into the mains to recharge the car's battery. The new idea is that a floor plate containing an induction coil is permanently connected to the mains and fitted in the garage where the car is parked. A second induction coil is fitted to the underside of the car and connected to the car's battery. Provided the car is parked in the correct position above the floor plate, the alternating magnetic field created by the plate can recharge the car's battery in about the same time as it would have taken if it had been directly plugged into the mains (Figure 2).

Figure 2 Wireless charging of a Nissan Leaf

8.1 ELECTROMAGNETIC INDUCTION

In 1823, Michael Faraday, the English physicist and chemist, made the discovery that electricity and magnetism could be combined to produce rotational movement – ultimately leading to the invention of the electric motor. For the next eight years, Faraday focused on his research in chemistry, but, returning to his work on electricity and magnetism, in 1831 he discovered a way of *producing* electricity from magnetism. Faraday had wrapped two wires around an iron ring (Figure 3), connected a battery and switch to the left hand coil and a galvanometer to the right hand coil. (The term 'galvanometer' was the historical name given to a moving-coil device used for detecting electric current.) Faraday observed that the galvanometer needle was deflected away from zero when the switch was closed and then returned to zero. On opening the switch, the galvanometer needle was deflected in the opposite direction and then returned to zero.

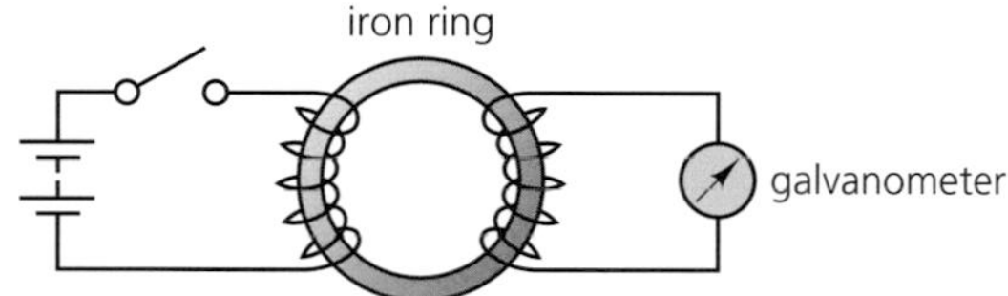

Figure 3 *Faraday's experiment led to the discovery of electromagnetic induction.*

Faraday described the induced voltage, which caused the brief flow of current in the right hand coil, as an electromotive force (emf). He explained that the action of opening and closing the switch caused a change in magnetic flux through the iron ring, and therefore through the right hand coil, and that it was this flux change that induced the emf. He described the effect as **electromagnetic induction**.

Faraday immediately set about trying to find other examples of electromagnetic induction. He discovered that moving a magnet into or out of a coil also induced an emf (Figure 4). If the magnet was held stationary, no effect was observed. He also observed that electromagnetic induction occurred if the magnet was held stationary and the coil was moved towards or away from the magnet.

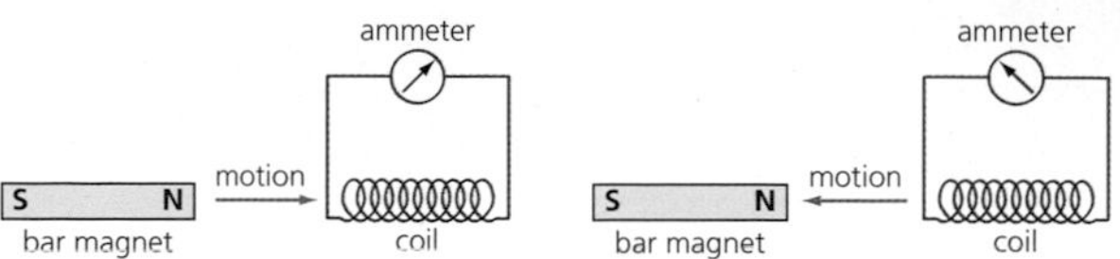

Figure 4 *Electromagnetic induction caused by relative movement between a coil and a magnet*

Faraday further discovered that emfs could be generated without the use of a coil. He found that an emf could be induced in a metal disc that was rotated in a magnetic field. His research led him to invent the Faraday disc generator (Figure 5), which was the world's first electric dynamo. Electrical contacts are made to the centre and the edge of the disc.

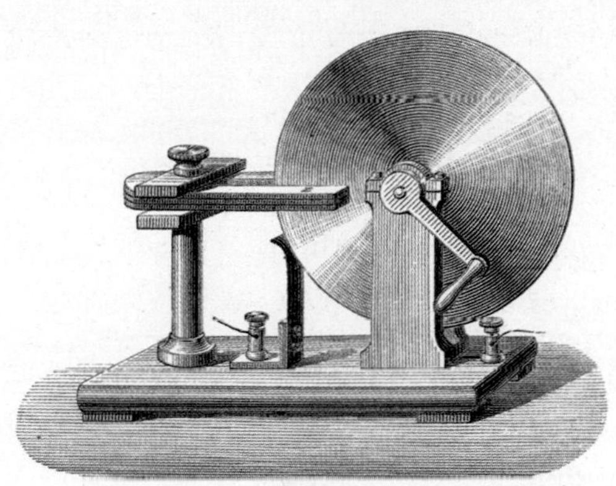

Figure 5 *A Faraday disc generator. As the handle is turned, the copper disc passes through the poles of the horseshoe magnet and an emf is generated between the two contacts, one in contact with the rim of the disc and one connected to the centre of the disc.*

Faraday also found that simply moving a wire through a magnetic field would generate an emf. Consider the wire in Figure 6 being moved upwards through a magnetic field. Since the wire is a metal, it contains vast numbers of free electrons. When the wire is moved upwards, essentially large numbers of electrons are moved upwards through the field. Charged particles moving in a magnetic field experience a magnetic force at right angles to both the field and the direction of motion (see section 7.3 of Chapter 7). The effect of the magnetic force is to push electrons away from one end of the wire and towards the other end. Since one end of the wire now has an excess of electrons, it becomes negative, while the other end has a deficit of electrons and becomes positive, creating an induced emf. Since there is a complete circuit, the induced emf drives an electric current. Note that the emf only exists while the wire is being moved.

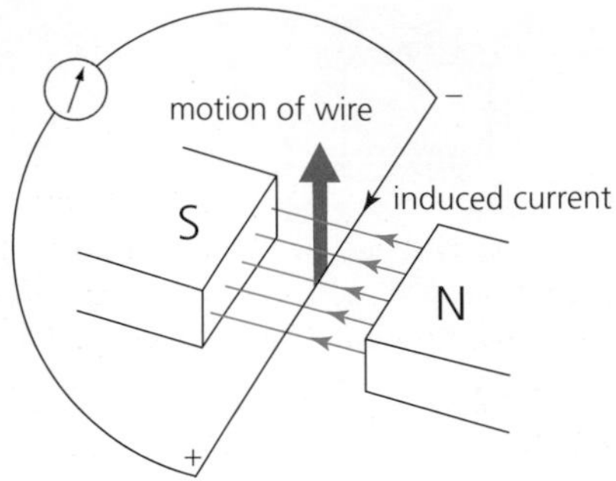

Figure 6 *An emf is induced in a wire moved upwards through a magnetic field.*

It is possible to predict which way the induced current flows using **Fleming's right hand rule** (Figure 7). The right hand rule, showing the direction of induced current when a conductor moves in a magnetic field, and the left hand rule, for the force on a current-carrying conductor in a magnetic field (section 7.1), are named after John Ambrose Fleming, a British electrical engineer and physicist.

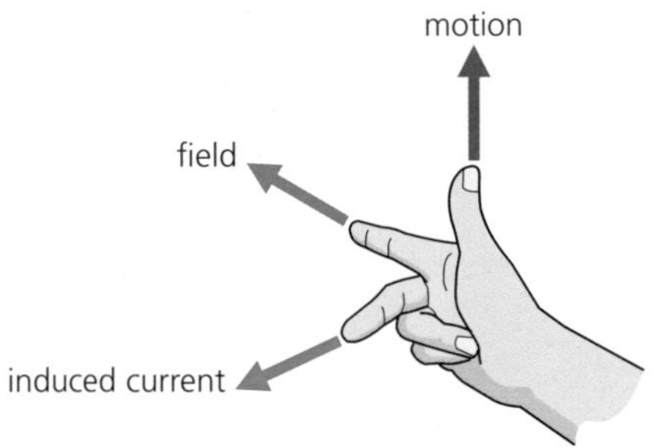

Figure 7 *Fleming's right hand rule*

The first finger points in the direction of the magnetic field. The thumb indicates the direction the wire is being moved. The middle finger then shows the direction of the induced current – which is opposite to the direction of flow of the electrons. (An alternative description is that the middle finger points to the end of the wire that the induced emf makes positive.)

QUESTIONS

1. In Figure 8, which end of the wire, A or B, becomes positive when the wire is being moved downwards through the field?

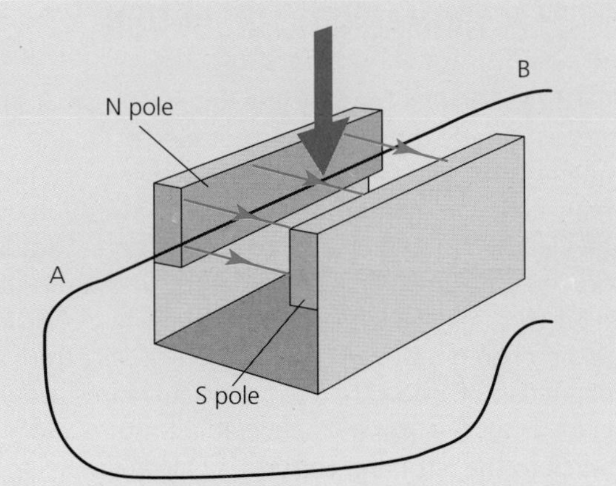

Figure 8 *Wire moved downwards through a magnetic field*

2. Figure 9 shows a Faraday disc generator. Which way does the induced current flow – from the rim of the copper disc to the centre, or the other way? [*Reminder*: The direction of a magnetic field is from N pole to S pole.]

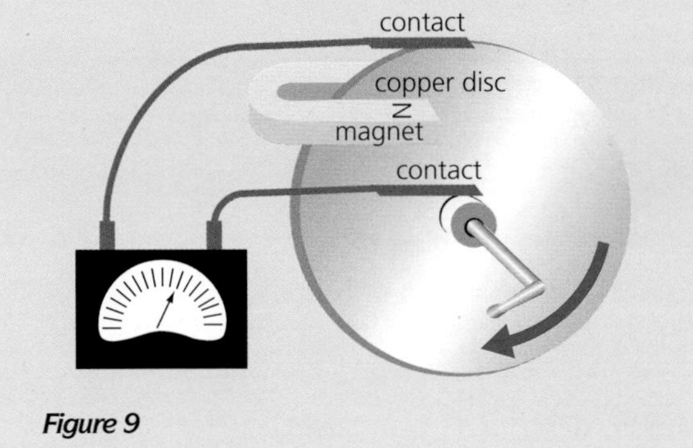

Figure 9

Faraday's law

Faraday established that any change in the magnetic environment near a conductor, whether it is a coil, a metal disc, or a single wire, would induce an emf in the conductor. He observed that the size of that emf increased when the change took place more quickly. His experiments led him to the following conclusions:

- When a conductor is moved through a magnetic field, the magnitude of the induced emf is equal to the rate at which magnetic flux is cut through by the conductor.
- When a magnet is moved near a coil, the magnitude of the induced emf is equal to the rate of change of flux multiplied by the number of turns on the coil, in other words, the rate of change of *flux linkage* $N\Phi$.

Faraday's conclusions are summarised by the following equation, known as **Faraday's law**:

$$\mathcal{E} = N\frac{\Delta\Phi}{\Delta t}$$

where $\mathcal{E}$ is the magnitude of the induced emf, $\Delta\Phi$ is the flux change that occurs in time Δt and N is the number of turns on the coil, which in the case of a straight conductor is equal to 1.

Worked example 1

A wire is moved upwards through a magnetic field of flux density of 500 mT (Figure 10).

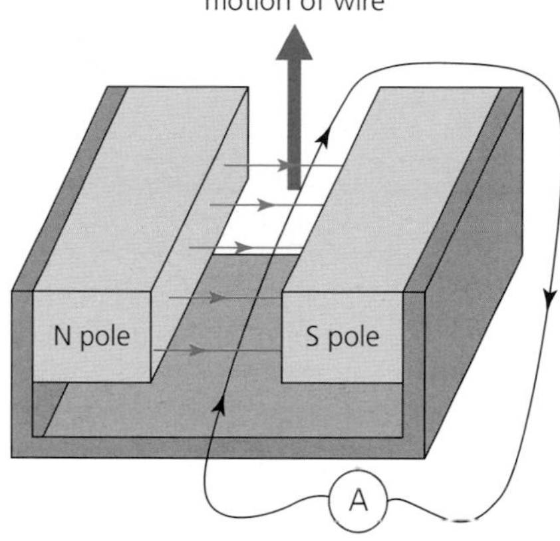

Figure 10

The length of wire that cuts through the magnetic flux is 5.0 cm and it moves through a distance of 2.0 cm in 0.50 s within the field.

Determine

a. the area of the magnetic field that the wire cuts through

b. the amount of flux that is cut by the wire

c. the size of the emf induced in the wire

d. the current that flows in the wire if the combined resistance of the wire and the ammeter is 0.45 Ω.

a. Area cut through by the wire = 0.05×0.02 $= 1.0 \times 10^{-3}\,\text{m}^2$

b. Magnetic flux cut by the wire = $\Phi = BA$ $= 0.5 \times 1.0 \times 10^{-3} = 5.0 \times 10^{-4}\,\text{Wb}$

c. Size (magnitude) of induced emf

$$= E = N\frac{\Delta\Phi}{\Delta t} = \frac{\Delta\Phi}{\Delta t} = \frac{5.0\times10^{-4}}{0.5} = 1.0\times10^{-3}\,\text{V}$$

d. Induced current

$$I = \frac{V}{R} = \frac{1.0\times10^{-3}}{0.45} = 2.2\times10^{-3}\,\text{A}$$

QUESTIONS

3. In a set-up like that in Figure 10, a wire is moved vertically *downwards* through a magnetic field of flux density 300 mT. The poles of the magnets producing the field have dimensions 4.5 cm by 2.0 cm and the wire passes between the poles in 0.25 s. Calculate

   a. the size of the emf induced in the wire

   b. the current induced in the wire if the combined resistance of the wire and the ammeter is 0.50 Ω.

4. An aircraft flying horizontally, due east at a speed of $200\,\text{m s}^{-1}$, is in a location where the flux density of the Earth's magnetic field is 1.7×10^{-4} T at an angle of 60° to the horizontal (Figure 11).

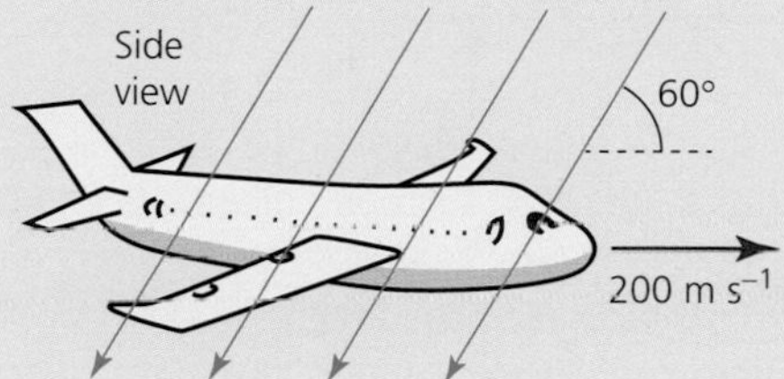

Figure 11

The wing span of the aircraft is 12 m. The wings cut through only the vertical component of the Earth's magnetic field. Calculate

   a. the vertical component of the Earth's magnetic field in the location of the aircraft

   b. the area swept out by the wings of the aircraft in one second

   c. the emf induced between the tips of the wings.

Worked example 2

A search coil (a small circular coil on a mounting, used to determine magnetic flux density) is made up of 5000 turns of radius of 6.5 mm. It is positioned in a magnetic field of known flux density 550 mT (Figure 12). Determine the average emf induced between the contacts to the coil if it is removed from the magnetic field in 0.25 s.

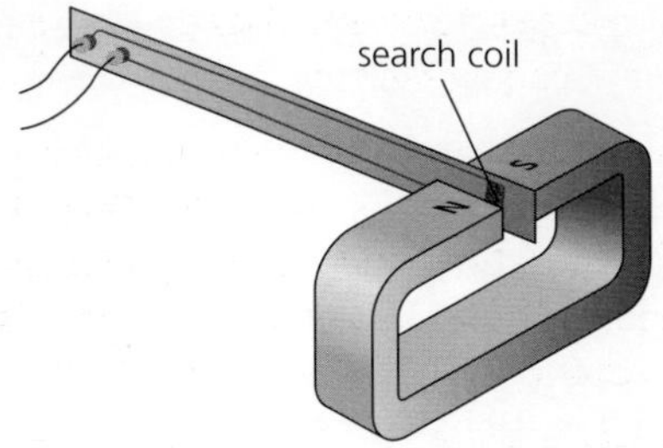

Figure 12

Area of the search coil is $A = \pi(6.5 \times 10^{-3})^2$
$= 1.327 \times 10^{-4}\ m^2 = 1.3 \times 10^{-4} m^2$

Induced emf is

$$\mathcal{E} = N\frac{\Delta\Phi}{\Delta t} = N\frac{BA}{\Delta t}$$

$$= 5000 \times \frac{550 \times 10^{-3} \times 1.327 \times 10^{-4}}{0.25} = 1.5V$$

QUESTIONS

5. A small circular coil of radius 2.0 cm and consisting of 20 turns is placed close to the N pole of a bar magnet so that its plane is perpendicular to the magnetic field lines (Figure 13). The magnetic flux density in the vicinity of the coil is 40 mT. Determine the average emf induced in the coil if it is rotated in 0.2 s so that it is edge-on to the field.

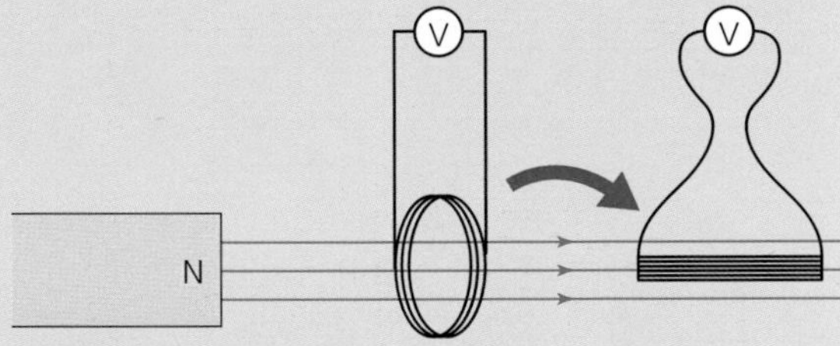

Figure 13

6. The electric current in the large coil in Figure 14 creates a magnetic field of flux density 1.2 mT along its horizontal axis. A search coil of radius 5.0 mm and 5000 turns is positioned inside the large coil with its plane at right angles to the magnetic field. The variable resistor is then adjusted steadily so that over a time interval of 4.0 s the current in the large coil is reduced to zero. Determine the emf induced in the search coil during the 4.0 s that the current in the large coil is reduced to zero.

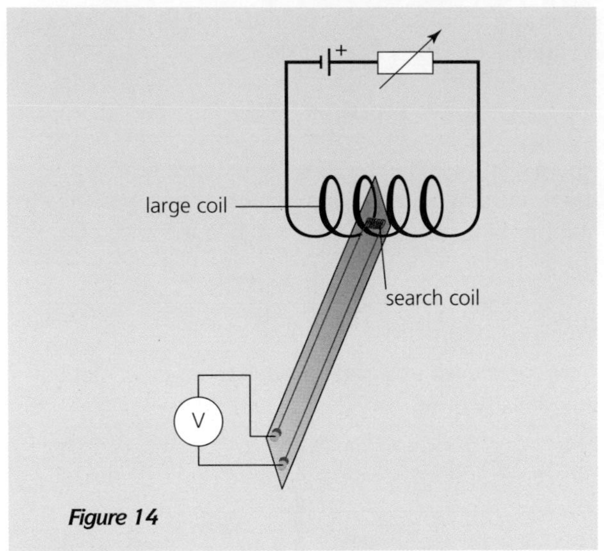

Figure 14

Lenz's law

In 1834, Heinrich Lenz, a Russian physicist, was studying electromagnetic induction and made an observation now summarised as Lenz's law in his honour. Lenz observed that, when a change in magnetic flux occurred resulting in an induced emf, the current driven by that emf created its own magnetic field which opposed the original flux change.

Lenz's law states that:

> The direction of an induced current opposes the change of magnetic flux that produces it.

Alternatively, Lenz's law can be quoted with reference to the direction of the induced emf rather than the induced current. But it is important to be aware that there has to be an induced current for any actual opposition to the flux change to occur. For example, when a magnet is moved towards a solenoid, the increase in flux linkage through the coil induces an emf, which drives a current in the solenoid (Figure 15). A current flowing in a solenoid creates a magnetic field similar in shape to that of a bar magnet. Lenz observed that the magnetic field created by the induced current in the solenoid created a N pole at the end of the solenoid to repel the approaching N pole of the magnet. Similarly, if a magnet already placed inside the solenoid is pulled out, the induced current in the solenoid creates a S pole to attract the receding N pole of the magnet.

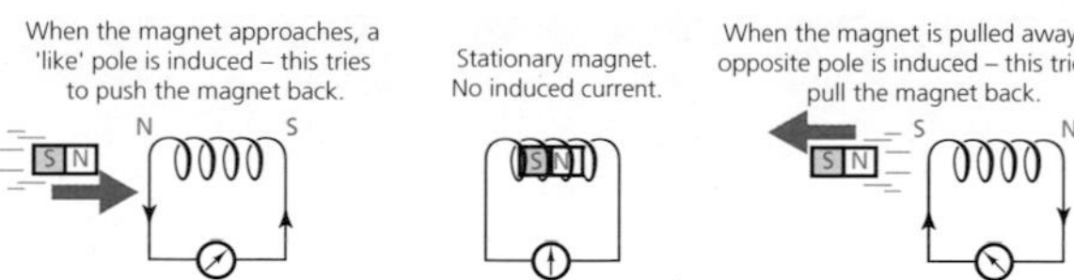

Figure 15 *Demonstrating Lenz's law*

The effect of Lenz's law can be described by considering energy conservation. Figure 16 shows two aluminium rings attached to an aluminium strip that is pivoted at its centre so that the arrangement is free to rotate in a horizontal circle. The ring on the right is complete but the ring on the left has a slit and is therefore not a complete circle. As the magnet (N pole first) is moved towards the right hand ring, a magnetic force is exerted on the electrons in the ring, creating an induced current. The direction of the induced current in the ring is anticlockwise, creating a N pole on the side facing the approaching magnet and producing repulsion between the ring and the magnet. To continue to move the magnet towards the ring requires work to be done in overcoming the repulsive force. When work is done, energy is transferred equal to the amount of work that is done. The mechanical work done in moving the magnet towards the ring against the repulsive force provides the energy needed to drive the induced current in the ring. Since the ring is free to rotate about the pivot, the repulsion from the magnet pushes the ring backwards away from the approaching magnet.

Figure 16 *Apparatus for observing Lenz's law*

If the magnet is then moved away from the right hand ring, the induced current flows in a clockwise direction, creating an S pole on the side of the ring facing the receding magnet. To continue to move the magnet away from ring requires work to be done to overcome the attraction between the ring and the N pole of the magnet. This work transfers the energy needed to drive the induced current. The arrangement starts to rotate in the opposite direction as the ring is attracted towards the receding magnet.

Movement of the magnet either towards or away from the left hand split ring has no visible effect. An emf is still induced in the ring but the lack of a complete circuit means that no current flows so there is no actual opposition to the flux change.

Another effect of Lenz's law can be demonstrated using a rotating aluminium disc and a strong magnet, set up as in Figure 17. As the disc is rotated, it cuts through the magnetic flux of the magnetic field, inducing emfs in the disc. Since the disc is solid metal, the induced emfs generate induced currents, which follow circular paths and are known as **eddy currents**. At any instant, the induced current in the section of the disc starting to enter the permanent magnetic field creates its own magnetic field such that the upper disc surface at that point acts as an N pole. The magnet's N pole repels this incoming induced N pole, slowing the rotation of the disc. Similarly, the induced current in the section of the disc in the process of leaving the permanent field creates an S pole at the upper disc surface. The attraction between this induced S pole and the magnet's N pole tries to prevent this part of the disc from leaving, and so slows the rotation of the disc. In this way, the eddy currents oppose the motion of the disc effectively, acting as an electromagnetic brake. Electromagnetic braking systems are used on rollercoasters, power equipment such as circular saws, rowing machines and other gym machines where extra resistance is required.

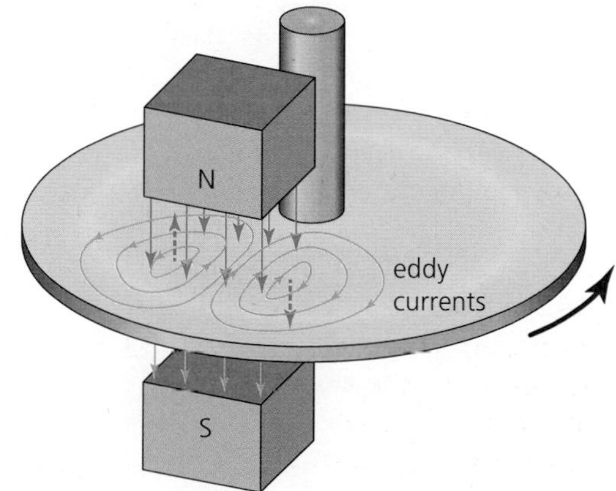

Figure 17 *A metal disc rotating in a magnetic field acts as a brake, opposing the rotating motion according to Lenz's law.*

QUESTIONS

7. The strongest permanent magnets are made from an alloy of the metal neodymium. Explain why a magnetised neodymium pellet takes 2 s to fall through a 30 cm long copper tube whereas an identical but unmagnetised neodymium pellet takes less than 0.3 s (Figure 18).

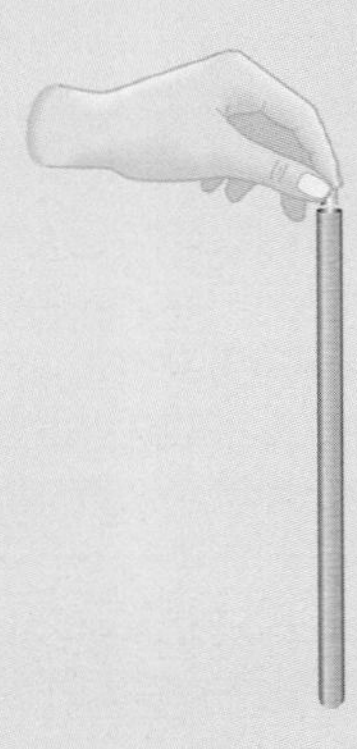

Figure 18

KEY IDEAS

- Faraday's law states that the magnitude of the induced emf in a conductor is equal to the rate at which the conductor cuts through magnetic flux, or the rate of change of flux linkage through a coil of wire.
- Faraday's law is summarised by the equation:

$$\mathcal{E} = N\frac{\Delta\Phi}{\Delta t}$$

- Lenz's law states that the direction of an induced current is such as to oppose the change causing it.

ASSIGNMENT 1: DEMONSTRATING FARADAY'S AND LENZ'S LAWS

(MS 3.1, MS 3.8, MS 3.9)

This assignment is an investigation designed to demonstrate Faraday's and Lenz's laws by using data logging equipment to monitor the effect of a magnet falling through a coil.

A solenoid, clamped so that its axis is vertical, is connected to a voltage sensor, which is connected to a data logger (Figure A1). The voltage sensor is an electronic device that measures the voltage across a component. It contains a micro-controller that ensures a high level of accuracy and consistency of readings. The input resistance of the sensor is about 1 MΩ, so negligible current flows through it. The magnet is allowed to fall through the solenoid, and the emf induced in the solenoid is measured and recorded by the sensor and data logger.

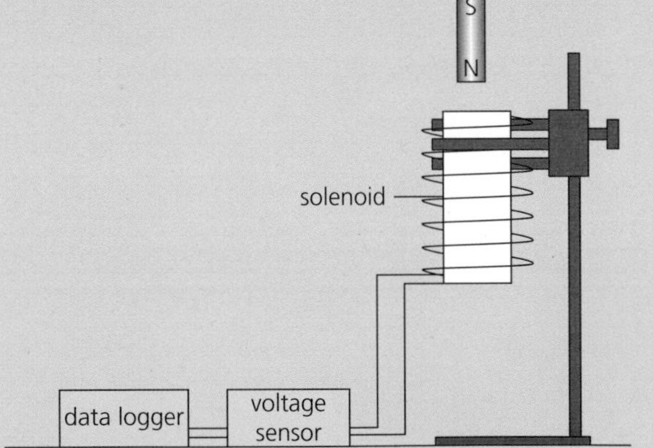

Figure A1 *Data logging equipment being used to monitor an induced emf*

When the data logger's measurements are analysed with appropriate software, a graph of voltage versus time (Figure A2) is produced.

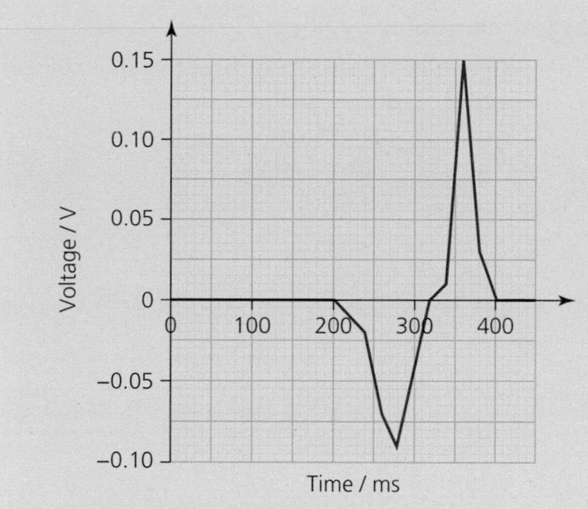

Figure A2 Voltage versus time graph

Questions

A1 a. Describe and explain the feature of the graph that demonstrates Lenz's law.

b. State what effect, if any, Lenz's law has on the acceleration of the falling magnet.

A2 Describe and explain the feature of the graph that demonstrates Faraday's law.

A3 Given that the equation for Faraday's law is $\mathcal{E} = N\frac{\Delta\Phi}{\Delta t}$, identify the significance of the area between the graph line and the time axis, and state whether the area enclosed by the negative peak should be equal to the area enclosed by the positive peak, explaining your answer.

A4 A student repeats the experiment but connects the solenoid to a moving-coil ammeter instead of the voltage sensor. The resistance of an ammeter is very small, so now the induced emf created when the magnet falls through the solenoid can drive a significant current in the solenoid. State and explain how the acceleration of the falling magnet might be affected as it passes through the solenoid.

8.2 THE ALTERNATING CURRENT (AC) GENERATOR

A simple alternating current (ac) **generator** (Figure 19) is constructed from a rectangular coil that is made to rotate in a magnetic field by turning a handle attached to the coil. As the coil is rotated, the flux linkage through the coil changes from a maximum, when the normal to the plane of the coil is parallel to the magnetic field, to zero, when the normal to the plane of the coil is at 90° to the magnetic field.

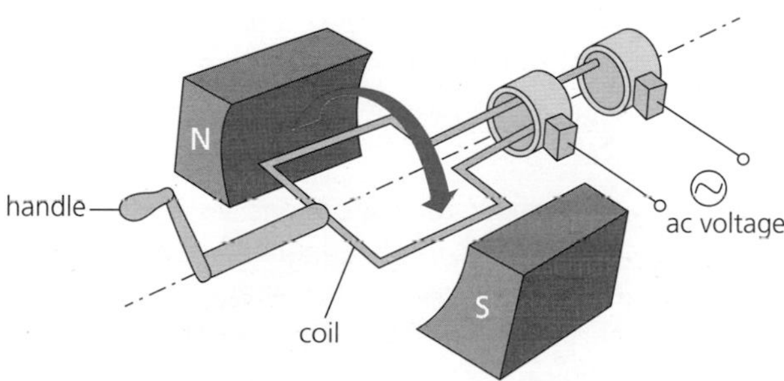

Figure 19 A simplified ac generator

The flux linkage, $N\Phi$, when the normal to the plane of the coil is at an angle θ to the magnetic field is given (see section 7.4 in Chapter 7) by

$$N\Phi = BAN\cos\theta$$

If the coil is rotated with angular speed ω, then angle $\theta = \omega t$ and the equation for the flux linkage becomes

$$N\Phi = BAN\cos\omega t$$

Faraday's law, $\mathcal{E} = N\frac{\Delta\Phi}{\Delta t}$, states that the induced emf in the coil is equal to the rate of change of flux linkage, which can be obtained mathematically by differentiating the expression for flux linkage with respect to time. Doing this gives this magnitude of the induced emf, $\mathcal{E}$, as

$$\mathcal{E} = BAN\,\omega\sin\omega t$$

where B is the magnetic flux density, A is the area of the coil, N is the number of turns on the coil and $\omega = 2\pi f$ is the angular speed of rotation, where f is the frequency of rotation of the coil (see Chapter 1).

Increasing any of the quantities, B, A, N or f, will increase the maximum value of the induced emf.

The variation of flux linkage and induced emf with time for one complete rotation of the coil is shown in Figure 20. The induced emf varies as a sine curve. Note that the emf is at its maximum when the flux linkage is zero, because, at this stage in the rotation, the flux linkage is *changing* at its greatest rate.

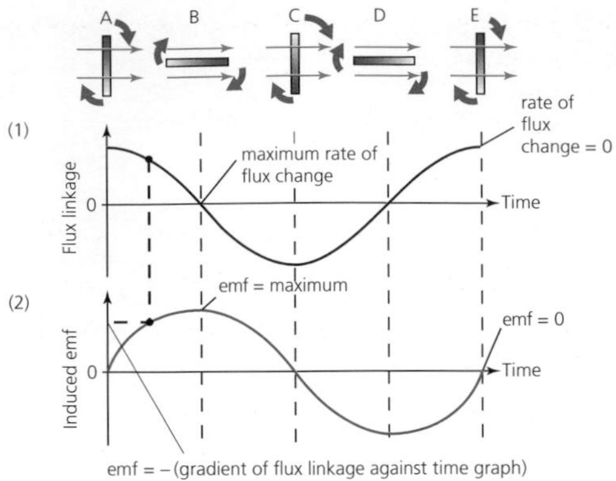

Positions A, B, C, D and E show the coil as viewed from the side for one complete rotation.

A: The coil is face-on to the field, the flux linkage is maximum and the emf is zero.

B: The coil is edge-on to the field, the flux linkage is zero and the emf is a maximum.

C: The coil is face-on to the field, the flux linkage is maximum and the emf is zero.

D: The coil is edge-on to the field, the flux linkage is zero and the emf is a maximum.

E: The coil is face-on to the field, the flux linkage is maximum and the emf is zero.

Figure 20 *Induction of an alternating emf with changing flux linkage in a rotating coil*

When the output of the ac generator is connected to the y-input of an oscilloscope, and the coil of the generator is rotated, a waveform in the shape of a sine curve is produced (Figure 21). Since the induced emf in the generator is given by $\mathcal{E} = BAN\,\omega \sin \omega t$, the maximum value of the induced emf is equal to $BAN\,\omega$, which is shown on the oscilloscope waveform as the **peak voltage** V_0. The **peak-to-peak voltage** $2V_0$ is also shown. The number of rotations completed by the coil of the generator in one second gives the frequency f of the alternating voltage output from the generator. Frequency $f = 1/T$, where T is the time period, the time for one complete rotation of the generator coil.

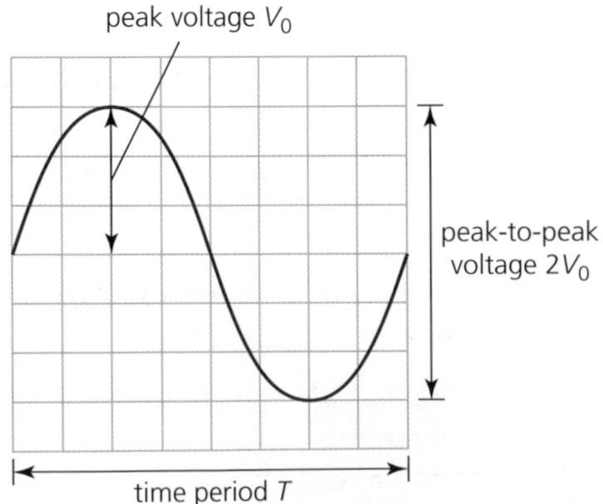

Figure 21 *Output voltage from an ac generator*

Worked example

A small ac generator made from a rectangular coil of 200 turns of wire rotates in a magnetic field of flux density 50 mT. The dimensions of the coil are 2.0 cm × 5.0 cm. The output from the generator is connected to the y-input of an oscilloscope and the peak voltage is measured as 2.0 V. Determine the frequency of rotation of the coil.

The peak voltage is equal to the maximum emf = $\mathcal{E} = BAN\,\omega$

Since angular frequency $\omega = 2\pi f$, maximum emf is $\mathcal{E} = BAN \times 2\pi f$, which rearranges to give

$$f = \frac{\mathcal{E}}{2\pi BAN}$$

$$= \frac{2}{2\pi \times 50 \times 10^{-3} \times 0.02 \times 0.05 \times 200} = 32\,\text{Hz}$$

QUESTIONS

8. The rectangular coil of an ac generator has 100 turns and dimensions 6.0 cm × 3.5 cm. It is rotated at a constant frequency of 25 Hz in a magnetic field of flux density 80 mT. Calculate
   a. the maximum flux linkage
   b. the maximum induced emf.
9. The coil of an ac generator has dimensions 10 cm × 5 cm and is made up of 100 turns. The coil is made to rotate at a frequency of 25 Hz in a magnetic field of 200 mT.
   a. Determine the peak output voltage from the generator.
   b. Sketch a graph of output voltage versus time, showing two cycles of the output voltage, and indicating appropriate scales on both axes.
10. The coil of an ac generator consists of N turns and rotates in a magnetic field of flux density B. The peak output voltage of the ac generator is V_0. What is the new peak output voltage if the frequency of rotation is halved, the flux density of the magnetic field is increased by a factor of 4 and the number of turns is doubled?

    A V_0 **B** $2V_0$ **C** $4V_0$ **D** $8V_0$

Stretch and challenge

11. An electric motor is made up of a coil positioned within a magnetic field. A power supply is used to drive an electric current through the coil so that the magnetic field exerts a force on the sides of the coil, causing it to rotate. Explain how Faraday's and Lenz's laws apply to the motor's rotating coil, and how the size of the current drawn from the power supply by the motor would be affected.

KEY IDEAS

- The emf induced in a coil rotating uniformly in a uniform magnetic field varies with time according to

$$\mathcal{E} = BAN\ \omega \sin \omega t$$

where B is the flux density, A is the area of the coil, N is the number of turns on the coil and ω is the angular speed of rotation.

- The sinusoidal emf gives rise to a sinusoidal voltage in an output circuit of frequency $\omega/2\pi$, peak value $BAN\omega$ and peak-to-peak value $2BAN\omega$.

8.3 ALTERNATING CURRENTS

Electricity, delivered around the UK by the National Grid, is produced by alternating current generators. Most of the UK's electricity is generated at power stations, where high-pressure steam-driven turbines rotate a large electromagnet, at a frequency of 50 Hz, inside a stationary coil. The emf generated at a power station varies as a sine curve, with a maximum emf, or peak voltage, typically about 35 kV. Since the mains voltage delivered to UK homes varies as a sine curve, any current that it drives also varies sinusoidally, which means that the current flowing in mains appliances continually reverses its direction. The usual circuit symbol for an alternating supply voltage is shown in Figure 22, although there are a number of alternative similar symbols.

Figure 22 *Alternating supply voltage symbol*

The sinusoidally varying mains alternating voltage and alternating current can be expressed as

$$\text{voltage: } V = V_0 \sin \omega t$$

$$\text{current: } I = I_0 \sin \omega t$$

where V_0 is the peak voltage, I_0 is the peak current and $\omega = 2\pi f$, where f is the alternating frequency.

The average value of a sine function is zero, regardless of the size of the peak value, so a special type of average is used for specifying the effective size of alternating voltages and currents. This is the **root mean square**, or **rms**, value.

The rms value of an alternating current is equal to the direct current that has the same heating effect.

The power P dissipated as heat in a resistor R is given by $P = I^2R$, so the power dissipated as heat in a resistor carrying an alternating current varies and is given by

$$P = (I_0 \sin \omega t)^2 R$$

The average value of any sine-squared function is $\frac{1}{2}$, so the average power dissipated as heat in a resistor by an alternating current is given by

$$P_{av} = \frac{1}{2}(I_0)^2 R$$

Comparing $P_{av} = \frac{1}{2}(I_0)^2 R$ with $P = I^2R$ shows that the direct current, I, that would have the same heating effect as an alternating current of peak value I_0 is given by

$$I = \sqrt{\tfrac{1}{2}(I_0)^2}$$

This is the root mean square value, I_{rms}. So

$$I_{rms} = \sqrt{\tfrac{1}{2}(I_0)^2} = \frac{I_0}{\sqrt{2}}$$

Similarly the root mean square value of an alternating voltage is

$$V_{rms} = \frac{V_0}{\sqrt{2}}$$

The average power dissipated in a resistor by an alternating current is

$$P_{av} = I_{rms}^2 R = V_{rms} I_{rms}$$

and the resistance in a circuit is

$$R = \frac{V_{rms}}{I_{rms}} = \frac{V_0}{I_0}$$

Worked example

The root mean square voltage of the mains in the UK is 230 V. Calculate the peak voltage, V_0, and the peak-to-peak voltage of the mains.

The equation for the rms value of an alternating voltage

$$V_{rms} = \frac{V_0}{\sqrt{2}}$$

can be rearranged to give the peak voltage

$$V_0 = \sqrt{2} \times V_{rms} = \sqrt{2} \times 230 = 325\text{V}$$

and thus the peak-to-peak voltage

$$2V_0 = 650\text{ V}$$

QUESTIONS

12. In the circuit shown in Figure 23, the peak voltage from the ac power supply unit is 20 V. Determine
    a. the rms voltage
    b. the rms current
    c. the average power output for the loud speaker.

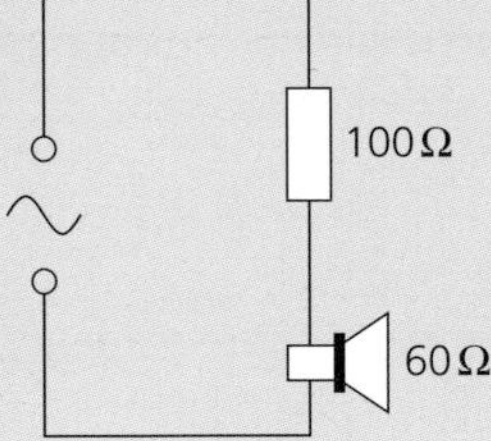

Figure 23

13. The root mean square voltage of the mains in the USA is 120 V. Calculate the peak voltage, V_0, and the peak-to-peak voltage of the US mains.

KEY IDEAS

- Sinusoidal alternating voltage and current are represented by $V = V_0 \sin \omega t$ and $I = I_0 \sin \omega t$, respectively, where $\omega = 2\pi f$ and f is the alternating frequency, and V_0 and I_0 are the peak voltage and peak current.
- The root mean square value of an alternating current is equal to the direct current that has the same heating effect, and is given by

$$I_{rms} = \frac{I_0}{\sqrt{2}}$$

- Similarly the root mean square value of an alternating voltage is

$$V_{rms} = \frac{V_0}{\sqrt{2}}$$

8.4 OSCILLOSCOPE APPLICATIONS

A cathode ray oscilloscope (Figure 24), or simply **oscilloscope**, is very useful as a dc and ac voltmeter, measuring time intervals and frequencies and also displaying alternating waveforms.

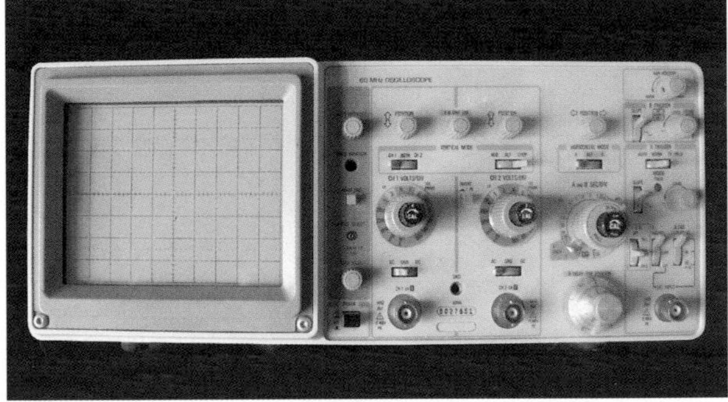

Figure 24 *A cathode ray oscilloscope of the type used in schools*

The screen of the oscilloscope can be considered as a graph of voltage on the y-axis versus time on the x-axis. The input voltage is connected to the y-inputs. The y-gain control (Figure 25a) enables the voltage value assigned to one division on the vertical voltage axis to be changed as required. The higher the y-gain value chosen, the greater the number of volts per grid division – so the *less* sensitive is the setting. If measurements are to be taken from the screen, a high sensitivity is needed, because the greater the vertical displacement of the waveform, the smaller the percentage uncertainty in the measured value.

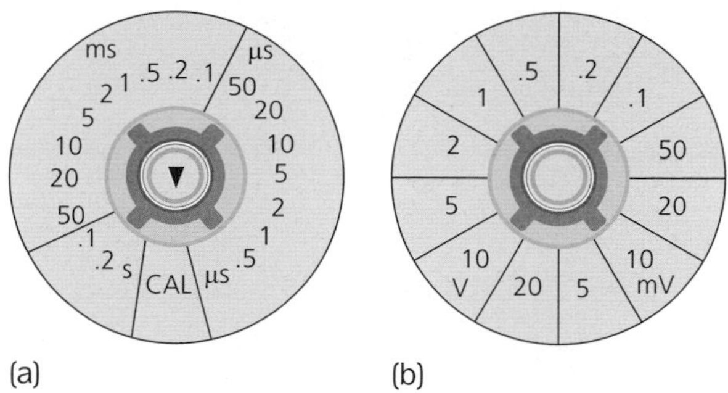

Figure 25 *(a) Oscilloscope y-gain control. (b) Time base control. The values on the controls only apply if the centre of the control knob is pointing in the direction indicated by 'CAL' (an abbreviation of 'calibrate').*

The **time base** control (Figure 25b) enables the time value assigned to one division on the horizontal time axis to be changed. With the time base switched on, but with no input voltage to the oscilloscope, either a horizontal line is produced on the screen or a dot of light is seen moving across the screen.

Measuring a dc voltage

The input control switch on the oscilloscope has three settings: DC, AC and GD. DC and AC obviously stand for dc and ac; GD stands for ground, and disconnects any input to the oscilloscope. It is helpful to first switch to GD and adjust the horizontal line produced so that it is along the central horizontal axis of the oscilloscope screen. The input control can then be switched to DC (or AC, depending on the type of input that is to be connected to the oscilloscope).

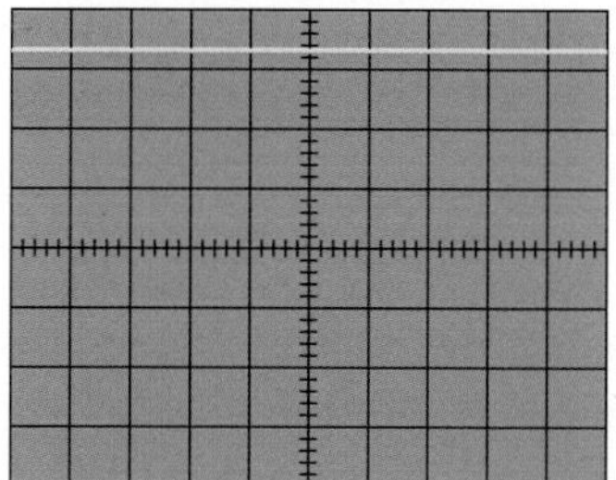

Figure 26 *Oscilloscope trace for a battery*

With the input control switch set to DC and a battery connected to the oscilloscope's *y*-input, a trace like that in Figure 26 is produced. If the intensity is kept low, the line is much finer and measurements can be made more precisely. If the *y*-gain control is set to 2.0 V/div, the emf $\mathcal{E}$ of the battery is given by

$$\mathcal{E} = \text{number of divisions} \times y\text{-gain} = 3.3 \times 2.0 = 6.6\,\text{V}$$

Since each division is divided into 5 subdivisions on the central axes, the uncertainty in the number of divisions measured is ±0.2, which gives a percentage uncertainty of $\frac{0.2}{3.3} \times 100 = \pm 6.1\%$. This corresponds to an uncertainty of ±0.4 V on the measured value of the battery's emf. Therefore the battery's emf is $\mathcal{E} = 6.6 \pm 0.4\,\text{V}$.

Measuring an ac frequency

If the oscilloscope input control is set to AC and a signal generator (or 'function generator'), set to its sine function, is connected to the *y*-input, a trace like the one in Figure 27a is produced. The time base setting for the waveform shown is 0.2 ms/div. The number of divisions corresponding to one cycle is measured along the central horizontal axis of the screen. The time for one cycle *T* can be determined from

$$T = \text{number of divisions for one cycle} \times \text{time base setting} = 6.6 \times 0.2 = 1.32\,\text{ms}$$

The uncertainty in the number of divisions is ±0.2 divisions, which corresponds to a percentage uncertainty of $\frac{0.2}{6.6} \times 100 = \pm 3.0\%$. The output frequency *f* from the signal generator is therefore

$$f = \frac{1}{T} = \frac{1}{1.32 \times 10^{-3}} = 758\,\text{Hz}$$

with a percentage uncertainty of ±3.0%, which is equivalent to ±23 Hz.

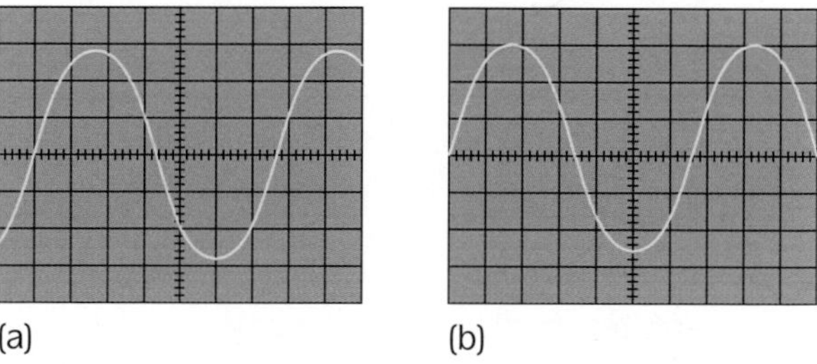

Figure 27 *Signal generator sine function output waveform. The waveform (b) is the same as the one in (a) but with its position shifted.*

Measuring ac peak-to-peak voltage

To obtain an accurate value for the peak to peak voltage, it can be helpful to use the *y* position control to move the positive peak so that it touches a horizontal grid line and then use the *x* position control so that the negative peak is aligned to the central vertical axis, as shown in Figure 27b. The *y*-gain setting for the trace shown is 0.1 V/div. The number of divisions corresponding to the peak-to-peak voltage is 5.6 div. The percentage uncertainty in the number of divisions for the peak to peak voltage is

$\frac{0.2}{5.6} \times 100 = \pm 3.6\%$. Therefore the peak-to-peak output voltage from the signal generator is $5.6 \times 0.1 = 0.56 \pm 0.02$ V, since 3.6% of 0.56 is 0.02 V.

Viewing other waveforms

A square wave alternating voltage (Figure 28) is fundamental to the operation of many electronic circuits and is usually one of the outputs provided by a signal generator. The vertical parts of the trace are dimmer than the horizontal sections, so to measure the frequency it may be necessary to adjust the intensity control on the oscilloscope to make the trace bright enough to make the vertical sections visible. To measure the peak-to-peak voltage, the trace should be made dimmer to make the horizontal sections thinner to enable more precise measurement.

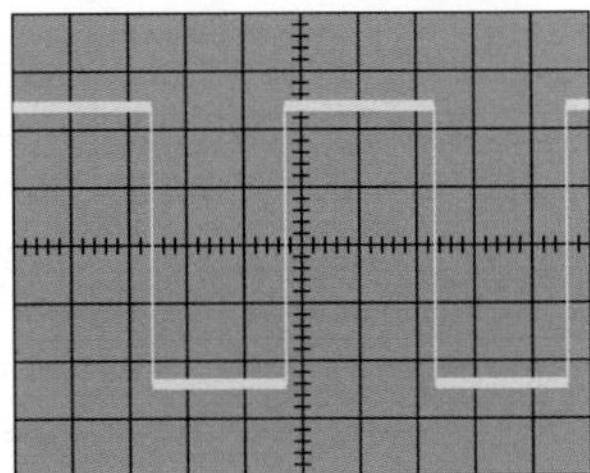

Figure 28 *Square wave output from a signal generator*

QUESTIONS

14. An oscilloscope is switched to GD (ground) and the horizontal line is positioned along the central axis. A cell is then connected to the *y*-input of the oscilloscope and the horizontal trace appears higher up the screen.

a. Describe how the emf of the cell can be determined from the trace shown.

b. Suggest any adjustments that could have been made that would have reduced the percentage uncertainty in the measured value of the cell's emf.

15. Figure 29 shows the waveform of the output of a signal generator on an oscilloscope screen with a *y*-gain setting of 0.5 V/div and a time base setting of 50 μs/div. Determine the following quantities, including an uncertainty with each value:

a. the peak-to-peak voltage

b. the rms voltage

c. the frequency of the output.

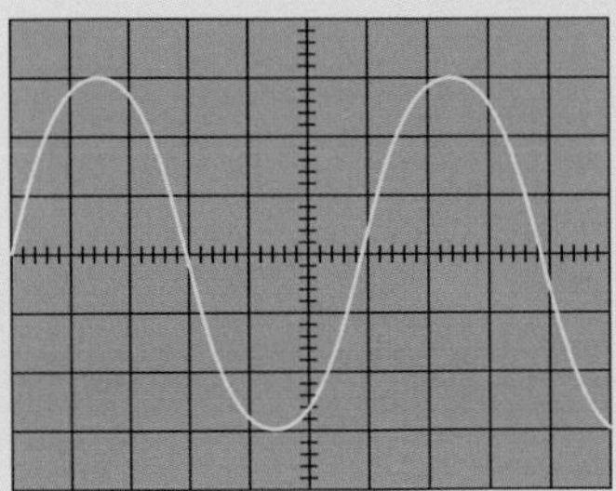

Figure 29

16. a. Determine the frequency of the square wave output shown in Figure 28, given that the time base is set to 1 ms/div.

b. Suggest an adjustment that should be made to the trace in Figure 28 to enable an accurate measurement of the square wave's peak-to-peak voltage.

17. The peak output voltage of a small ac generator is 2.6 V at a frequency of 20 Hz. Select the most appropriate *y*-gain and time base settings from Table 1 to display *two cycles* of the output from the generator on an oscilloscope screen, which has 10 horizontal divisions and 8 vertical divisions, ensuring that the waveform takes up as much of the screen as possible.

y-gain, volts/div	Time base, ms/div
0.2	2
0.5	5
1	10
2	20

Table 1

KEY IDEAS

- An oscilloscope screen is an adjustable voltage versus time graph.
- The sensitivity of the vertical axis of the oscilloscope screen can be adjusted using the *y*-gain control.
- The sensitivity of the horizontal axis of the oscilloscope screen can be adjusted using the time base control.

REQUIRED PRACTICAL: APPARATUS AND TECHNIQUES

Investigation (using a search coil and an oscilloscope) of the effect on magnetic flux linkage of varying the angle between a search coil and the magnetic field direction

This practical uses the phenomenon of electromagnetic induction to demonstrate how the magnetic flux linkage through a small search coil varies as its orientation within the magnetic field, created by a larger circular coil, is changed.

The practical gives you the opportunity to:

- use appropriate analogue apparatus to record a range of measurements (to include length/distance or angle) and to interpolate between scale markings
- use appropriate digital instruments, including electrical multimeters, to obtain a range of measurements (to include current and voltage)
- correctly construct circuits from circuit diagrams using dc power supplies, cells and a range of circuit components
- use an oscilloscope, including volts/division and time base.

Apparatus

This method uses a vertically mounted large circular coil. The coil shown in Figure P1 has a useful flat plastic surface through its centre. A signal generator is connected to this coil, so that alternating current flows in it. An axial search coil is connected to the y-input of an oscilloscope (Figure P2). A multimeter set as an ammeter is in series with the signal generator and the large coil. A half-metre rule and the search coil are positioned on the plastic surface (Figure P3).

Figure P1 *Vertically mounted circular coil*

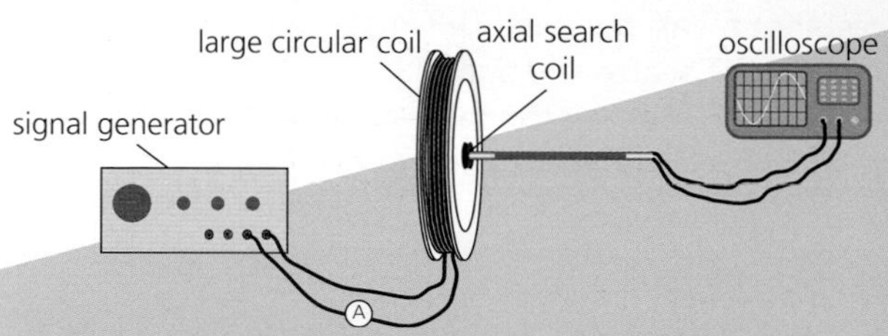

Figure P2 *Electrical connections*

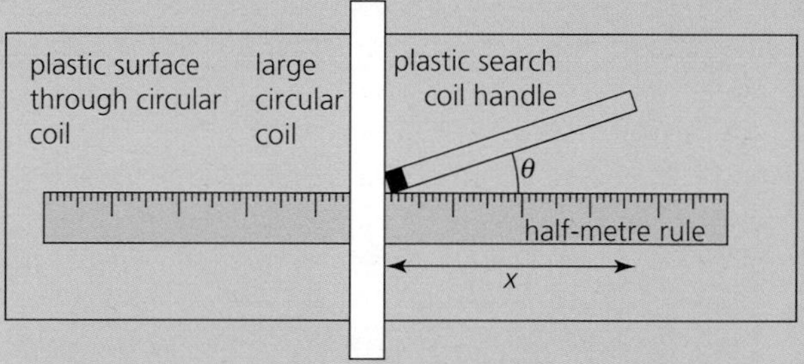

Figure P3 *Plan view showing position of the axial search coil and half-metre rule relative to the large circular coil*

Principle

The alternating current in the large coil creates an alternating magnetic field. The search coil is positioned close to the centre of the large coil, so that the flux from the large coil passes through the search coil. The alternating magnetic field causes the flux linking the search coil to change continuously. According to Faraday's law, an emf is induced in the search coil, and this will also be alternating. If the alternating current in the large coil is kept constant, the peak value of the emf induced in the search coil is directly proportional to the maximum flux linkage (BAN) – see section 8.2.

Techniques

The peak value and the frequency of the alternating current in the large circular coil are both kept constant. The multimeter, set as an ammeter, is used to monitor the current in the large coil to ensure that it remains constant.

The emf induced in the search coil is measured using the y-gain setting on the oscilloscope.

Angle θ in Figure P3 is the angle between the direction of the magnetic field created by the large coil and the normal to the plane of the axial search coil. Angle θ is therefore also equal to the angle between the half-metre rule and the plastic handle of the search coil, which can be measured with a protractor or by applying trigonometry to the appropriate distance measurements. The angle can be varied by moving the search coil but taking care to ensure that the centre of the search coil remains at the same point relative to the large circular coil.

QUESTIONS

P1 State the independent, the dependent and the control variables in the experiment.

P2 A student sets up the experiment and adjusts the frequency and amplitude of the signal generator along with the *y*-gain and time base of the oscilloscope until at least one full cycle of the induced alternating emf is displayed on the oscilloscope screen (Figure P4). The *y*-gain setting on the oscilloscope is 5 mV/div. Determine the peak-to-peak value of the emf induced in the search coil, and its uncertainty.

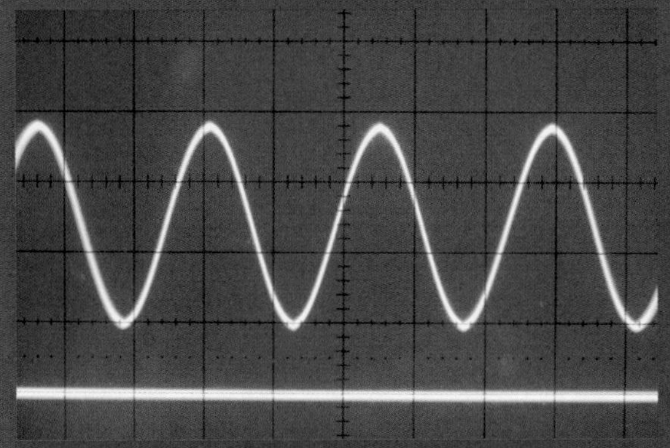

Figure P4 *The induced emf displayed on an oscilloscope screen*

P2 a. Figure P3 shows that distance x can be measured to determine a value for angle θ for a particular orientation of the search coil relative to the large circular coil. Distance x is measured along the half-metre rule from a point in line with the centre of the search coil to a point in line with the end of the search coil's plastic handle. Suggest how to ensure that distance x is measured accurately.

b. A student undertakes the experiment and measures distance x as 21.2 cm and the distance from the centre of the search coil to the end of the plastic handle as 22.8 cm. Determine a value for θ with an uncertainty.

c. How does the method involving the measurement of distance x compare with the use of a protractor, with the smallest division of 1°, for determining a value for θ?

8.5 THE OPERATION OF A TRANSFORMER

A major advantage of producing electrical energy using an alternating current generator is that it is relatively easy to change the size of an alternating voltage using a **transformer**. The root mean square voltage generated at a power station is typically 25 kV but can be changed to 400 kV for efficient transmission around the country via the National Grid (see section 8.6), and then changed to much lower values for users. On a smaller scale, a transformer is used to change the mains supply at 230 V to 12 V for a low-voltage power supply unit used in a school laboratory.

A transformer is made from two coils wound separately on a soft iron core (Figure 30a). The circuit symbol for a transformer is shown in Figure 30b.

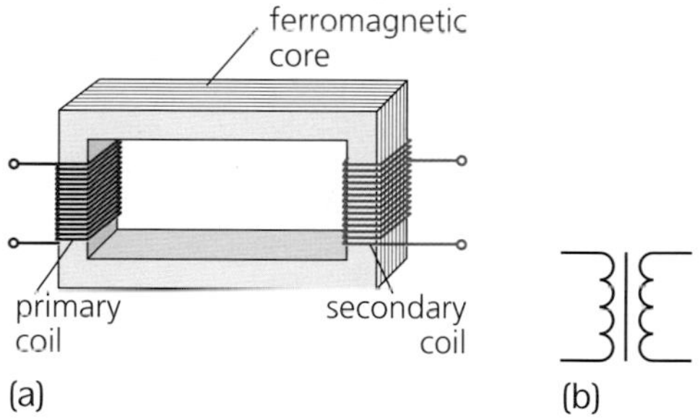

Figure 30 *A transformer and its circuit symbol. In a circuit the convention is that the coil on the left is the primary (input) and the coil on the right is the secondary (output).*

The alternating voltage to be changed is applied to the primary coil, causing an alternating current to flow in that coil (Figure 31). The alternating current in the primary coil creates an alternating magnetic field in the iron core, which in turn creates a changing magnetic flux through the secondary coil. The changing magnetic flux induces an emf in the secondary coil in accordance with Faraday's law. It is important to understand that there is no electrical connection between the primary and secondary coils. The only link between the two coils is the magnetic field in the iron core. The shape of the transformer is designed to ensure that as much of the magnetic flux from the primary coil as possible passes through the secondary coil.

If there are fewer turns on the secondary than on the primary, the output emf is smaller than the input voltage, and the transformer is said to be a **step-down transformer**. If there are more turns on the secondary than on the primary, the output voltage is greater than the input voltage, and the transformer is said to be a **step-up transformer**.

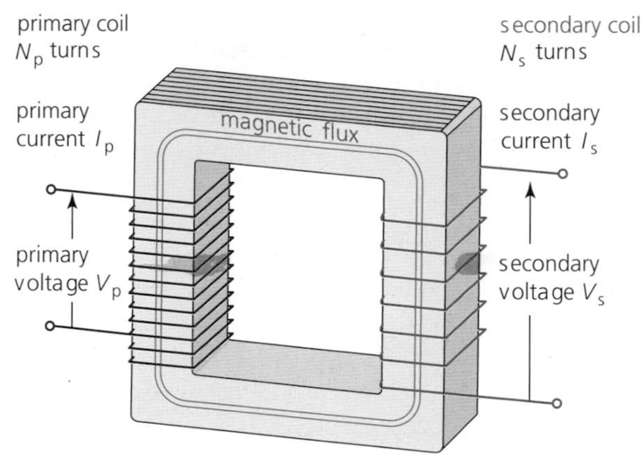

Figure 31 *Transformer operation*

The ratio of the voltages across the primary and secondary coils is equal to the ratio of the number of turns.

The **transformer equation** is

$$\frac{N_s}{N_p} = \frac{V_s}{V_p}$$

where N_p and N_s are the number of turns on the primary and secondary coils, respectively, V_p is the voltage applied to the primary coil and V_s is the emf induced in the secondary coil. The equation can be rearranged and written as

$$\frac{N_p}{N_s} = \frac{V_p}{V_s}$$

Efficiency of a transformer

The input power to the transformer is $P_{in} = I_pV_p$, and the output power is $P_{out} = I_sV_s$. No transformer is 100% efficient – the input power, P_{in}, is always greater than the output power, P_{out}.

The **efficiency** of as transformer as a fraction is given by

$$\text{efficiency} = \frac{\text{output power}}{\text{input power}} = \frac{I_sV_s}{I_pV_p}$$

The efficiency as a percentage is given by

$$\text{efficiency} = \frac{I_sV_s}{I_pV_p} \times 100\%$$

An **ideal transformer** is an imaginary transformer that does not lose any electrical energy, so in theory its input power is equal to its output power: $I_pV_p = I_sV_s$. Although such a transformer does not actually exist, the equation does serve to illustrate the following:

A step-up transformer steps up the voltage and steps down the current. A step-down transformer steps down the voltage and steps up the current.

There are a number of reasons why transformers are less than 100% efficient.

- Energy is dissipated as heat in the primary and secondary coils. To minimise this energy loss, the windings of the coils are made from a low-resistance wire, typically a thick wire of low resistivity such a copper.
- The alternating magnetic flux through the iron core induces emfs in the core that drive **eddy currents**. These eddy currents generate heat in the core, resulting in a loss of energy. This energy loss is minimised by using a laminated iron core rather than a solid iron core (see Figure 32). The laminations are alternate layers of iron and insulator. The layers of high-resistance insulator prevent the flow of eddy currents.
- Since the current in the primary coil is an alternating current, the iron core is repeatedly magnetised in one direction, then demagnetised, then magnetised in the other direction. The repeated magnetisation and demagnetisation of the core generates heat, which results in a loss of energy. This energy loss is minimised by using a core made from a magnetically soft material, such as soft (or annealed) iron, which can be magnetised and demagnetised easily.
- Some of the magnetic flux created by the primary coil does not pass through the secondary coil. To achieve maximum flux linkage, the transformer is designed to keep the two coils as close together as possible. For example, the coils can be wound on top of each other around the same section of the iron core.

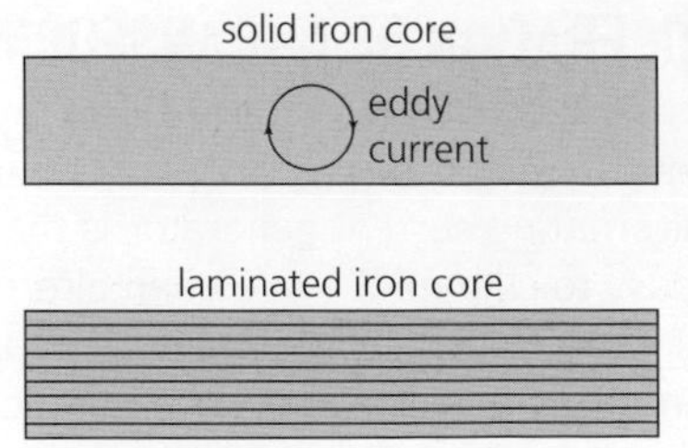

Figure 32 *Laminations in the iron core reduce eddy currents.*

Worked example

The generator at a power station generates electricity at an rms voltage of 25 kV, and a transformer steps up this voltage to 400 kV for transmission around the country via the National Grid. Given that the primary current is 40 A and the transformer is 85% efficient, determine

a. the turns ratio of the transformer

b. the secondary current.

a. Turns ratio is $\frac{N_s}{N_p} = \frac{V_s}{V_p} = \frac{400}{25} = 16$.

b. Efficiency is $\frac{I_sV_s}{I_pV_p} = 0.85$, which gives

$$I_s = 0.85 \times \frac{I_pV_p}{V_s} = \frac{0.85 \times 40 \times 25}{400} = 2.1\text{ A}$$

QUESTIONS

18. a. Calculate the number of turns on the secondary coil of a step-down transformer that would enable a 12 V bulb to be used with the 230 V mains supply, if there are 480 turns on the primary.

b. Calculate the secondary current, given that the primary current is 0.20 A and the transformer efficiency is 80%.

19. Explain why a transformer works using alternating current but not using direct current.

ASSIGNMENT 2: TESTING THE TRANSFORMER EQUATION AND MEASURING TRANSFORMER EFFICIENCY

(MS 1.5, PS 1.1, PS 1.2, PS 2.3, PS 2.4, PS 3.3)

An experiment is set up to make measurements to test the equation

$$\frac{N_s}{N_p} = \frac{V_s}{V_p}$$

and to determine the efficiency of a demountable step-down transformer.

A low-voltage variable ac power supply unit is used to deliver an alternating current to the 400-turn primary coil of a transformer. Multimeters set to appropriate ac ranges are used to measure the primary current and voltage. The 200-turn secondary coil delivers an alternating current to an 18 Ω resistor, again with multimeters set to measure the secondary current and voltage. A suitable circuit diagram is shown in Figure A1.

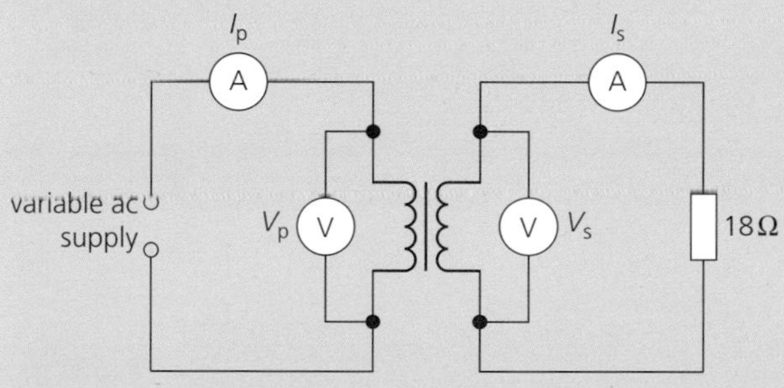

Figure A1

The power supply unit is varied to produce eight data sets for primary voltages varying from 1 V to 8 V with all four multimeter readings taken for each data set.

Questions

A1 Which of the quantities labelled on the circuit diagram is the independent variable?

A2 Describe and explain what type of graph line would be obtained if V_s were plotted on the y-axis and V_p on the x-axis, using data obtained from making measurements using the circuit shown in Figure A1.

A3 A student makes measurements (Table A1) for one data set using the circuit shown in Figure A1. Assuming that the accuracy of a multimeter is reflected by the number of decimal places recorded for each value, determine the efficiency of the transformer and its uncertainty.

Primary voltage V_p / V	Primary current I_p / mA	Secondary voltage V_s / V	Secondary current I_s / mA
4.00	71	2.01	112

Table A1

A4 Suppose a student undertaking the experiment mistakenly used the 400-turn coil as the secondary rather than the primary.

a. If the maximum output voltage from the variable low-voltage supply was 12 V, what is the maximum possible value for V_s?

b. Explain why using coils of 200 and 400 turns makes the experiment much safer than coils of 200 and 1200 turns.

Induction heating

Induction heating involves heating a metal object using the phenomenon of electromagnetic induction to induce electric currents in the object to be heated. The effect can be demonstrated using the jumping ring experiment (Figure 33). The arrangement is essentially a step-down transformer with the aluminium ring equivalent to a single-turn secondary coil. A step-down transformer steps up the current and, with typically 1000 turns on the primary, the turns ratio is considerable, resulting in a very large current being induced in the aluminium ring, making it very hot. When the alternating voltage is applied to the coil, the resulting current creates an alternating magnetic field in the iron core, which becomes weaker with increasing distance from the coil. The induced current in the ring creates its own magnetic field, which opposes the field in the iron core, causing the ring to jump upwards and then levitate at the position where its weight is balanced by the magnetic force. The technique is used in an induction furnace where temperatures in excess of 2000 °C are achieved.

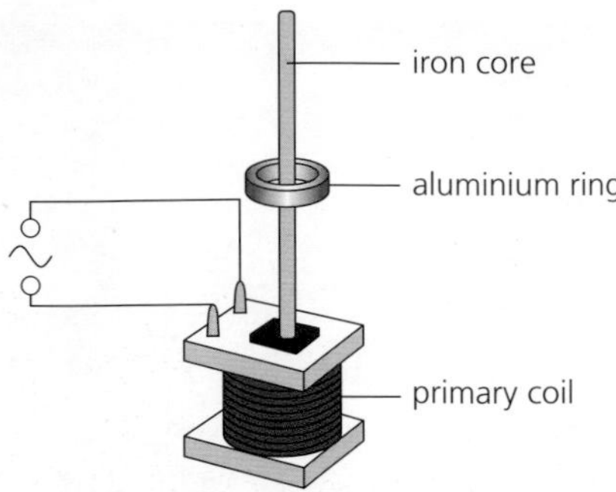

Figure 33 *'Jumping ring' experiment*

QUESTIONS

20. What would happen to the aluminium ring in Figure 33 if the primary coil current was increased?

21. A student suggests that the ring in Figure 33 is placed in liquid nitrogen before being slotted over the iron core. Suggest what effect this would have when the ac power supply is switched on.

KEY IDEAS

- A transformer can be used to step up or step down voltages.
- The transformer equation is

$$\frac{N_s}{N_p} = \frac{V_s}{V_p}$$

- Low-resistance windings, a soft iron core and laminations in the core reduce energy losses in a transformer.
- Since power $P = IV$, and, neglecting energy losses, the output power is equal to the input power, stepping up the voltage steps down the current, and vice versa.

8.6 TRANSMISSION OF ELECTRICAL POWER

Early power stations were relatively small and built close to the factories requiring the electricity. In the 1920s, planning started on creating the National Grid designed to distribute electricity all over the UK. The use of transformers is crucial to the transmission of electrical energy over long distances because the stepping up of voltage for transmission enables the energy lost as heat in the cables to be kept to a minimum.

The power P dissipated as heat in a cable of resistance R carrying current I is given by

$$P = I^2R$$

Since the current I is squared, then, if the size of the current is reduced by a factor of 10, the power lost as heat decreases by a factor of 100. A step-up transformer at a power station is therefore used to step up the 25 kV produced by the generator to 400 kV used for transmission, resulting in the current being stepped down and the power loss during transmission being significantly reduced.

Step-down transformers are then used at substations to reduce the voltage to values suitable for use by consumers. The voltage is changed to 33 kV and 11 kV for heavy and light industry, respectively, and finally to 230 V for use in the home (Figure 34).

The National Grid operates very efficiently, with electrical heating losses in generators, transformers and cables being less than about 3% of peak electricity demand.

Worked example

A power station is connected to a town by cables of total resistance 10 Ω, delivering 50 MW.

Calculate the power losses in the cables if the power is supplied to the town

a. directly from the generator at 25 kV

b. after having been stepped up to 400 kV.

a. To deliver 50 MW at 25 kV requires an electric current I given by

$$I = \frac{P}{V} = \frac{50 \times 10^6}{25000} = 2000\,\text{A}$$

Power loss in cables is

$$P = I^2R = 2000^2 \times 10 = 40\,\text{MW}$$

b. To deliver 50 MW at 400 kV requires an electric current I given by

$$I = \frac{P}{V} = \frac{50 \times 10^6}{400 \times 10^3} = 125\,\text{A}$$

Power loss in cables is

$$P = I^2R = 125^2 \times 10 = 0.16\,\text{MW}$$

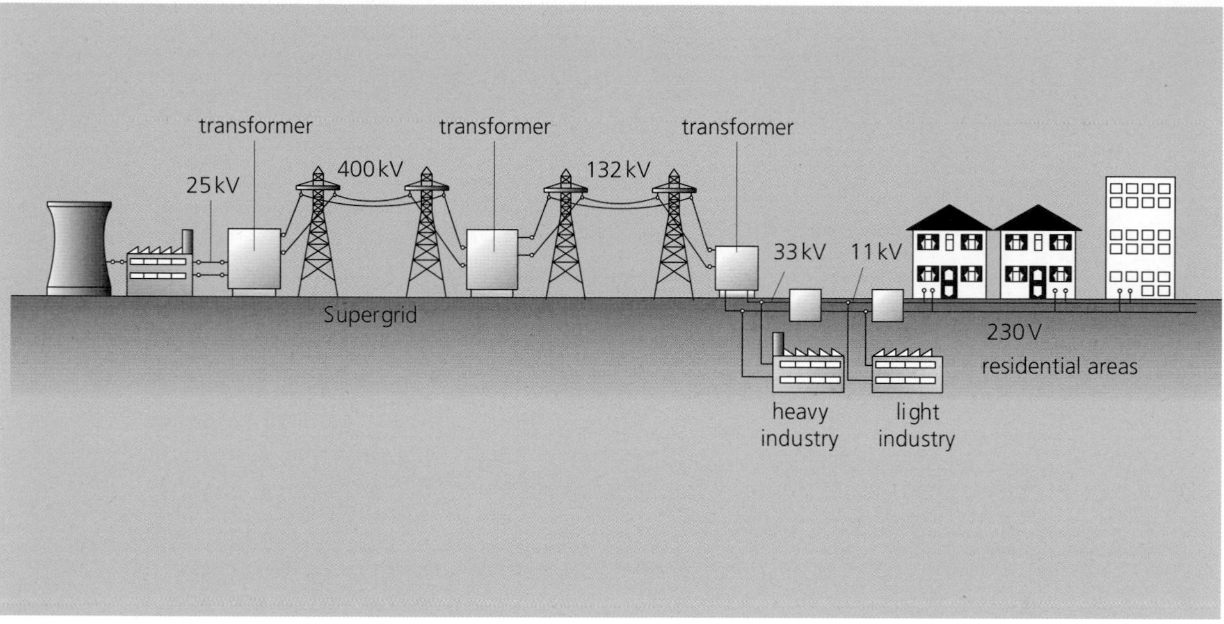

Figure 34 The National Grid

QUESTIONS

22. A power cable of total resistance 20 Ω is used to deliver 2.0 MW at 33 kV to a heavy industrial plant. Determine the power wasted in the cable.

23. A portable diesel-powered generator is capable of producing 4 kW of electrical power at a voltage of 240 V for use on a remote farm that is not connected to the National Grid. If cables of total resistance 1.2 Ω deliver the electricity to the farmhouse, determine

    a. the power lost as heat in the cables when the generator is running at 4 kW

    b. the minimum size of the voltage available in the farmhouse.

KEY IDEAS

- Transmission of electricity at high voltage and low current significantly reduces energy losses.
- The power P dissipated as heat in a cable of resistance R carrying current I is given by

$$P = I^2R$$

PRACTICE QUESTIONS

1. Figure Q1 shows an end view of a simple electrical generator. A rectangular coil is rotated in a uniform magnetic field with the axle at right angles to the field direction. When in the position shown in Figure Q1 the angle between the direction of the magnetic field and the normal to the plane of the coil is θ.

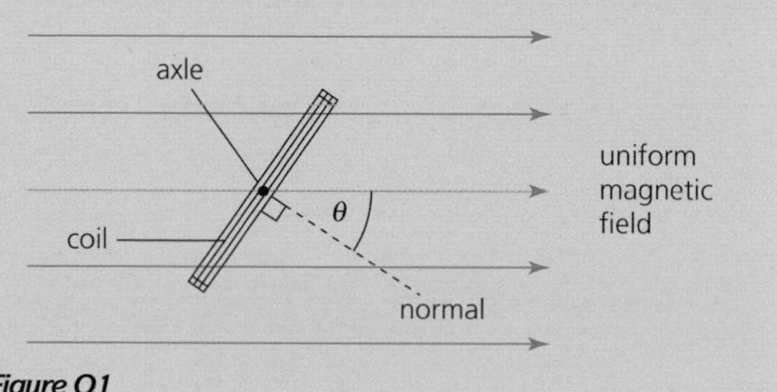

Figure Q1

a. The coil has 50 turns and an area of $1.9 \times 10^{-3}\,m^2$. The flux density of the magnetic field is 2.8×10^{-2} T. Calculate the flux linkage (in Wb turns) for the coil when θ is 35°, expressing your answer to the appropriate number of significant figures.

b. The coil is rotated at constant speed, causing an emf to be induced.

i. Sketch a graph on axes like those in Figure Q2 to show how the induced emf varies with angle θ during one complete rotation of the coil, starting when $\theta = 0$. Values are not required on the emf axis of the graph.

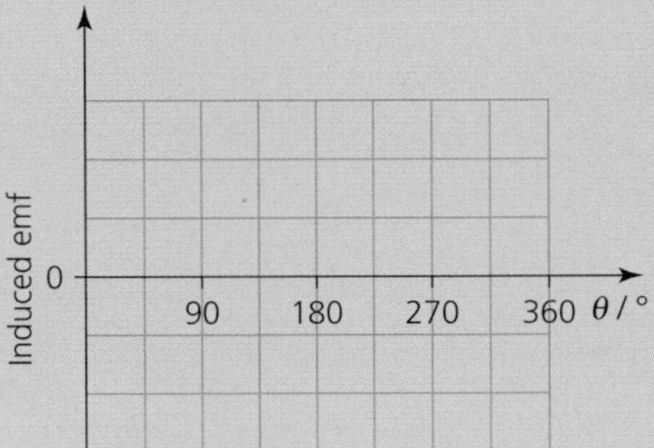

Figure Q2

ii. Give the value of the flux linkage for the coil (in Wb turns) at the positions where the emf has its greatest values.

iii. Explain why the magnitude of the emf is greatest at the values of θ shown in your answer to **b i**.

AQA June 2010 Unit 4 Section B Q5

2. Figure Q3 shows how a sinusoidal alternating voltage varies with time when connected across a resistor, *R*.

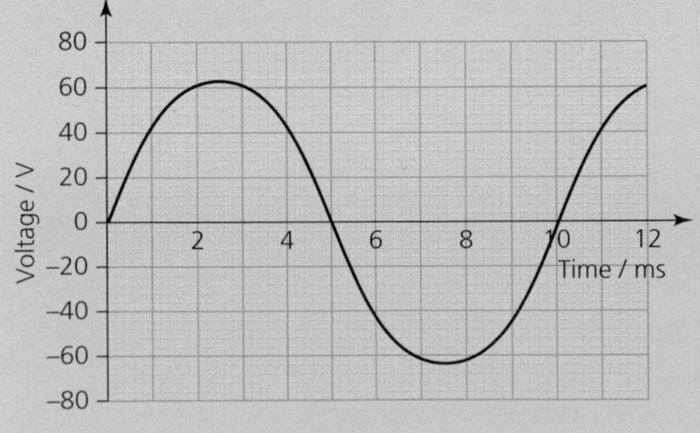

Figure Q3

a. i. State the peak-to-peak voltage.

ii. State the peak voltage.

iii. Calculate the root mean square (rms) value of the alternating voltage.

iv. Calculate the frequency of the alternating voltage. State an appropriate unit.

b. An oscilloscope has a screen of eight vertical and ten horizontal divisions. Describe how you would use the oscilloscope to display the alternating waveform in Figure Q3 so that two complete cycles are visible.

AQA June 2013 Unit 1 Q5 part

3. a. i. Outline the essential features of a step-down transformer when in operation.

ii. Describe **two** causes of the energy losses in a transformer and discuss how these energy losses may be reduced by suitable design and choice of materials. The quality of your written communication will be assessed in this question.

b. Electronic equipment, such as a TV set, may usually be left in 'standby' mode so that it is available for instant use when needed. Equipment left in standby mode continues to consume a small amount of power. The internal circuits operate at low voltage, supplied from a transformer. The transformer is disconnected from the mains supply only when the power switch on the equipment is turned off. This arrangement is outlined in Figure Q4.

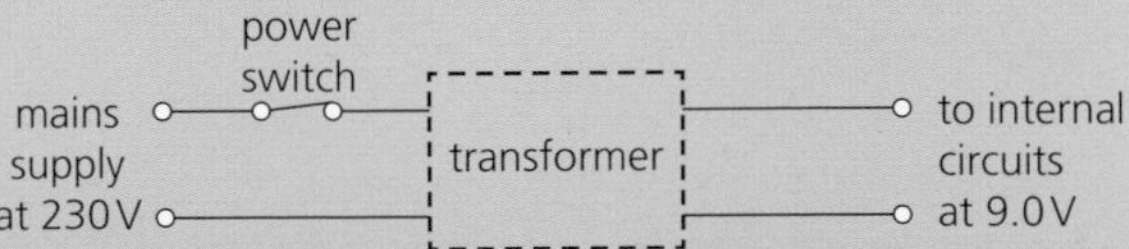

Figure Q4

When in standby mode, the transformer supplies an output current of 300 mA at 9.0 V to the internal circuits of the TV set.

i. Calculate the power wasted in the internal circuits when the TV set is left in standby mode.

ii. If the efficiency of the transformer is 0.90, show that the current supplied by the 230 V mains supply under these conditions is 13 mA.

iii. The TV set is left in standby mode for 80% of the time. Calculate the amount of energy, in J, that is wasted in one year through the use of standby mode. [1 year = 3.15×10^7 s]

iv. Show that the cost of this wasted energy will be about £4, if electrical energy is charged at 20p per kWh.

c. The power consumption of an inactive desktop computer is typically double that of a TV set in standby mode. This waste of energy may be avoided by switching off the computer every time it is not in use. Discuss **one** advantage and **one** disadvantage of doing this.

AQA January 2010 Unit 4 Section B Q4

4. The magnetic flux through a coil of 5 turns changes uniformly from 15×10^{-3} Wb to 7.0×10^{-3} Wb in 0.50 s. What is the magnitude of the emf induced in the coil due to this change of flux?

A 14 mV **B** 16 mV **C** 30 mV **D** 80 mV

AQA January 2013 Unit 4 Section A Q23

5. A rectangular coil of area *A* has *N* turns of wire. The coil is in a uniform magnetic field, as shown in Figure Q5. When the coil is rotated at a constant frequency *f* about its axis XY, an alternating emf of peak value $\mathcal{E}_0$ is induced in it.

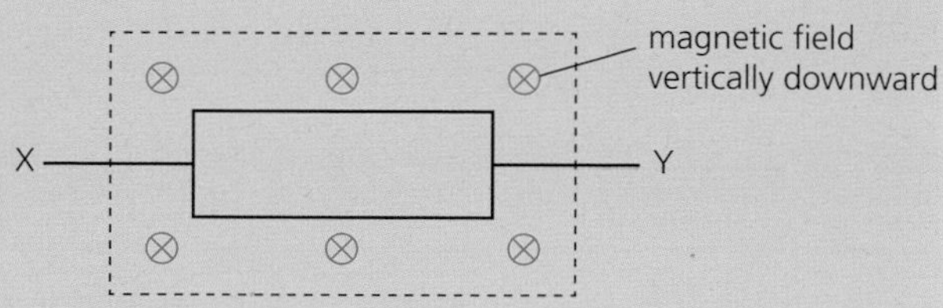

Figure Q5

What is the maximum value of the magnetic flux linkage through the coil?

A $\frac{\mathcal{E}_0}{2\pi f}$ **B** $\frac{\mathcal{E}_0}{\pi f}$ **C** $\pi f \mathcal{E}_0$ **D** $2\pi f \mathcal{E}_0$

AQA June 2014 Unit 4 Section A Q24

6. The graph in Figure Q6 shows how the magnetic flux linkage, $N\Phi$, through a coil changes when the coil is moved into a magnetic field.

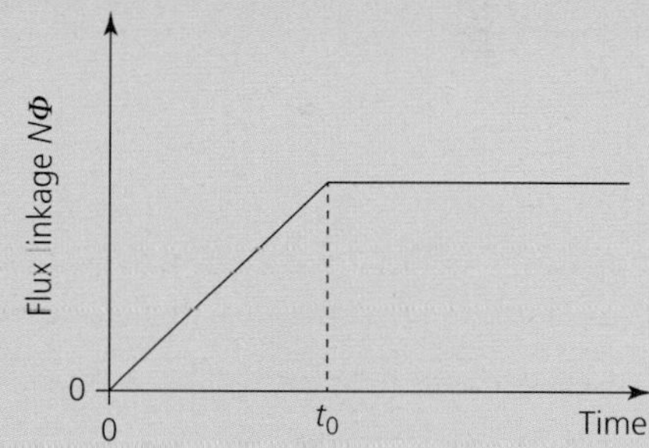

Figure Q6

The emf induced in the coil

A increases then becomes constant after time t_0

B is constant then becomes zero after time t_0

C is zero then increases after time t_0

D decreases then becomes zero after time t_0

AQA June 2011 Unit 4 Section A Q22

9 RADIOACTIVITY

PRIOR KNOWLEDGE

You will already be familiar with the structure and constituents of the atom and know something of how these were discovered. You will have used the electronvolt as a unit of energy for subatomic particles. You will know (*from Chapter 2 of Year 1 Student Book*) about the emission of alpha, beta and gamma radiation from the decay of radioactive isotopes, and may be aware of the use of radioisotopes in medical diagnosis. From Chapters 5 and 6 of this book you will be familiar with the concept of electric potential and have experience of the maths involved in an exponential decay.

LEARNING OUTCOMES

In this chapter you will gain an in-depth understanding of Rutherford's alpha particle scattering experiment and how, from it, Rutherford made an estimation of the size of an atom's nucleus. You will develop further appreciation of how knowledge of atomic structure developed over time, with each scientist building on the ideas of another. You will learn of the measuring techniques involved in investigating radioactive decay, along with the mathematical techniques used in the analysis of decay data. You will also gain knowledge of the many useful applications of ionising radiation.

(Specification: 3.8.1.1 to 3.8.1.3, 3.8.1.5 part)

NASA's *New Horizons* spacecraft, launched in 2006, flew past Jupiter in 2007, past Pluto in 2015 and is expected to continue on its journey (Figure 1) through the outer Solar System, gathering data and transmitting information back to Earth for possibly another 20 years. The *New Horizons* spacecraft has no solar cells and no batteries, so what will continue to provide the electrical power for the spacecraft's scientific instruments for so many years?

Appropriately, on its journey to Pluto, heat and electricity for *New Horizons* was supplied by plutonium. The radioactive isotope plutonium-238 decays by emitting alpha particles with 5.6 MeV of kinetic energy.

Figure 1 *An artist's impression of NASA's New Horizons spacecraft*

The alpha particles transfer their kinetic energy to the atoms of plutonium by a series of collisions, heating the plutonium pellet to over 500 °C (see Figure 2). Onboard *New Horizons* is a radioisotope power system (RPS), which transforms the heat generated by the plutonium directly into electricity using solid-state thermoelectric converters. An RPS is a reliable source of both electricity and heat, which can work in the harsh environments of deep space where solar panels to charge batteries would be totally ineffective. Since the half-life of plutonium-238 is 88 years, an RPS can continue working effectively for many years, and has been used as the energy source for several of NASA's missions, including the *Mars Exploration Rovers*, *Cassini* and *Voyagers 1* and *2*.

Figure 2 *A red hot pellet of plutonium heated by the radioactive decay of its own nuclei.*

9.1 THE STRUCTURE OF THE NUCLEUS

The idea of the atom was first suggested by the ancient Greeks, but it was not until the early 19th century that modern atomic theory started to develop, with the experimental work of John Dalton, the English chemist and physicist. Dalton proposed that each element was made up of its own atoms, which were identical, and that compounds were formed by a combination of atoms of different elements. He thought that atoms were both indivisible and indestructible. However, in 1897, Joseph John ('J. J.') Thomson, the English physicist and engineer, identified the electron as a negatively charged particle with mass, energy and momentum. He proposed that the electron could be a fundamental constituent of every atom. He also suggested that the observation that atoms appeared to be electrically neutral showed that, as well as electrons, atoms must have some sort of positive charge to balance the electrons' negative charge. In the light of his discovery of the electron, Thomson revised Dalton's model of the atom to include his own theory that atoms could be subdivided into smaller parts. Thomson proposed his 'plum pudding model' of the atom (*see section 2.1 in Chapter 2 in Year 1 Student Book*), in which he suggested that most of the mass of the atom and its positive charge were uniformly distributed, and within this positive charge the electrons were embedded but free to move around.

Crucial to the further development of the atomic model was the discovery of radioactivity as an atomic phenomenon by French physicist Henri Becquerel in 1896. Research by Marie and Pierre Curie identified the radioactive elements polonium and radium, and the Curies, along with Becquerel, were jointly awarded the 1903 Nobel Prize in Physics in recognition of their work on radioactivity.

In 1899, New Zealand-born physicist Ernest Rutherford, working at McGill University in Canada, identified two different types of radiation from radioactive materials, which he named alpha and beta. In measuring the charge-to-mass ratio of alpha radiation, Rutherford was able to identify it as consisting of positively charged particles. His technique for detecting the position of alpha particles involved the use of photographic film. During his experiments, he noticed that the images produced by alpha particles on the film were blurred if they had first passed through a thin sheet of mica. Rutherford concluded that something about the mica was causing the alpha particles to be scattered through small angles.

Rutherford scattering

In 1907, Rutherford returned to the UK to take up the post of Chair of Physics at the University of Manchester, where he continued his work on the scattering of alpha particles. He also worked with Thomas Royds, an English physicist, with whom he was able to identify the alpha particle as a doubly ionised helium atom (*see section 2.4 in Chapter 2 of Year 1 Student Book*).

Rutherford designed – and in 1909 Hans Geiger, a German physicist, with his research assistant Ernest Marsden, conducted – the now famous **Rutherford scattering** experiment (Figures 3 and 4) in which alpha particles were fired at a thin piece of gold foil (*see section 2.2 in Chapter 2 in Year 1 Student Book*). The gold foil had to be very thin to make it most likely that a particular alpha particle would only experience one scattering and the experiment had to take place in an evacuated container because alpha particles can only travel a few centimetres in air. A zinc sulfide screen attached to the end of a rotatable microscope was used to detect the alpha particles at various angles to the beam's original direction. When an alpha particle hit the zinc sulfide screen, a flash of light, known as a **scintillation**, was produced, and the number of flashes in a given time at a particular scattering angle were counted by Geiger and Marsden.

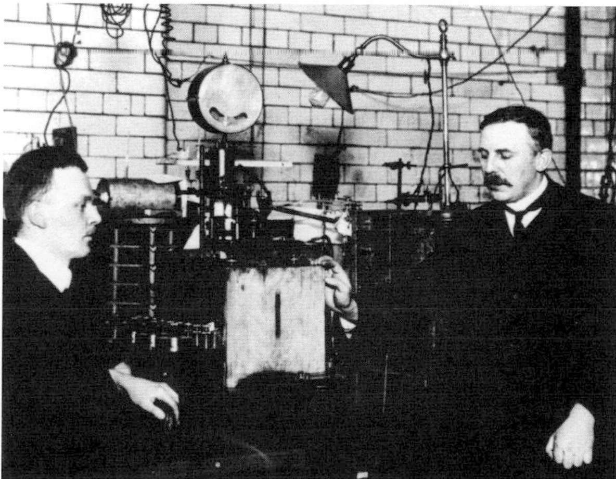

Figure 3 *Rutherford (right) and Geiger with alpha particle detection equipment in their Manchester laboratory*

Crucial to the experiment was that the source of the alpha particles was designed to produce a narrow parallel beam in order to define the precise location where the beam hit the foil and enable the scattering angle θ to be accurately measured (Figure 5). Also the scattering angle, θ, would have been influenced by the speed of the alpha particles, with a slower alpha particle being scattered by a

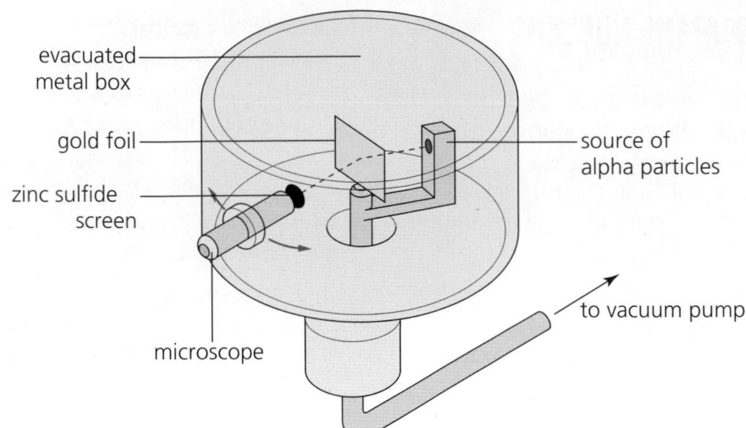

Figure 4 *Rutherford's alpha particle scattering apparatus*

larger angle than a faster-moving alpha particle. It was therefore necessary that the alpha particles were monoenergetic. Fortunately, the alpha particles emitted from a particular radioactive source all have the same kinetic energy.

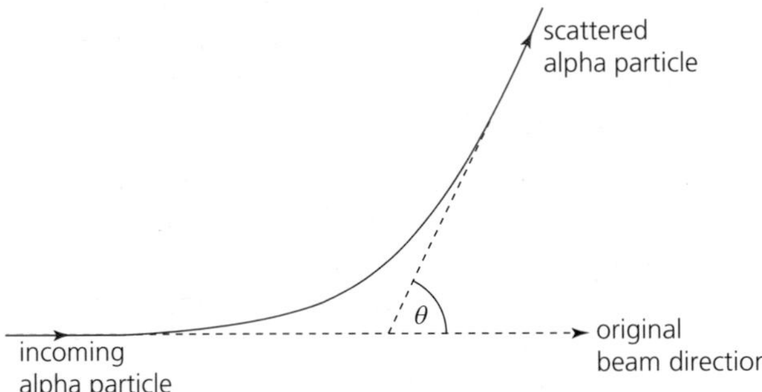

Figure 5 *Path of a scattered alpha particle*

Geiger and Marsden observed that about 99.95% of the alpha particles passed straight through the foil without being scattered. Approximately 1 in 2000 of the alpha particles were scattered, and about 1 in 8000 were scattered through an angle θ greater than 90° – that is, bounced backwards towards the alpha source. When Geiger and Marsden reported their observation of back-scattered alpha particles to Rutherford, he reportedly described it as the most incredible event of his life.

Applying conservation of momentum to an elastic collision between a moving object and a stationary object shows that the moving object only bounces backwards if its mass is very much smaller than the mass of the stationary object. After analysing Geiger and Marsden's data, Rutherford concluded that the pattern of large angles of alpha particle scattering could only occur if the alpha particle was colliding with a positively charged, dense object with a mass much larger than the alpha particle's mass.

The nuclear model

Rutherford concluded that Thomson's plum pudding model, in which the positive charge was distributed throughout the atom, must be incorrect, as it predicted that only small scattering angles would occur. He proposed the nuclear model of the atom, in which a tiny central nucleus containing most of the atom's mass and all of its positive charge was surrounded by the atom's orbiting electrons (Figure 6). He suggested that most of the atom was in fact empty space, which accounted for the observation that the vast majority of the alpha particles were not scattered at all, and that the large scattering angles occurred only when the alpha particles passed close to the massive positive nucleus.

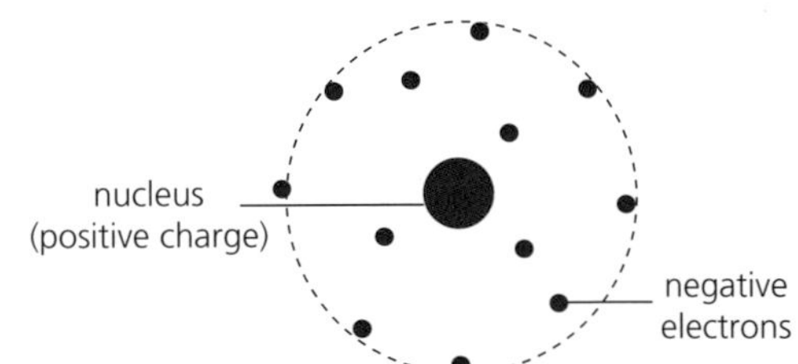

Figure 6 *Rutherford's model of the nuclear atom*

The backward scattering of some of the alpha particles enabled Rutherford to predict the size of the central nucleus – he calculated a value for an upper limit for the radius of the gold nucleus. He considered an alpha particle making a head-on collision with a gold nucleus (Figure 7) and calculated a value for the distance of closest approach, d, assuming that the repulsion between the alpha particle and the gold nucleus was due to the **Coulomb repulsion** between two positive charges (see section 5.1 of Chapter 5).

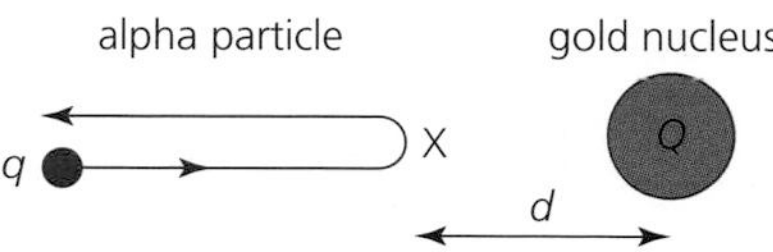

Figure 7 *A head-on collision between an alpha particle and a gold nucleus*

The equation for electrical potential at a distance r from a charge Q is

$$V = \frac{Q}{4\pi\varepsilon_0 r}$$

(see section 5.3). Therefore, at the point of closest approach, labelled X on Figure 7, the electrical potential V_X due to the positive charge Q of the gold nucleus is given by

$$V_X = \frac{Q}{4\pi\varepsilon_0 d}$$

where d, the distance of closest approach, is the distance from X to the centre of the gold nucleus. The electric potential energy E_p of the alpha particle of charge q at point X is therefore given by

$$E_p = \frac{Q}{4\pi\varepsilon_0 d} \times q$$

since electric potential energy is equal to the electric potential at a point multiplied by the size of the charge moved from infinity to that point.

However, when the alpha particle reaches X, it is momentarily stationary and all its original kinetic energy, E_k, has been changed to electrical potential energy. Therefore $E_p = E_k$ so

$$E_k = \frac{Q}{4\pi\varepsilon_0 d} \times q$$

Rutherford knew that the radioactive source used in the experiment emitted alpha particles of energy 5 MeV and that the charges on the alpha particle (which he knew to be a doubly ionised helium atom) and the gold nucleus were $2e$ and $79e$, respectively. Rearranging the above equation gives an equation for the distance of closest approach, d:

$$d = \frac{Q}{4\pi\varepsilon_0 E_k} \times q$$

Substituting the data gives the value for the distance of closest approach, d, of the alpha particle:

$$d = \frac{79 \times 1.6 \times 10^{-19}}{4\pi \times 8.85 \times 10^{-12} \times 5 \times 10^{6} \times 1.6 \times 10^{-19}} \times 2 \times 1.6 \times 10^{-19}$$

$$= 4.5 \times 10^{-14}\,\text{m}$$

Rutherford assumed that his value for the distance of closest approach of the alpha and the gold nucleus provided an upper limit for the radius of the gold nucleus. He had no way of knowing just how much smaller the radius of a gold nucleus was compared with his value of d. However, it was known at the time that atomic diameters were in the range of 10^{-9} to 10^{-10} m, so Rutherford's estimate of an upper limit for the radius of the nucleus confirmed that most of an atom is empty space.

There are a number of factors that limited the accuracy of Rutherford's value for the upper limit for the radius of the gold nucleus:

- Firstly, a look at the apparatus used (Figure 4) shows that the location of the alpha source makes it impossible to detect a head-on collision with a scattering angle of 180°.
- The assumption that the force between the alpha particle and the gold nucleus is just Coulomb (electrostatic) repulsion is not correct, since alpha particles, being hadrons, will also interact with the nucleus via the strong nuclear force (*see section 4.1 in Chapter 4 in Year 1 Student Book*), although at the distances involved the strong nuclear force is likely to be significantly smaller than the Coulomb repulsion.
- Also, the finite size of the alpha particles introduces an uncertainty in the calculation of the distance of closest approach, and the recoil of the gold nucleus, which would take place during the collision with the alpha particle, would cause a slight change in the size of the distance of closest approach.

QUESTIONS

1. In Rutherford's scattering experiment, alpha particles were fired at gold foil and the number of alpha particles scattered at various angles were measured.

a. Why is it essential that the alpha particles fired at the gold foil had the same kinetic energy?

b. Suggest why

   i. the beam of alpha particles had to be both narrow and parallel

   ii. the gold foil had to be very thin.

c. Which force, other than Coulomb repulsion, could have affected the scattering pattern of the alpha particles by gold nuclei?

d. In Rutherford's scattering experiment, how were the scattered alpha particles detected?

2. Express, as an order of magnitude, how many times greater the radius of an atom is compared with the radius of the nucleus, using Rutherford's upper limit for the radius of the nucleus and an estimate of atomic size as 10^{-10} m.

3. Which aspect of the alpha particle scattering results supported Rutherford's model of the atom over Thomson's plum pudding model?

4. From his analysis of the alpha particle scattering results, Rutherford established that the path taken by a scattered alpha particle was a hyperbola. He also identified a quantity that he called the impact parameter, *b*, which influenced the angle of scattering (Figure 8). Rutherford was able to calculate the value of *b* corresponding to different angles of scattering.

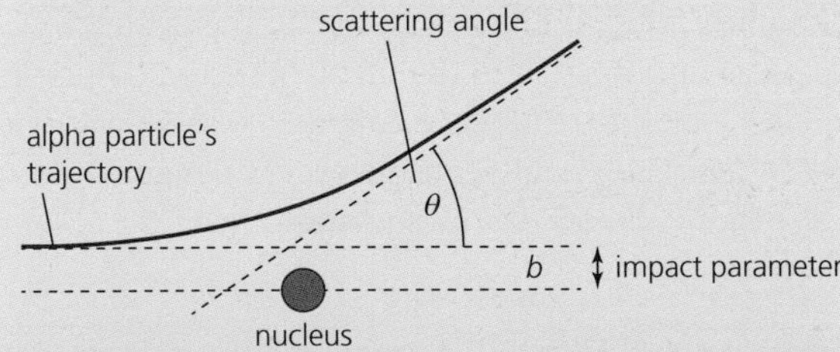

Figure 8 *Hyperbolic path of scattered alpha particle*

State how the size of the scattering angle θ would be affected if

a. the alpha particle's trajectory had a greater impact parameter *b*

b. the alpha particle had been emitted from a higher-energy source but had the same impact parameter.

The nature of the nucleus

The understanding of the structure of the atom continued to evolve over time, with one scientist building on the work of another scientist, and so on. Just as Thomson revised Dalton's model, and Rutherford revised Thomson's plum pudding model, atomic models have continued to be refined to fit with new discoveries.

- In 1913, the Rutherford model was replaced by the Bohr model (*see section 8.2 in Chapter 8 in Year 1 Student Book*), which included Niels Bohr's modification that electrons could only orbit the nucleus at discrete distances associated with specific energies.
- In 1917, another of Rutherford's experiments, in this case knocking hydrogen nuclei out of nitrogen atoms using alpha particles, provided evidence that the nuclei of all elements were made up of hydrogen nuclei, which were named **protons** in 1920.
- In 1926, Erwin Schrödinger developed the electron cloud model, which was further refined by Werner Heisenberg one year later.
- In 1932, James Chadwick discovered the **neutron** (*see section 2.3 in Chapter 2 in Year 1 Student Book*).

The ideas of all these scientists led to the modern nuclear model of the atom shown in Figure 9.

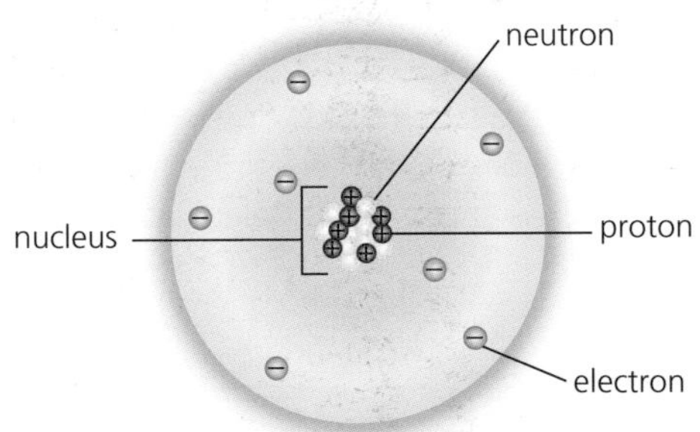

Figure 9 *The modern nuclear model of the atom*

In general, an atom is represented by the symbol:

$$^{A}_{Z}X$$

where X is the chemical symbol of the element, *A* is the **nucleon number**, which is the sum of the numbers of protons and neutrons, and *Z* is the **proton number**. A particular element is made up of atoms containing a specific number of protons characteristic of that element. **Isotopes** of an element have the same number of protons but different numbers of neutrons. In an electrically neutral atom, the number of protons is equal to the number of electrons.

Isotopes can also be written with the name of the corresponding element followed by the nucleon number. For example, $^{40}_{20}Ca$ is also written as calcium-40.

The nucleon number expressed as a mass in grams approximates to the mass of one mole of that isotope (that is, the isotope's molar mass).

This tells us that the mass of 1 mole of calcium-40 can be taken as 40 g, and this is reasonably accurate, given that the actual mass of 1 mole of calcium-40 is 39.96 g to four significant figures. The relationship between the number of nucleons and the mass of 1 mole arises because of the definition of the mole:

One **mole** of substance contains the same number of entities as there are atoms in 12 g of pure carbon-12, that number being the **Avogadro constant**, $6.02 \times 10^{23}\ mol^{-1}$.

So 1 mole of carbon-12 has by definition a mass of *exactly* 12 g. Since nuclei are all made up of the same stuff (neutrons and protons), and ignoring the mass of the orbiting electrons, then, to about three significant figures, the mass of 1 mole of an isotope is equal to its nucleon number in grams.

Worked example

Determine the number of atoms in 1 g of gold-197.

The mass of 1 mole of gold-197 can be taken as 197 g. Since the number of atoms in 1 mole is equal to the Avogadro constant, 6.02×10^{23}, the number of atoms n in 1 g of gold-197 can be calculated from:

$$n = \frac{1}{197} \times 6.02 \times 10^{23} = 3.06 \times 10^{21}.$$

QUESTIONS

5. a. Determine the number of protons and the number of neutrons in the following isotopes of iron:
   i. $^{56}_{26}Fe$
   ii. $^{59}_{26}Fe$

   b. Calculate the number of atoms in 1 g of iron-56.

KEY IDEAS

- Scientific knowledge evolves over time, with one scientist building on the ideas of another.
- Rutherford's alpha particle scattering experiment involved firing alpha particles at gold foil and then detecting the scattered alpha particles.
- The results of the alpha particle scattering experiment showed that most of the mass and all of the positive charge of the atom was concentrated in a tiny nucleus.
- Rutherford was able to estimate an upper limit ($\approx 10^{-14}$ m) for the size of a gold nucleus by equating the original kinetic energy of an alpha particle to the electric potential energy of the alpha particle at its distance of closest approach to the gold nucleus.
- Isotopes of an element have the same number of protons but different numbers of neutrons.
- The nucleon number expressed as a mass in grams gives the value of the isotope's molar mass.

9.2 ALPHA (α), BETA (β) AND GAMMA (γ) RADIATION

Identification of α and β by absorption experiments

Rutherford's initial identification of two types of ionising radiation, alpha and beta, was based on the extent to which these radiations were absorbed by other materials. He found that one, which he called beta radiation, was much more penetrating than the other. Figure 10 shows apparatus that can be used to determine the penetrating ability of ionising radiation. The **Geiger–Müller (GM) tube** is a device that registers a pulse of electricity each time an ionising particle, for example an alpha or beta particle, enters the tube. The GM tube is connected to a digital counter, which keeps count of the number of ionising particles entering the tube (see Figure P1 in the Required practical later in this chapter).

When investigating the radiation from a radioactive source using a GM tube and counter, it is necessary to record the number of counts during a fixed period (measured using a stopwatch) with no radioactive source present. This is required because the

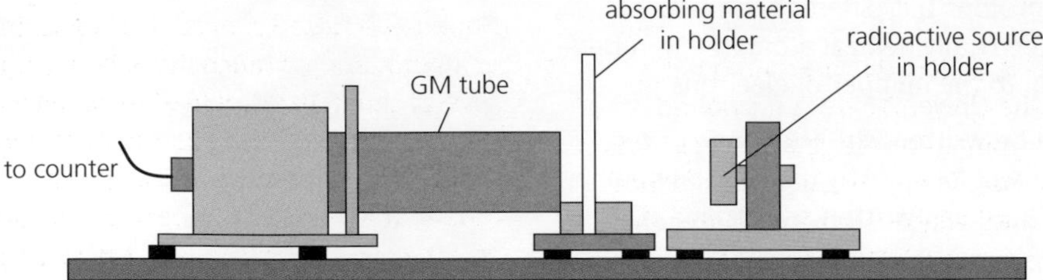

Figure 10 *Measuring the absorption of ionising radiation*

GM tube detects **background radiation** from the surroundings (see section 9.4). The 'background count rate' in counts s^{-1} is found by dividing the number of counts recorded by the measured time interval.

To investigate the absorption of beta radiation, a beta source is positioned as in Figure 10 and the number of counts is recorded over a fixed period of time for various thicknesses of aluminium positioned in the absorbing material holder.

When investigating the absorption of alpha radiation, the absorbing material holder is removed, and the number of counts is recorded over a fixed period of time for various distances between the source and GM tube, effectively using air as the absorber.

The 'corrected count rate' values in counts s^{-1} due only to the source are found firstly by dividing the number of counts by the measured time interval, then subtracting the background count rate:

corrected count rate =
measured count rate – background count rate

Essential safety precautions include handling the alpha and beta sources using long tongs and minimising the time for which the sources are removed from their lead-lined boxes.

The data for the results of the absorption experiments show that the corrected count rate decreases to zero for beta with an aluminium absorber of thickness of about 3 mm, and for alpha falls rapidly with an air gap of about 5 cm between the source and GM tube.

Nature of α and β and relative risks

Alpha radiation consists of particles each made up of two protons and two neutrons held together by the strong nuclear force – in effect, helium nuclei. Their short range in air, typically a few centimetres, is due to their intense ionisation of the air molecules. For example, to knock an electron out of a molecule of air requires about 34 eV of energy, so a 5 MeV alpha particle with a range in air of 5 cm produces about 30 000 ion pairs per centimetre of its path. Alpha particles are easily stopped by a sheet of paper and by skin, so they are not usually a significant risk to human health, provided they are outside the body, because all their energy is dissipated in the outer (dead) layers of skin. Alpha emitters are, however, very dangerous when inside the human body – which can occur when radioactive gas or dust is ingested – because all their energy is deposited in a small volume, causing significant damage to cells.

Beta radiation consists of electrons (or positrons, in the case of beta-plus decay), some of which travel at speeds close to the speed of light. Their range in air is typically 2–3 m. Since electrons are much smaller than alpha particles and have half the charge, they are less likely to interact with matter, therefore produce less ionisation and consequently are more penetrating. Sources of beta radiation present a greater risk to us than alpha particles (of the same activity level) outside the human body because they can penetrate sufficiently to damage surface tissues and the eyes. But they are less of a risk inside the body because their energy is distributed over a greater volume.

QUESTIONS

6. Which radiation, alpha or beta, is more harmful to humans if the source has been ingested, assuming that sources of similar activity are being compared? Explain your answer.
7. If an alpha particle and a beta particle are emitted with the same energy, which would have the greater range in air? Explain your answer
8. a. Why is it essential to handle a beta source with long tongs and for the shortest time possible?
   b. Why is it essential that the source is kept in a lead-lined box?

Gamma radiation

In 1903, Marie Curie (Figure 11) presented her doctoral thesis at the Université de la Sorbonne, becoming the first woman to receive a doctoral degree in the history of France. In her thesis, she described the technique that had enabled her to discover the element radium. She reported that she had shown that the radium sample emitted alpha, beta *and* a third type, gamma radiation, by applying a magnetic field to the emitted radiation. The alpha and beta radiation, being made up of charged particles, are deflected by the magnetic field, but gamma radiation consists of high-energy photons of electromagnetic radiation and passes through the magnetic field without deflection (*see section 2.4 in Chapter 2 in Year 1 Student Book*).

Figure 11 *Marie Curie, having jointly received the 1903 Nobel Prize for Physics with Becquerel and her husband Pierre, went on to receive the 1911 Nobel Prize for Chemistry 'in recognition of her services to the advancement of chemistry by the discovery of the elements radium and polonium, by the isolation of radium and the study of the nature and compounds of this remarkable element.'*

QUESTIONS

9. In Figure 12, the radioactive emissions are in a uniform magnetic field. In which direction is the field?

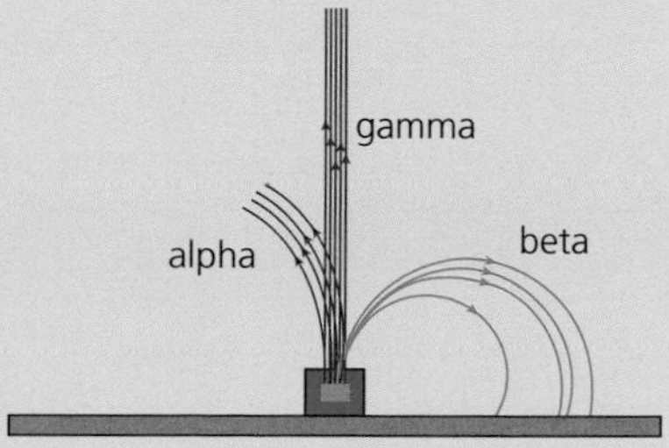

gamma

alpha

beta

Figure 12 *A magnetic field acting in the region of radiation emitted from radium*

Gamma radiation is much less ionising than alpha and beta, and so much more penetrating. It requires several centimetres of lead as an effective shield. Since the absorption of gamma radiation by air is negligible, the relationship between the intensity of gamma radiation and the distance from its source follows an **inverse square law** as the radiation spreads outwards. This assumes that, while the gamma intensity is being measured, the source activity is unchanged, so only holds for sources with a long half-life (see section 9.5).

The **intensity** I of electromagnetic radiation is defined as the radiation energy passing per second through an area of 1 m^2 normal to the radiation.

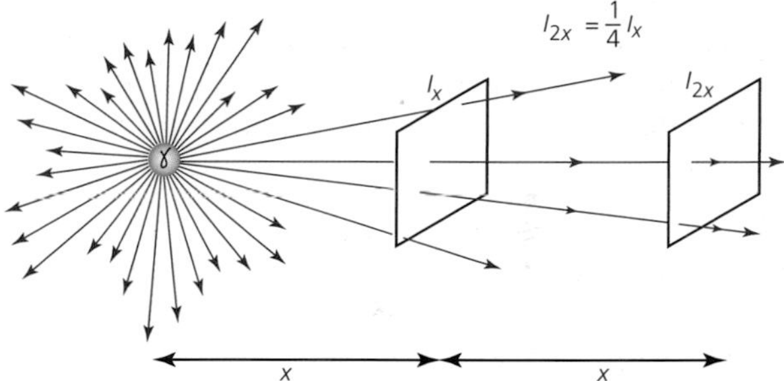

Figure 13 *Inverse square law for gamma radiation*

For a point gamma source (Figure 13) emitting N gamma photons per second each of energy hf, the radiation energy emitted per second is equal to Nhf. At distance x from the source, all the photons from the source pass through a total area of $4\pi x^2$, which is the surface area of a sphere of radius x. The intensity of the radiation at distance x is therefore given by

$$I = \frac{\text{energy emitted per second}}{\text{total area}} = \frac{Nhf}{4\pi x^2}$$

Therefore the intensity of gamma radiation is

$$I = \frac{k}{x^2} \quad \text{where} \quad k = \frac{Nhf}{4\pi}$$

which is a constant provided the activity of the source remains constant.

If the intensity of the source at distance x is I_x, then the intensity at a distance $2x$ from the source is I_{2x}, where

$$I_{2x} = \frac{k}{(2x)^2} = \frac{k}{4x^2} = \frac{1}{4} I_x$$

which shows that doubling the distance from the gamma source reduces the intensity by a factor of 4, illustrating an inverse square law relation.

If a GM tube and counter are being used to detect gamma radiation, the corrected count rate at a particular distance from the source is directly proportional to the gamma intensity at that distance. A graph of corrected count rate versus distance between a gamma source and GM tube is shown in Figure 14. Increasing the separation of the GM tube and source from 10 cm to 20 cm reduces the corrected count rate to $\frac{1}{4}$, from 1000 to 250. Increasing the separation further to 30 cm reduces the corrected count rate to $\frac{1}{9}$ of its value at 10 cm. The graph illustrates the steep fall in gamma intensity with increasing distance from the source due to the inverse square law. It also shows the importance of *always using long-handled tongs to manipulate a gamma source* and always storing the source well away from people.

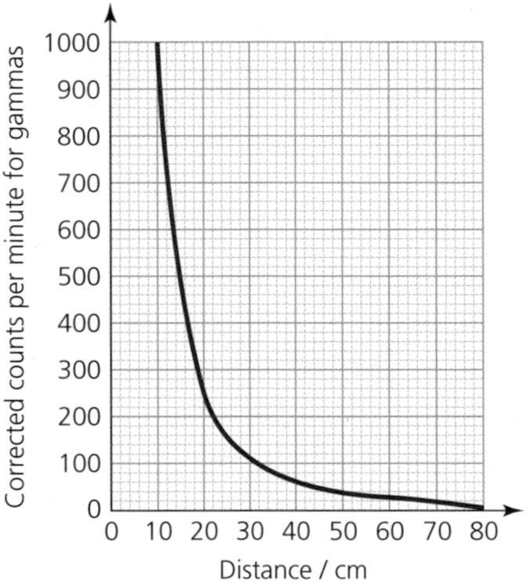

Figure 14 *Graph of corrected count rate versus distance for gamma radiation from a point source*

The considerable penetrating power of gamma radiation means that it presents a greater danger to human health when the source is outside of the human body than either of alpha or beta.

Worked example

A point source of gamma radiation with a long half-life produces a corrected count rate of 14.2 counts s^{-1} when placed 200 mm from a GM tube connected to a counter.

a. Determine the corrected count rate when the source is place 270 mm from the GM tube.

b. Determine how far from the GM tube the source would have to be placed to produce a corrected count rate of 1.0 counts s^{-1}.

a. The intensity I of gamma radiation varies inversely with the square of the distance x from a point source: $I = \frac{k}{x^2}$. Since the corrected count rate C is directly proportional to the radiation intensity then

$$C = \frac{\text{constant}}{x^2}$$

and therefore

$$C \times x^2 = \text{constant}$$

Therefore

$$14.2 \times 200^2 = C_{new} \times 270^2$$

which gives the new count rate $C_{new} = 7.8$ counts s^{-1}. (Note that it is not necessary to convert the distance measurements to metres, as the conversion factors on both sides of the equation would have cancelled.)

b. Again using $C \times x^2 = \text{constant}$:

$$14.2 \times 200^2 = 1.0 \times x^2$$

which gives $x = 754$ mm.

QUESTIONS

10. The corrected count rate at a distance of 30.0 cm from a gamma source is 12.9 counts s^{-1}. At what distance from the source would the corrected count rate fall to 8.7 counts s^{-1}?

11. The corrected count rate at a distance of 25.0 cm from a gamma source is 15.4 counts s^{-1} as measured by a GM tube and counter.
    a. Given that the diameter of the GM tube is 20 mm, determine the number of gamma photons emitted from the source every second, assuming that 5% of the gamma photons entering the tube are detected.
    b. Given that the photon energy is 100 keV, determine the intensity of the gamma radiation at a distance of 2.0 m from the source.

12. Why does the intensity of beta radiation in air not follow an inverse square law with distance from the source?

Summary of the properties of alpha, beta and gamma radiation

Table 1 and Figure 15 summarise what we have covered so far in this chapter. In Table 1, the asterisks (*) indicate that the actual values depend on the initial energy with which the radiation is emitted.

QUESTIONS

13. Absorption tests on a radioactive sample of unknown origin yield the following results. When the source is held close to a GM tube, the count rate is high, but it drops suddenly when the GM tube is moved a few centimetres from the sample. With the GM tube and sample positioned 10 cm apart, a 3 mm thick sheet of aluminium inserted between the sample and the GM tube does not change the count rate, which remains well above background. Determine what type or types of radiation are emitted from the sample, explaining your answer.

14. State precautions that should be taken when undertaking any experiment using a source of ionising radiation.

Radiation	Nature	Penetrating power	Range in air	Ionising effect	Behaviour in electric and magnetic fields
Alpha	Two protons and two neutrons (helium nucleus)	Easily stopped, e.g. by a sheet of paper or the outer layer of (dead) skin cells	A few centimetres	Intensely ionising: an alpha particle will cause about 10^4 to 10^5 ion pairs per centimetre in air	Positively charged, so deflected by electric and magnetic fields; but relatively massive, so deflected less than beta particles
Beta	An electron	Stopped by thin (a few millimetres)* metal sheet	Several metres	Less intensely ionising than alpha: a beta particle will cause about 1000 ion pairs per centimetre*	Negatively charged, so deflected in opposite direction to alpha; deflected more than alpha, as the mass is much less
Gamma	High-frequency electromagnetic radiation	Reduced in intensity by half by about 5 cm of concrete or 1 cm of lead*	Ten to hundreds of metres*	Weakly ionising: about 10 ion pairs per centimetre*	Not charged, so undeviated by a magnetic field or an electric field

Table 1 *Alpha, beta and gamma radiation*

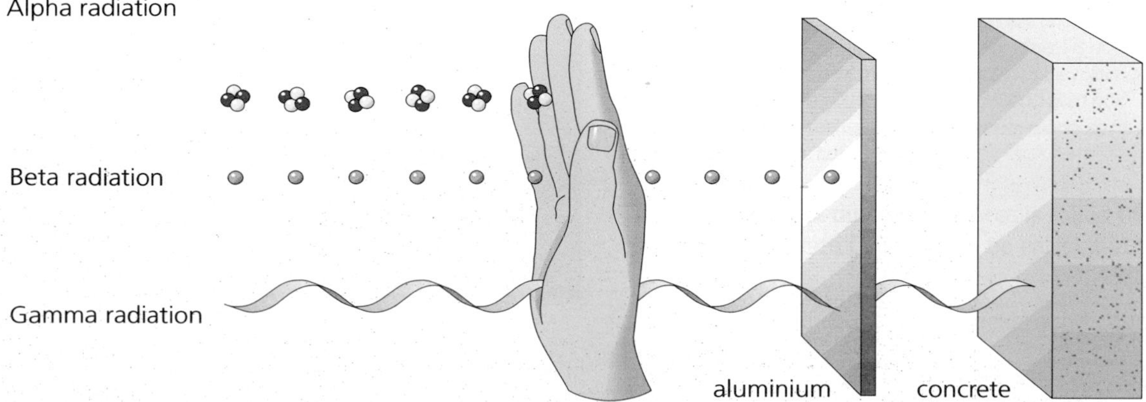

Figure 15 *The penetrating power of alpha, beta and gamma radiation*

A statistical analysis of repeat readings of a large number of counts generated by a source of radiation shows that the uncertainty in a total count of N is $\pm\sqrt{N}$. For example, the uncertainty in a count of 500 is $\pm\sqrt{500} = \pm 22$, which corresponds to a percentage uncertainty of ±4.4%. Consider a count of 100. The uncertainty would be $\pm\sqrt{100} = \pm 10$, which corresponds to a percentage uncertainty of ±10%. Therefore, the uncertainty in the number of counts can be determined by simply taking the square root of the number of counts recorded.

Analysis

Theory predicts that the intensity I of gamma radiation is related to distance x by the relation $I = \frac{k}{x^2}$. Since the corrected count rate, C, is directly proportional to the intensity, $C = \frac{\text{constant}}{x^2}$. Therefore, a graph of $\frac{1}{\sqrt{C}}$ on the y-axis versus distance x on the x-axis being a straight line through the origin would demonstrate the inverse square law. However, there are two complications that would make it likely that the straight-line graph would reveal a systematic error. Firstly the actual radioactive source is inside the source container, not at the end, from which the distance measurements are made. The second complication concerns the operation of the GM tube. The sensitive area of the GM tube – where ionisation takes place – is actually within the tube but the measurement of x is actually made to the front of the GM tube, creating an additional uncertainty in the distance measurement. There is therefore a possible systematic error, e, in the distance measurement x, as shown in Figure P3.

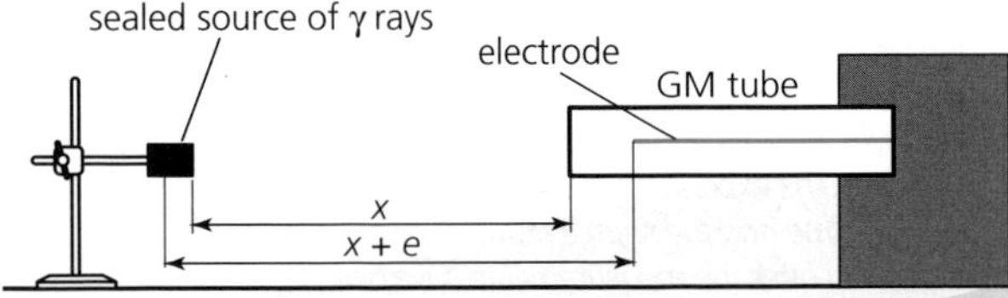

Figure P3 *The distance from source to detector is x + e*

QUESTIONS

P1 State the dependent, the independent and control variables.

P2 A measurement of background radiation in a laboratory gives a count of 480 in a time of 20.0 minutes. Determine the background count rate in counts per second.

P3 A student undertakes the experiment and obtains values for the counts in 60 s at various distances between a gamma source and GM tube, as shown in Table P1. It can be assumed that the activity of the source is constant for the duration of the experiment.

Distance x / mm	Number of counts in 60 s
100	2101
200	725
300	378
400	238
500	169
600	125

Table P1

a. Determine the percentage uncertainty in the number of counts recorded when distance $x = 100$ mm.

b. For each data set in Table P1, determine

i. the number of counts per second

ii. the corrected count rate C in counts s^{-1}.

c. i. Plot a graph of $\frac{1}{\sqrt{C}}$ on the y-axis versus distance x on the x-axis.

ii. Determine a value for the systematic error, e.

d. Use the graph to predict the measured count rate when $x = 450$ mm.

9.3 APPLICATIONS OF RADIOISOTOPES

Despite the hazards of radioactive emissions, some radioactive isotopes, or **radioisotopes**, are used for medical treatments and diagnosis. The radioactive isotopes used for medical purposes are carefully selected so that their benefits outweigh the associated risks.

- Tiny quantities of the alpha-emitting isotope radium-233 are injected into tumorous tissue to directly kill cancer cells.
- The beta-emitting isotope iodine-131 is used for the treatment of thyroid cancer.
- Beta-plus emitters, such as carbon-11, nitrogen-13 and fluorine-18, are chemically added to compounds used by body processes to act as diagnostic **tracers** in conjunction with a PET scanner (*see Assignment 1 at the end of section 3.3 of Chapter 3 in Year 1 Student Book*).
- Low-energy gamma sources are used as tracers to produce images of internal organs. They are combined with a compound that is taken up by a certain part of the body, and the gamma radiation emitted is detected by a gamma camera to track the movement of the compound.
- Gamma radiation is also used to sterilise medical equipment – it can easily kill bacteria and viruses.

There are also many industrial uses of radioactive isotopes.

- Polonium-210, an alpha emitter, is used as a static eliminator to neutralise static electricity in industrial processes such as the production of paper, plastics and textiles.
- The beta source strontium-90 is used in the manufacture of materials requiring a specific thickness, such as paper, plastic films, aluminium foil and thin sheets of steel. The thickness of the sheet of material is controlled by measuring the amount of radiation that passes through it (Figure 16). The detector is connected to a computer, which controls the pressure of the rollers to maintain the correct thickness.
- Gamma-emitting isotopes are often used as industrial tracers. For example, a small amount of radioactive gas can be introduced into a pipeline and the gamma intensity above the ground can be measured to determine the position of a leak.

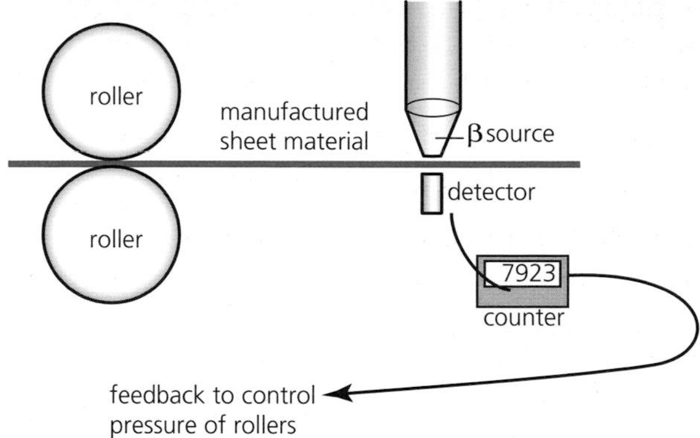

Figure 16 *Beta radiation is used to monitor sheet thickness.*

People who work with ionising radiation wear a radiation **dosimeter badge** (Figure 17) for monitoring the cumulative amount of radiation that their bodies have absorbed. The measurement of the dose, in sievert (Sv), takes account of the relative biological effects of the ionising radiation.

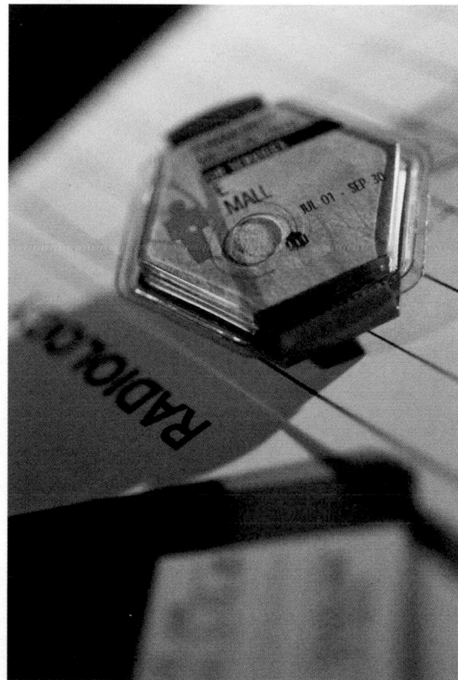

Figure 17 *Radiation dosimeter badge*

QUESTIONS

15. Suggest an occupation that would require the wearing of a radiation dosimeter badge.

9.4 SOURCES OF BACKGROUND RADIATION

Every day we are exposed to ionising radiation from the air we breathe, the rocks we walk on and the food we eat. This radiation is called **background radiation** and the amount of background radiation absorbed by a human is usually expressed in millisievert (mSv) per year. The major sources of this background radiation are illustrated in Figure 18.

- Some rocks contain unstable isotopes, for example, potassium, uranium and thorium, which decay to other radioactive products, some of which are gases that can seep out of the ground. Granite is one such rock that is very common in the UK. Its uranium content includes radioactive radon gas in its decay products. The level of radon gas depends on where you live (Figure 19).
- Radioactive isotopes in the ground that dissolve in water are taken up by plants and then animals, which then enter the food chain.
- The Earth is continually being bombarded by cosmic rays, which include particles and gamma radiation from the Sun and from other sources outside of the Solar System.

Along with these naturally occurring sources of background radiation, there are also artificial sources, such as those used in medical applications, waste from the nuclear power industry, and fallout from nuclear weapons testing.

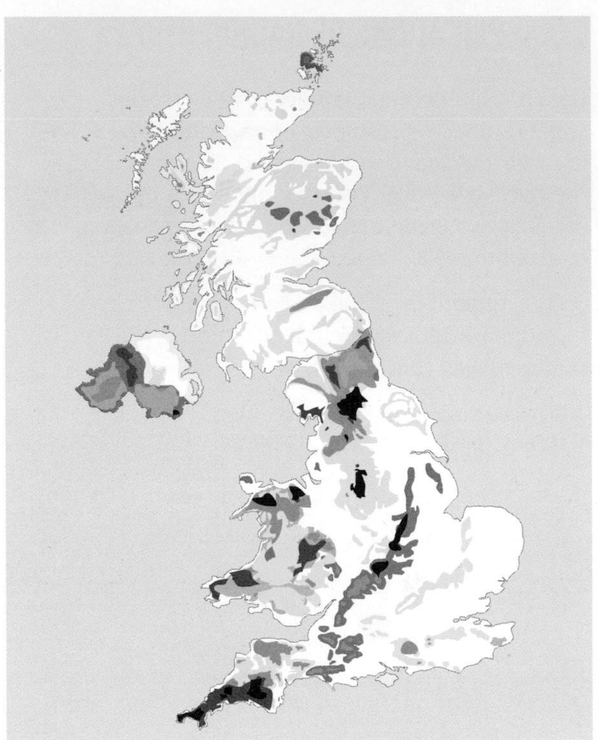

Figure 19 *UK radon level map: the darker the colour, the greater the chance of a higher level of radioactivity from radon.*

QUESTIONS

16. Suggest two features of planet Earth that provide us with some protection from cosmic rays.

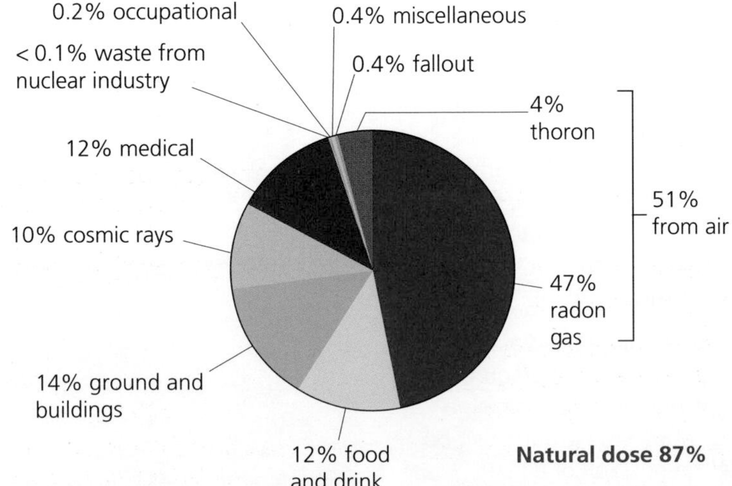

Figure 18 *Sources of background radiation*

9.5 RADIOACTIVE DECAY

Radioactive decay is a process by which an unstable nucleus loses energy by emitting radiation. The decay of an individual nucleus is described as a *random event* because it is not possible to predict when the nucleus will undergo a decay. However, if a large enough number of atoms of a particular isotope is studied, it is possible to predict the most likely proportion of them that will decay in a given time.

A more familiar random event is the tossing of a coin. Since there are two possible outcomes to the tossing of a coin, heads or tails, the **probability** of getting heads is $\frac{1}{2}$. Probability is a number between 0 and 1 that gives an estimate of the likelihood that a random event will produce a particular outcome. So, while the outcome of a single coin toss cannot be predicted, if 1000 coins were thrown into the air, it is possible to successfully predict that more or less half would land heads upwards.

Probability gives the fraction of a large sample of random events that would show a particular outcome.

Another example of a random event is the throwing of a dice. If a six-sided dice is thrown, the probability of throwing a '1' is $\frac{1}{6}$. Therefore, if 600 six-sided dice were thrown, it would be reasonable to predict that approximately 100 of them will show a '1'. Again, the probability, in this case $\frac{1}{6}$, gives the fraction of the large number of dice being thrown showing a '1'.

An analogy can be made between the decay of an unstable nucleus and the throwing of a dice. Instead of a large number of dice, we have a sample of a radioactive isotope, which contains a large number of atoms. One throw of the large number of dice is analogous to a specific period of time, 1 s for example, during the decay of the isotope, and the probability of throwing a '1' is equivalent to the probability of an individual nucleus decaying per second.

If a radioactive sample contained N atoms then, by analogy:

$$\text{number of decays per second} = \text{probability of decay per second} \times N$$

To take the analogy a step further, consider other types of dice other than those that are six-sided. For example, the probability of showing a '1' when a 20-sided dice (Figure 20) is thrown is $\frac{1}{20}$. The various types of dice are analogous to the various types of radioisotopes, which have different values for the probability of an individual nucleus decaying per second.

Figure 20 *A 20-sided dice*

The probability of an individual nucleus of a particular radioisotope decaying per second is called the **decay constant**, λ, which has the unit s^{-1}. The decay constant is also equal to the fraction of a sample decaying per second. Therefore, if there are N atoms in a sample:

$$\text{number of decays per second} = \lambda \times N$$

The number of decays per second is defined as the **activity**, A, and is measured in **becquerel** (Bq), where 1 Bq is equal to one decay per second. Therefore:

$$\text{activity } A = \lambda N$$

The activity of a sample of a specific isotope decreases in proportion with the decreasing number of radioactive atoms present.

The **half-life**, $T_{1/2}$, of a radioactive isotope is defined as the time for the activity of a sample of that isotope to halve. The activity of an isotope with a large decay constant falls more rapidly and therefore has a short half-life. The gamma emitter used to demonstrate the inverse square law (see the Required practical) has to have a half-life that is long enough so that its activity does not change during the course of the experiment.

Worked example 1

Determine the number of decays per second of a mole of the isotope radium-226 given that the value of its decay constant is $1.4 \times 10^{-11}\,s^{-1}$. [1 mole contains 6.02×10^{23} atoms, equal to the Avogadro constant]

Number of decays per second = activity $A = \lambda N$
$= 1.4 \times 10^{-11} \times 6.02 \times 10^{23} = 8.4 \times 10^{12}$ Bq

QUESTIONS

17. Determine the activity of a sample of polonium-210 of mass 1.0 g, given that its decay constant is $5.8 \times 10^{-8}\,s^{-1}$. [*Reminder*: Since the nucleon number of polonium is 210, the molar mass can be taken as 210 g.]

ASSIGNMENT 1: MODELLING RADIOACTIVE DECAY

(MS 0.3, MS 1.2, MS 1.3, PS 2.1)

The aim of this assignment is to model radioactive decay using plastic cubes to represent the atoms of a sample of a radioactive isotope (Figure A1). The cubes are thrown gently onto a table, and those cubes which land painted side upwards, representing the atoms that have decayed, are removed and counted. The process is repeated until there are no more cubes left to throw. A table, with the headings shown in Table A1, is suitable for recording the data. A graph of the number of cubes remaining on the *y*-axis versus throw number on the *x*-axis can be plotted to illustrate the pattern of the decay.

Figure A1 *Painted cubes used to model radioactive decay*

Throw number	Number decayed	Number remaining
0	0	500
1		

Table A1

The graph (Figure A2) can be used to determine an average value for the number of throws required to halve the number of cubes remaining. This value would be analogous to **half-life**, $T_{1/2}$, in actual radioactive decay.

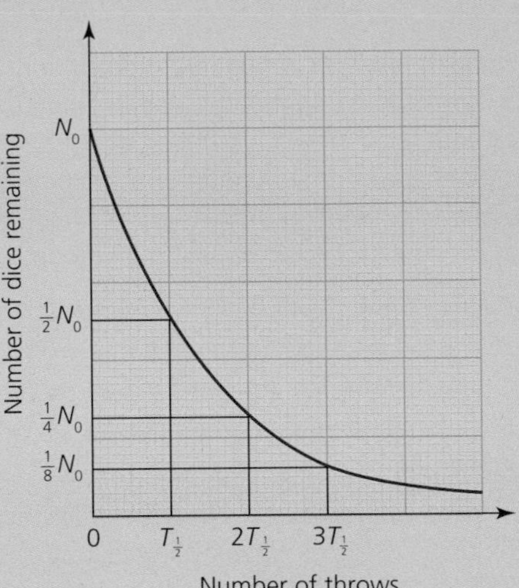

Figure A2 *Graph of number of dice remaining versus number of throws*

QUESTIONS

A1 A student undertakes the experiment and plots a graph of number remaining versus number of throws for the data obtained to determine the half-life of the dice 'decay'.

A drop from 500 to 250 cubes takes 3.7 throws.

A drop from 400 to 200 cubes takes 3.9 throws.

A drop from 300 to 150 cubes takes 3.8 throws.

A drop from 200 to 100 cubes takes 3.8 throws.

Obtain an average value for the constant increase in the number of throws that reduces the number of cubes remaining to half, along with its percentage uncertainty.

A2 Suppose that, instead of using 500 painted cubes, the student uses 500 20-sided dice. A dice showing a '1' represents a decay. How would you expect the number of throws needed for the number of dice remaining to halve to change? Explain your answer.

A3 Another student decides to use 500 20-sided dice to model the radioactive decay of a particularly active sample. For each throw of the dice, he removes those showing an even number. How many throws would be needed to reduce the number of dice remaining to less than 50?

A4 The throwing of six-sided dice can be useful for modelling radioactive decay and can help to introduce the idea of half-life. However, there are some flaws in the model. Suggest how the model may be failing.

Exponential decay

Consider a sample of a radioactive isotope consisting of N nuclei. If, during a very small time interval, Δt, the number of nuclei that decay is ΔN, then the number of decays per second is equal to $\frac{\Delta N}{\Delta t}$. However, since the number of decays per second is the activity, A, of the sample and $A = \lambda N$, then we have

$$\frac{\Delta N}{\Delta t} = -\lambda N$$

The minus sign is necessary because ΔN represents a decrease in the value of N. Activity A is the magnitude of $\frac{\Delta N}{\Delta t}$.

The solution to the above equation (see question 21) is

$$N = N_0 e^{-\lambda t}$$

where N_0 is the number of radioactive nuclei in the sample at time $t = 0$.

This equation shows that radioactivity is an example of **exponential decay** – the number of atoms of the original radioactive isotope decreases exponentially as time passes. The decrease occurs more quickly the larger the value of the decay constant λ (Figure 21).

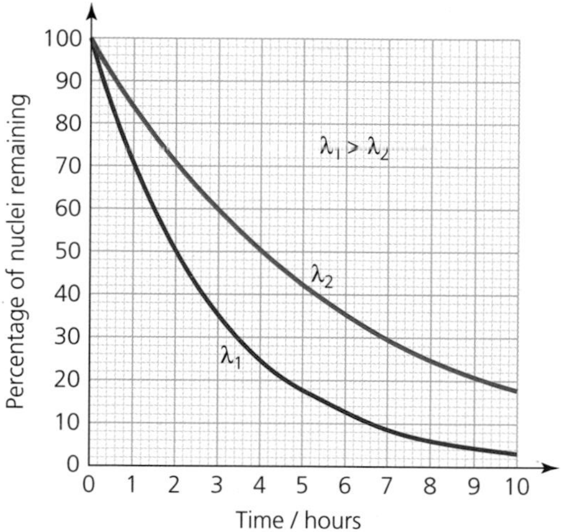

Figure 21 *Exponential decay of the percentage of undecayed nuclei remaining*

Since $A = \lambda N$, then activity $A = \lambda N_0 e^{-\lambda t}$, which can be written as

$$A = A_0 e^{-\lambda t}$$

where $A_0 = \lambda N_0$ is the initial activity of the sample.

The activity of a sample decreases exponentially with time.

The corrected count rate, C, due to a sample of a radioactive isotope, as measured by a GM tube and counter (see section 9.2), is directly proportional to the sample's activity. Therefore, it can be assumed that the corrected count rate also decreases exponentially, so

$$C = C_0 e^{-\lambda t}.$$

The half-life $T_{1/2}$ of a radioactive isotope is therefore the time for each of the following to halve:

- the number of radioactive atoms, N
- the activity, A
- the corrected count rate, C.

Since the decay of a sample of a radioactive isotope can be monitored using a GM tube and counter, it is convenient to determine the half-life of the isotope from a graph of corrected count rate, C, versus time.

Since the time for N, A and C to halve has a constant value ($T_{1/2}$), the following pattern results:

- after 1 half-life, N, A and C are all $\frac{1}{2}$ of their original values
- after 2 half-lives, N, A and C are all $\frac{1}{4}$ of their original values
- after 3 half-lives, N, A and C are all $\frac{1}{8}$ of their original values ...
- after n half-lives, N, A and C are $\frac{1}{2^n}$ of their original values.

Worked example 2

A graph of corrected count rate versus time for a sample of a particular radioactive isotope is shown in Figure 22. Use the graph to determine the half-life of the radioactive isotope.

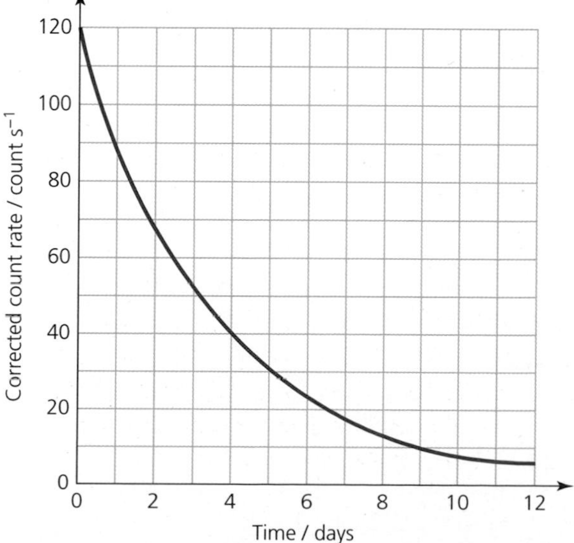

Figure 22

The time for the count rate to halve from 120 counts s^{-1} to 60 counts s^{-1} is 2.5 days.

The time for the count rate to halve from 60 counts s^{-1} to 30 counts s^{-1} is 2.5 days.

The time for the count rate to halve from 30 counts s^{-1} to 15 counts s^{-1} is 2.5 days.

It can be concluded that the half-life of the isotope is 2.5 days.

Worked example 3

A radioactive isotope has a half-life of 28 years. Determine the fraction of the isotope's atoms remaining in a sample after 112 years.

A time of 112 years corresponds to 4 half-lives, so the fraction of the isotope's atoms remaining $= \frac{1}{2^4} = \frac{1}{16}$.

QUESTIONS

18. A graph of corrected count rate versus time for a sample of a particular isotope is shown in Figure 23. Use the graph to determine the isotope's half-life.

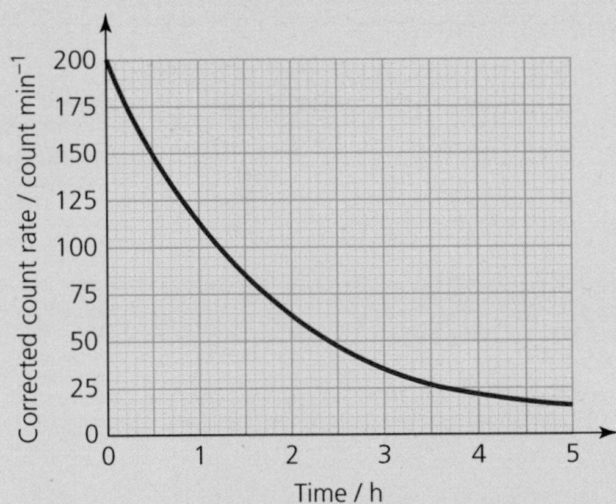

Figure 23

19. A sample of radioactive isotope has an activity of 1.2×10^{16} Bq. Given that the half-life of the isotope is 5 years, determine the activity of the sample after 25 years have elapsed.

20. A freshly prepared sample of iodine-131 has a mass of 1.0 g. Given that the decay constant of iodine-131 is $1.0 \times 10^{-6}\,s^{-1}$, determine the activity of the sample 24 hours after its preparation.

Relation between decay constant and half-life

Consider the equation $N = N_0 e^{-\lambda t}$. When the time that has elapsed is equal to $T_{1/2}$, the number of radioactive atoms still present is $\frac{1}{2}N_0$. Substituting into $N = N_0 e^{-\lambda t}$ gives

$$\frac{1}{2}N_0 = N_0 e^{-\lambda T_{1/2}}$$

which gives

$$\frac{1}{2} = e^{-\lambda T_{1/2}}$$

and therefore

$$2 = e^{\lambda T_{1/2}}$$

Taking natural logarithms gives

$$\ln 2 = \lambda T_{1/2}$$

which rearranges to give

$$T_{1/2} = \frac{\ln 2}{\lambda}$$

The half-life of a particular isotope is inversely proportional to its decay constant.

Worked example 4

A freshly prepared sample of the radioisotope fluorine-18 has a mass 0.010 g. The decay constant of fluorine-18 is $1.05 \times 10^{-4}\,s^{-1}$. Determine

a. the initial activity of the sample

b. the activity after 2 hours

c. the half-life of the isotope.

a. The mass of 1 mole is 18 g.

Initial number of fluorine nuclei in sample

$$= \frac{0.01}{18} \times 6.02 \times 10^{23} = 3.344 \times 10^{20}$$

Initial activity $A_0 = \lambda N_0$
$= 1.05 \times 10^{-4} \times 3.344 \times 10^{20} = 3.512 \times 10^{16}$
$= 3.5 \times 10^{16}$ Bq

b. Activity after 2 hours $A = A_0 e^{-\lambda t}$

$= 3.512 \times 10^{16} \times e^{-1.05 \times 10^{-4} \times 2 \times 3600}$

$= 1.6 \times 10^{16}$ Bq

c. Half-life $T_{1/2} = \frac{\ln 2}{\lambda} = \frac{\ln 2}{1.05 \times 10^{-4}}$

$= 6.6 \times 10^3$ s $= 110$ minutes

QUESTIONS

21. A sample of a radioactive isotope is produced that contains 2.0×10^{16} atoms. The isotope has a half-life of 25 minutes. Calculate
    a. the decay constant
    b. the number of nuclei that decay in the first 15 minutes
    c. the energy released, in joules, from the sample in the first 15 minutes, given that each decay results in the emission of an alpha particle of energy 5.0 MeV.

Stretch and challenge

22. The equation $\frac{\Delta N}{\Delta t} = -\lambda N$ can be rewritten as the differential equation

$$\frac{dN}{dt} = -\lambda N$$

and rearranged to give

$$\frac{dN}{N} = -\lambda \, dt$$

Integrate this last equation between time $t = 0$ (when the number of radioactive atoms is N_0) and time t (when the number of radioactive atoms is N). Take inverse natural logarithms of the resulting equation to show that $N = N_0 e^{-\lambda t}$.

Using logarithms to analyse decay data

Applying logarithms to the radioactive decay equation $A = A_0 e^{-\lambda t}$ gives

$$\ln A = \ln A_0 + \ln e^{-\lambda t}$$

which can be written

$$\ln A = \ln A_0 - \lambda t$$

and rearranged to give

$$\ln A = -\lambda t + \ln A_0$$

Comparing this with the equation of a straight line, $y = mx + c$, shows that a graph of $\ln A$ on the y-axis versus time t on the x-axis will be a straight line with a gradient $-\lambda$. A graph of $\ln A$ versus t is described as a log–linear graph.

Worked example 5

Strontium-90 is one of the radioactive isotopes that are produced as waste products in nuclear fission reactors. The activity of a small sample of strontium-90 was measured over a 10 year period and results are shown in Table 2. Plot a suitable graph to determine the half-life of strontium-90.

Time / years	Activity / MBq
0	200.0
1	195.1
2	190.3
3	185.7
4	181.1
5	176.7
6	172.4
7	168.2
8	164.1
9	160.1
10	156.1

Table 2 *Decay data for strontium-90*

It can be seen from the data in the table that, during the 10 year period, the activity has not even halved, and so a graph of activity versus time would not enable the half-life to be determined. Therefore, it is necessary to plot $\ln A$ versus time (Figure 24) and obtain the gradient, which is equal to $-\lambda$, and from this the half-life can be calculated.

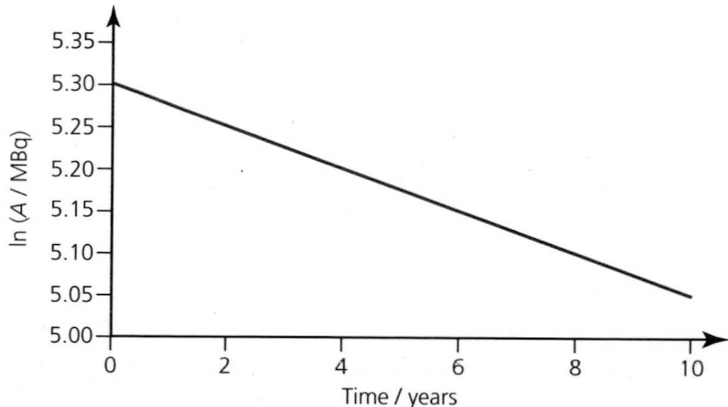

Figure 24

The gradient is equal to -0.025 years^{-1}.

$$\text{Half-life } T_{1/2} = \frac{\ln 2}{\lambda} = \frac{\ln 2}{0.025} = 28 \text{ years}$$

QUESTIONS

23. The graph of $\ln(C / \text{counts s}^{-1})$ versus time in years for a sample of a particular radioactive isotope is shown in Figure 25. Determine the half-life of the isotope.

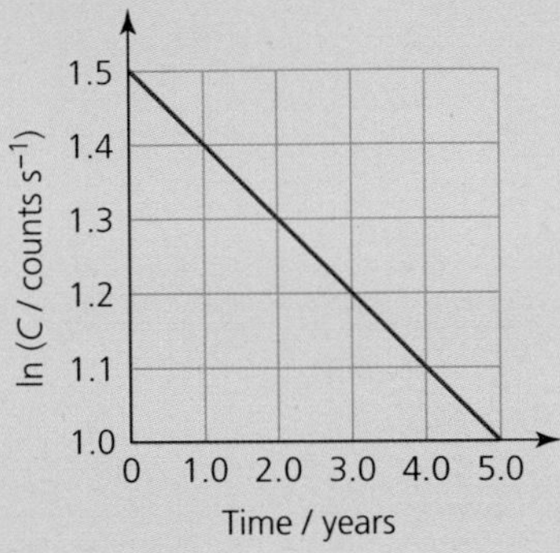

Figure 25

KEY IDEAS

- Radioactive decay of a particular nucleus is a random event, which can be modelled by other random events such as the throwing of a dice.
- The activity (A) of a sample of a radioisotope is the number of decays per second, in becquerel (Bq); 1 Bq = 1 decay per second.
- The activity of a sample of a radioisotope decreases in proportion with the decreasing number of radioactive atoms present:
  $\frac{\Delta N}{\Delta t} = -\lambda N$ or $A = \lambda N$ (since activity A is the magnitude of $\frac{\Delta N}{\Delta t}$)
  where λ is the decay constant in s^{-1}.
- The number of atoms of a particular radioactive isotope in a sample decreases exponentially with time,
  $$N = N_0 e^{-\lambda t}$$
  as does the activity,
  $$A = A_0 e^{-\lambda t}$$
  and the corrected count rate,
  $$C = C_0 e^{-\lambda t}$$
- The half-life ($T_{1/2}$) of a radioactive isotope is the time for the activity to halve.
- Half-life $T_{1/2}$ and decay constant λ are related by
  $$T_{1/2} = \frac{\ln 2}{\lambda}$$
- The half-life can be obtained from a graph of corrected count rate versus time (unless one half-life has not elapsed).
- The magnitude of the gradient of a graph of $\ln A$ versus time t gives the decay constant λ, since
  $$\ln A = -\lambda t + \ln A_0$$

9.6 IMPLICATIONS OF HALF-LIFE VALUE

Choice of half-life in applications of radioisotopes

The beta emitter used to monitor an industrial process producing sheets of paper of a specific thickness (see section 9.3) requires a long half-life so that its activity remains constant for an extended period of time. On the other hand, the radioactive isotopes chosen for use as tracers in medical diagnosis need to have short half-lives so that the time for which they are active inside the human body is kept to a minimum.

Consideration of half-life of radioactive waste

The approximate half-lives of some of the radioactive isotopes constituting the radioactive waste from a nuclear reactor are shown in Table 3. Since activity $A = \lambda N$, and the shorter the half-life, the larger the decay constant λ, then, on a per mole basis, the waste isotopes with shorter half-lives have the much higher initial activity ($A_0 = \lambda N_0$) and are therefore very dangerous and require both remote-controlled handling and cooling.

Radioisotope	Half-life
Barium-140	13 days
Cerium-144	280 days
Krypton-85	11 years
Strontium-90	28 years
Caesium-137	30 years
Plutonium-239	24000 years
Caesium-135	2.3 million years
Iodine-129	15.7 million years

Table 3 *Half-lives of some constituents of radioactive waste*

The requirement for cooling occurs as a result of the 'decay heat' generated by radioactive decay – the energy of the ionising radiation emitted from these isotopes is converted to the thermal energy of the surrounding atoms. However, the activity of the radioactive isotopes with shorter half-lives falls relatively quickly, and the safety issue then becomes one of long-term safe storage of the isotopes with longer half-lives.

Half-life in radioactive dating

Radioactive dating is a technique used to determine the age of an object based on the amount of a naturally occurring radioactive isotope present in the material from which the object is made.

Potassium–argon dating, for example, enables determination of how long ago a rock was formed, based on the ratio of argon-40 to potassium-40 contained within the rock. Potassium is a common element found in material from which rocks are formed. The isotope potassium-40 is radioactive and has a half-life of 1250 million years, with two possible modes of decay, forming either argon-40 or calcium-40, the decay to calcium-40 being eight times the more likely. Originally, the material from which the rock was made was molten and the argon, formed from the decay of potassium-40, escaped as a gas. However, once the rock solidified, the argon became trapped within the rock. Using mass spectrometry, the ratio of the potassium-40 atoms to argon-40 atoms in a sample can be determined, and from this the time elapsed since the rock's formation can be calculated (see Figure 26).

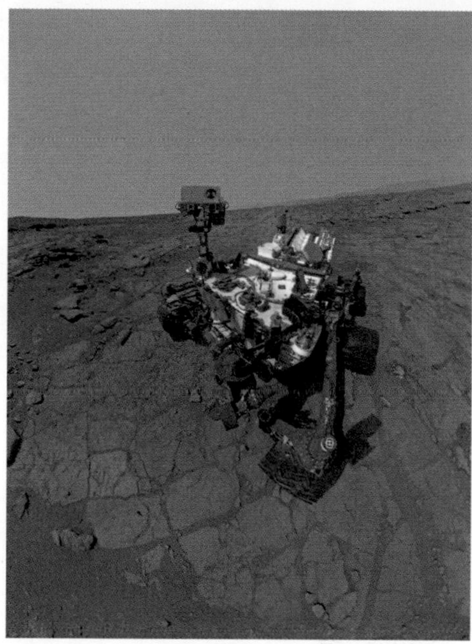

Figure 26 *The first geological dating of a rock on a planet other than Earth took place on Mars in December 2013. NASA's Mars rover Curiosity drilled into the surface of the planet to extract a rock sample and used the potassium–argon technique to date the sample's age at between 3.85 and 4.56 billion years.*

Carbon dating is a method used to date artefacts that are made of organic materials, based on the amount of carbon-14 present. There are three naturally occurring isotopes of carbon: carbon-12 and carbon-13, which are stable, and carbon-14, which decays to nitrogen-14 by the emission of beta radiation with a half-life of 5730 years. Carbon-14 is created in the Earth's atmosphere by the action of cosmic rays. Plants take up atmospheric carbon dioxide by photosynthesis and therefore contain traces of carbon-14 with typically one carbon atom in every 10^{12} being carbon-14. The plants are then ingested by animals, so every living thing is constantly exchanging carbon-14 with its environment for as long as it lives. Once the plant or animal dies, however, this exchange stops, and the amount of carbon-14 gradually decreases. The time elapsed since the plant or animal died can be determined by comparing the amount of carbon-14 still present with the amount of carbon-14 likely to have been present when the animal was living.

Originally, carbon dating involved measuring the activity of the sample under investigation. A modern carbon dating method uses the accelerator mass spectrometry (AMS) technique, which measures the ratio of the carbon-14 to the carbon-12 content in the sample directly. Since the AMS technique counts the carbon atoms present rather than just counting the beta particles emitted from the sample, it is possible to date very small samples accurately.

Carbon dating is less accurate for samples that are less than 200 years old, because the change in the ratio of carbon-14 to carbon-12 is too small to be measured accurately. Also, for samples older than about 60 000 years there are too few carbon-14 atoms remaining to be measured accurately.

Worked example

Carbon dating is to be used to date a piece of ancient timber. The activity of a sample of the wood is measured at 0.33 Bq. The activity of an equal mass of living wood is measured at 1.20 Bq. Determine the time that has elapsed since the tree from which the timber originally came was felled, stating any assumptions you may make. [Half-life of carbon-14 = 5730 years]

The decay constant of carbon-14 is

$$\lambda = \frac{\ln 2}{T_{1/2}} = \frac{\ln 2}{5730 \times 365 \times 24 \times 3600}$$

$$= 3.836 \times 10^{-12} \text{s}^{-1}$$

Assuming that the ratio of carbon-12 to carbon-14 in living wood when the tree was felled is the same as it is today, and assuming that the sample has not been contaminated, then using $A = A_0\, e^{-\lambda t}$ gives

$$0.33 = 1.2 \times e^{-3.836 \times 10^{-12} \times t}$$

which gives

$$0.275 = e^{-3.836 \times 10^{-12} \times t}$$

Taking natural logarithms gives

$$-1.291 = -3.836 \times 10^{-12} \times t$$

and the time elapsed $t = 3.37 \times 10^{11}$ s $= 10\,700$ years.

QUESTIONS

24. A sample from a piece of bone is found to have 0.392 times as many carbon-14 atoms as a piece of living bone of the same mass. Calculate the age of the bone in years, given that the decay constant for carbon-14 is $3.84 \times 10^{-12}\text{s}^{-1}$.

PRACTICE QUESTIONS

1. a. i. A radioactive source is positioned 10 cm from a Geiger–Müller (GM) tube that is connected to a counter (Figure Q1). When **no absorber** is present, the number of counts recorded in a 10 minute period is 13065. The uncertainty in a measured number of counts, N, can be taken as $\pm\sqrt{N}$. Determine the uncertainty in the recorded number of counts.

ii. An aluminium absorber, 3 mm thick, is then placed between the source and the GM tube, and the number of counts recorded in a 10 minute period is 12874. Determine the uncertainty in the recorded number of counts.

iii. A student studies the two sets of data and concludes that, since the number of counts recorded was reduced by the aluminium absorber, the source must be emitting beta radiation. However, another student suggests that the drop in the recorded number of counts is more likely to be due to the aluminium absorber stopping alpha particles from reaching the GM tube. Discuss the validity of these two conclusions and decide for yourself the nature of the radiation being emitted from the source, given that a reading of the background radiation in the laboratory taken over a period of 10 minutes was 243.

b. Describe the procedures that would ensure that the experiment was conducted safely.

2. a. In a radioactivity experiment, background radiation is taken into account when taking corrected count rate readings in a laboratory. One source of background radiation is the rocks on which the laboratory is built. Give **two** other sources of background radiation.

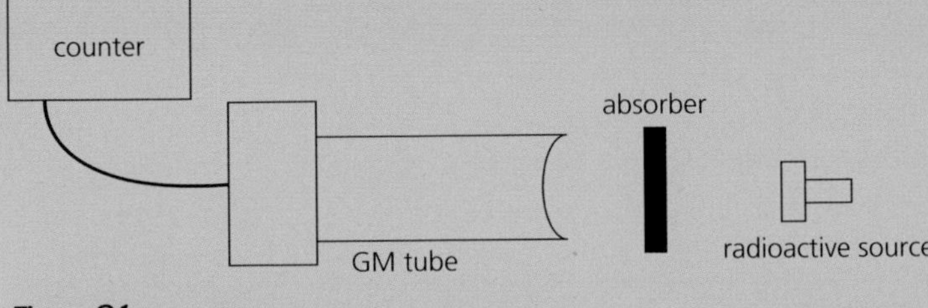

Figure Q1

b. A γ ray detector with a cross-sectional area of $1.5 \times 10^{-3}\,m^2$ when facing the source is placed 0.18 m from the source. A corrected count rate of 0.62 counts s^{-1} is recorded.

   i. Assume the source emits γ rays uniformly in all directions. Show that the ratio

   $$\frac{\text{number of } \gamma \text{ photons incident on detector}}{\text{number of } \gamma \text{ photons produced by source}}$$

   is about 4×10^{-3}.

   ii. The γ ray detector detects 1 in 400 of the γ photons incident on the facing surface of the detector. Calculate the activity of the source. State an appropriate unit.

c. Calculate the corrected count rate when the detector is moved 0.10 m further from the source.

AQA June 2012 Unit 5 Section 1 Q3

3. The carbon content of living trees includes a small proportion of carbon-14, which is a radioactive isotope. After a tree dies, the proportion of carbon-14 in it decreases due to radioactive decay.

   a. i. The half-life of carbon-14 is 5740 years. Calculate the decay constant in yr^{-1} of carbon-14.

   ii. A piece of wood taken from an axe handle found on an archaeological site has 0.375 times as many carbon-14 atoms as an equal mass of living wood. Calculate the age of the axe handle in years.

   b. Suggest why the method of carbon dating is likely to be unreliable if the sample is

   i. less than 200 years old

   ii. more than 60 000 years old.

AQA June 2014 Unit 5 Section 1 Q2

4. Which one of the following statements regarding the isotopes of a particular element is **incorrect**?

   A The isotopes have the same number of protons.

   B The isotopes have different nucleon numbers.

   C The isotopes have the same molar mass.

   D Neutral atoms of the isotopes have the same number of electrons.

5. What is the half-life of a radioactive sample whose activity falls to $\frac{1}{128}$ of its initial value in 100 days?

   A 20.0 days **B** 16.7 days
   C 14.3 days **D** 12.5 days

6. A gamma source with a long half-life emits 2500 gamma photons per second. How many gamma photons will pass through an area of $1.0\,m^2$ every second at a distance of 2.0 m from the source?

   A 400 **B** 200
   C 100 **D** 50

10 NUCLEAR ENERGY

PRIOR KNOWLEDGE

You will know from Chapter 9 about alpha, beta and gamma radiations as a result of nuclear decay, and about their associated hazards. You will be familiar with the idea of using beams of particles to probe the structure and size of nuclei.

LEARNING OUTCOMES

In this chapter you will learn how the wave behaviour of electrons can be used to measure nuclear radii. You will find out how Einstein's equation $E = mc^2$ can be used to calculate the energy released in nuclear decays, in fission and in fusion. You will gain an understanding of how knowledge of excited states within the nucleus can be applied in medical diagnoses, and learn the principles on which the operation of a nuclear reactor is based.

(Specification: 3.8.1.4 to 3.8.1.8)

JET, the Joint European Torus (Figure 1) is the world's largest 'magnetic-confinement plasma' experiment. Its aim is to investigate the potential of nuclear fusion as a virtually limitless energy source for future generations. JET, based on the Russian 'tokamak' design, is located at the Culham Science Centre in Oxfordshire and is a joint venture collectively used by more than 40 European laboratories.

The problems with nuclear fusion are firstly that extremely high temperatures are required. Hence the plasma (hot ionised gas) in which fusion occurs needs to be magnetically confined so that it does not make contact with the container walls, which

Figure 1 *The JET tokamak*

would cause the plasma to cool, stopping the fusion reactions, and lead to erosion of the surface of the wall. Secondly, in all attempts so far, more energy has been needed to create fusion reactions than has been generated. JET, which has been in operation since the mid-1980s, holds the current energy release record, but still has not been able to achieve 'plasma energy breakeven': the point at which a fusion device generates as much energy as it consumes. However, another much bigger fusion experiment, ITER, the International Thermonuclear Experimental Reactor, is currently under construction in southern France and is scheduled to start its plasma experiment in 2020. It is also based on the tokamak design and will have a plasma volume of 840 m^3. It is hoped that eventually ITER will be able to generate 500 MW, about 10 times more power than it will consume.

In the meantime, it has recently been reported that the US technology giant, Lockheed Martin, could be just 10 years away from constructing a compact fusion reactor that will be no bigger than a jet engine and capable of providing many times more energy than it consumes.

10.1 NUCLEAR INSTABILITY

A **nuclide** is a type of nucleus with a specific number of protons, a specific number of neutrons and a specific energy state. There are 339 naturally occurring nuclides, of which 254 have never been observed to decay and are therefore described as stable. Nuclides present on Earth today that were present when the Earth was formed 4.6 billion years ago are either stable or are radioactive with half-lives long enough to have survived those billion years in sufficient numbers to be still detectable. Other naturally occurring nuclides present today have been formed from the radioactive decay of unstable nuclides.

There are also around 2963 artificially produced nuclides, of which about 2400 have half-lives less than 1 hour. A graph of neutron number, N, versus proton number, Z, for stable nuclides is shown in Figure 2.

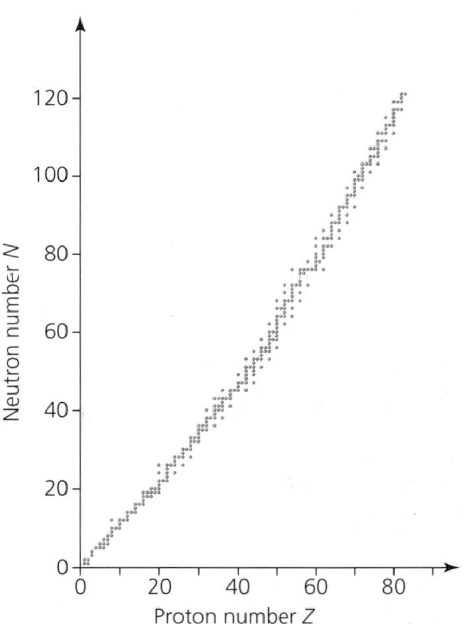

Figure 2 *Graph of neutron number versus proton number for stable nuclides*

For nuclides with proton number 20 or less, the graph follows an approximate straight line for which $N \approx Z$. For stable nuclei with proton numbers greater than 20, the ratio $\frac{N}{Z} > 1$ and continues to increase for the heavier nuclei. The heaviest stable nuclide is bismuth-209, which has 83 protons and 126 neutrons and a ratio $\frac{N}{Z} = 1.52$.

The naturally occurring nuclide uranium-238 is unstable and decays by alpha emission with a half-life of 4.5 billion years. A nucleus of uranium-238 has 92 protons and 146 neutrons and, if plotted on the graph in Figure 2, would be in the top right-hand corner, beyond and to the right of the stability line. A section of an N versus Z graph in that region is shown in Figure 3, with a point plotted for uranium-238 at top right, and illustrating its decay, initially by alpha emission, to eventually form stable lead. This is called a **decay series**.

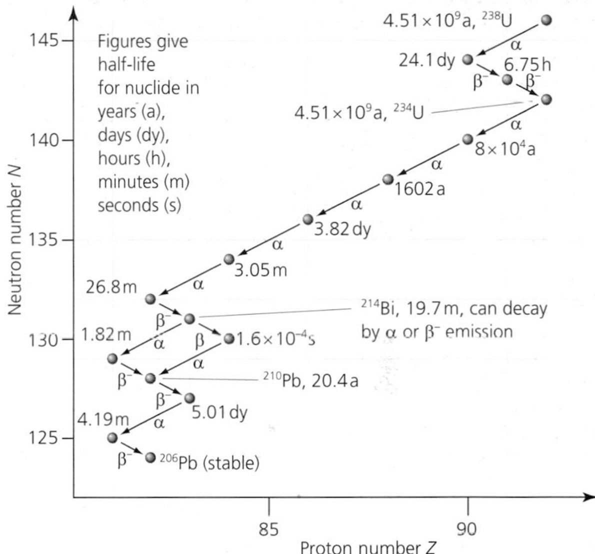

Figure 3 *Uranium-238 decay series*

Alpha decay

The decay of a uranium-238 nucleus by the emission of an alpha particle is represented in Figure 4. The decay equation is

$$^{238}_{92}\text{U} = \,^{234}_{90}\text{Th} + \,^{4}_{2}\text{He}$$

The alpha particle, which consists of 2 protons and 2 neutrons, is represented here by the symbol for a helium nucleus; it is also possible to use the symbol $^{4}_{2}\alpha$. The nuclide formed by the decay, the **daughter** nucleus, is thorium-234, which has a nucleon number 4 less than the **parent** nucleus, and a proton number 2 less than its parent.

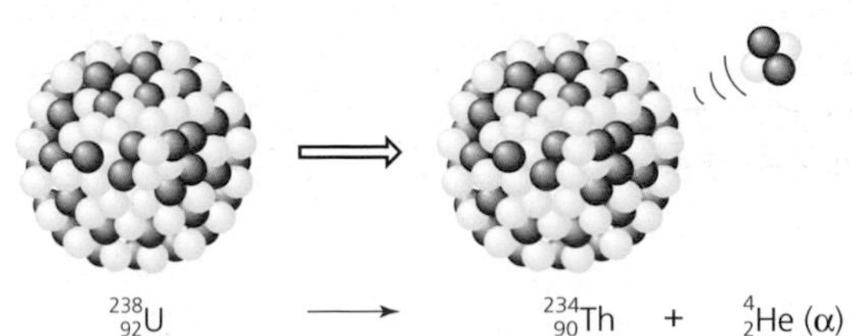

Figure 4 *Alpha decay of uranium-238*

The general equation for alpha decay is

$$^{A}_{Z}\text{X} \rightarrow \,^{A-4}_{Z-2}\text{Y} + \,^{4}_{2}\text{He}$$

In a decay equation, the nucleon numbers balance and the proton numbers balance.

Each radioactive decay results in an output of energy known as the **decay energy**. In pure alpha decay, the energy appears as kinetic energy of the daughter nucleus and of the alpha particle, and is shared in a fixed ratio, so that the alpha particles emitted from a sample of a particular radioactive isotope have a distinct energy (*see section 3.5 in Chapter 3 of Year 1 Student Book*) and are described as being **monoenergetic**.

Beta-minus decay

Thorium-234, the daughter nucleus of uranium-238, is itself unstable and decays with a half-life of 24.1 days (see Figure 3), emitting a beta-minus particle and an electron antineutrino (*see section 3.5 in Chapter 3 of Year 1 Student Book*). It forms protactinium-234. The equation for the decay is

$$^{234}_{90}\text{Th} \rightarrow {}^{234}_{91}\text{Pa} + {}^{0}_{-1}\text{e} + {}^{0}_{0}\bar{\nu}_e$$

The daughter nucleus, protactinium-234, has the same nucleon number as its parent but has a proton number that exceeds its parent by 1. The beta-minus particle is shown here by the symbol for an electron; alternatively the symbol $^{0}_{-1}\beta$ can be used. In beta-minus emission, a neutron, in a nucleus over-rich in neutrons, changes into a proton, an electron and an electron antineutrino. The proton remains inside the nucleus, but the electron and the antineutrino are emitted. The general equation for beta-minus decay is

$$^{A}_{Z}\text{X} \rightarrow {}^{A}_{Z+1}\text{Y} + {}^{0}_{-1}\text{e} + {}^{0}_{0}\bar{\nu}_e$$

The change within the nucleus is represented by

$$\text{n} \rightarrow \text{p} + \text{e}^- + \bar{\nu}_e$$

The decay energy is shared between the daughter nucleus, the electron and the electron antineutrino. The energy spectrum (Figure 5) of the emitted electrons shows that there is a specific value for their maximum kinetic energy, which is characteristic of the parent nucleus.

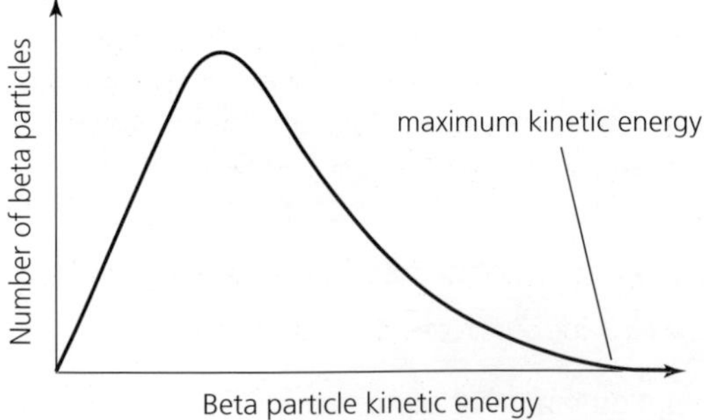

Figure 5 *Beta particle energy spectrum*

Protactinium-234, the daughter nucleus of thorium-234, is also a beta-minus emitter, decaying to form a nucleus of uranium-234:

$$^{234}_{91}\text{Pa} \rightarrow {}^{234}_{92}\text{U} + {}^{0}_{-1}\text{e} + {}^{0}_{0}\bar{\nu}_e$$

There are 11 more stages in this uranium-238 decay series until the stable isotope of $^{206}_{82}\text{Pb}$ is reached (see Figure 3).

QUESTIONS

1. Write a decay equation for each of the following decays, which form part of the uranium-238 decay series.
   a. Radium-226, $^{226}_{88}\text{Ra}$, decays by alpha emission to form radon (symbol Rn).
   b. Lead-214, $^{214}_{82}\text{Pb}$, decays by beta-minus emission to form bismuth (symbol Bi).
   c. Polonium-210, $^{210}_{84}\text{Po}$, decays by alpha emission to form a stable isotope of lead (symbol Pb).

Stretch and challenge

2. When a parent nucleus decays by alpha emission, the decay energy is released as kinetic energy of the alpha particle and daughter nucleus. Since the alpha particle and daughter are the only two products of the decay, the decay energy is shared in a fixed ratio, which always has the same value for a particular parent nuclide. Show that the ratio is

$$\frac{\text{alpha kinetic energy}}{\text{daughter kinetic energy}} = \frac{\text{mass of daughter}}{\text{mass of alpha}}$$

   assuming that the parent nucleus was stationary when the decay took place.

Beta-plus decay

Nuclei that are over-rich in protons can decay by beta-plus emission. Very few of these types of nuclide occur naturally, but there are many synthetic beta-plus emitters. For example, oxygen-15, frequently used in PET (positron emission tomography) imaging, is produced by bombarding nitrogen-14 with deuterons using a cyclotron. A nucleus of oxygen-15 decays with a half-life of about 2 hours to form a nucleus of nitrogen-15, emitting a positron and an electron neutrino in the process (*see section 4.5 in Chapter 4 of Year 1 Student Book*). The decay equation is

$$^{15}_{8}\text{O} \rightarrow ^{15}_{7}\text{N} + ^{0}_{+1}\text{e} + ^{0}_{0}\nu_e$$

The daughter nucleus, nitrogen-15, has the same nucleon number as its parent but has a proton number that is 1 less than its parent. The beta-plus particle can be shown by the symbol for a positron, or alternatively the symbol $^{0}_{+1}\beta$ can be used.

The general equation for beta-plus decay is

$$^{A}_{Z}\text{X} \rightarrow ^{A}_{Z-1}\text{Y} + ^{0}_{+1}\text{e} + ^{0}_{0}\nu_e$$

In beta-plus emission, a proton changes into a neutron, a positron and an electron neutrino, and is represented by

$$\text{p} \rightarrow \text{n} + \text{e}^+ + \nu_e$$

The neutron remains inside the nucleus but the positron and electron neutrino are emitted. The positrons are emitted with a range of energies up to a specific maximum value, as in beta-minus decay (Figure 5). Note that the change of a proton to form a neutron, a positron and an electron neutrino only occurs inside an unstable nucleus; a free proton is a stable particle.

Decay by electron capture

An alternative decay for a nucleus over-rich in protons is called **electron capture** (or sometimes K-capture). This is very rare in nature, but beryllium-7, which is naturally produced in the atmosphere by the action of cosmic rays, decays by electron capture with a half-life of 53 days. Nuclides have been produced synthetically by electron capture. The process involves the nucleus of an electrically neutral atom absorbing one of its inner orbital electrons, resulting in a proton changing into a neutron with the simultaneous emission of an electron neutrino. An outer orbital electron undergoes a transition to replace the absorbed electron, resulting in the emission of an X-ray photon. The decay equation for beryllium-7 is

$$^{7}_{4}\text{Be} + ^{0}_{-1}\text{e} \rightarrow ^{7}_{3}\text{Li} + ^{0}_{0}\nu_e$$

The daughter nucleus, lithium-7, has the same nucleon number as its parent but has a proton number that is 1 less than its parent. The general equation for decay by electron capture is

$$^{A}_{Z}\text{X} + ^{0}_{-1}\text{e} \rightarrow ^{A}_{Z-1}\text{Y} + ^{0}_{0}\nu_e$$

The change within the nucleus is represented by (*see section 4.5 in Chapter 4 of Year 1 Student Book*)

$$\text{p} + \text{e}^- \rightarrow \text{n} + \nu_e$$

Other decay mechanisms

Other less common types of decay include neutron emission, proton emission and spontaneous fission. The nuclide cobalt-53m has been observed to decay to iron-52 with the emission of a proton, and beryllium-13 decays to form beryllium-12 with the emission of a neutron. Very heavy nuclei such as uranium-235 and plutonium-239 have also been observed to spontaneously split into two smaller nuclei.

QUESTIONS

3. The synthetic radioisotope, fluorine-18, $^{18}_{9}\text{F}$, which is used in PET imaging, decays by beta-plus emission 97% of the time and by electron capture 3% of the time. Both types of decay produce the same stable isotope of oxygen. Write a decay equation for both types of decay.

Looking closer at instability

Consider the graph of neutron number, N, versus proton number, Z, shown in Figure 6. Points corresponding to stable nuclei lie in the central lilac area. Unstable nuclei decaying by alpha decay tend to be heavy nuclei with corresponding points on the N–Z graph being mostly beyond the stability region (dark red area). Points corresponding to beta-minus emitters tend to be to the left of the stability region in the green area, while points corresponding to nuclides decaying by beta-plus emission or electron capture tend to be to the right of the stability region, in the blue area. Stable nuclei that have a proton number less than 20 tend to lie on or close to the $N = Z$ reference line.

The strong nuclear force is responsible for binding together protons and neutrons to form atomic nuclei. The strong nuclear force is attractive between nucleons at separations of about 1 fm (1 fm = 10^{-15} m) and can overcome the Coulomb repulsion between adjacent protons. However, at distances around 3 fm the strong nuclear force becomes negligible, and the net force between two protons at 3 fm is repulsive (*see section 2.3 in Chapter 2 of Year 1 Student Book*).

Consequently, in order for increasingly heavier nuclei to exist, the ratio, $\frac{N}{Z}$, of the number of neutrons to the number of protons, must increase, reaching a value of about 1.6 for very heavy nuclei.

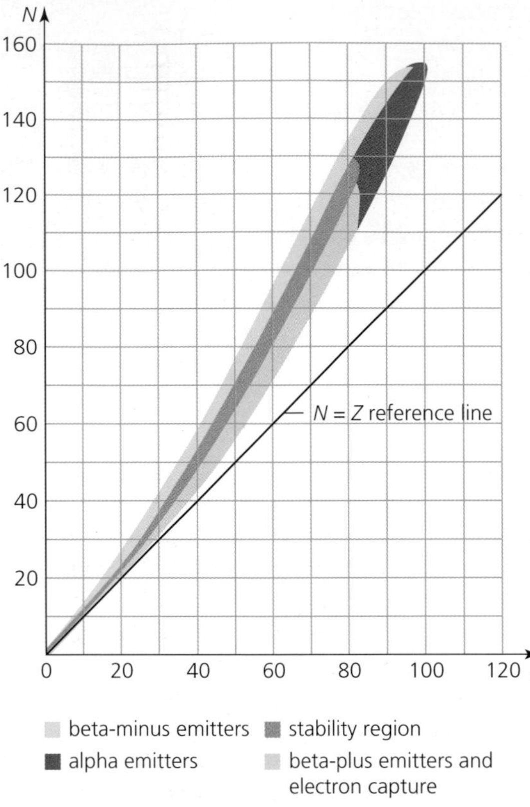

Figure 6 *Graph of N versus Z showing methods of decay of unstable nuclei*

An enlarged part of the lower left-hand section of the graph in Figure 6 is shown in Figure 7, showing some of the isotopes of carbon. Carbon-12 and carbon-13 can be seen plotted in the stability region.

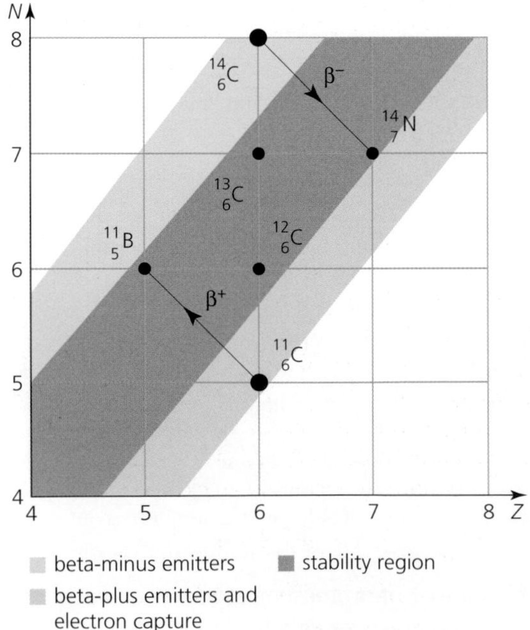

Figure 7 *A section of an N versus Z graph showing carbon isotopes*

The beta-minus emitter carbon-14 and the beta-plus emitter carbon-11 can be seen plotted outside of the stability region. Notice that the emission of a beta-minus particle transmutes the unstable carbon-14 nucleus to form nitrogen-14, which lies in the stability region. Similarly, the emission of a beta-plus particle transmutes the unstable carbon-11 nucleus to form the stable nucleus boron-11.

Worked example

Thorium-232, $^{232}_{90}\text{Th}$, decays to form a stable isotope of lead, $^{208}_{82}\text{Pb}$, by a series of alpha and beta-minus decays. In the series of decays, alpha decay occurs six times and beta-minus occurs n times. Determine the value of n.

Each alpha decay changes the nucleon number by 4 and the proton number by 2. Therefore, six alpha decays result in a new value for the nucleon number $= 232 - (6 \times 4) = 208$, and a new proton number $= 90 - (6 \times 2) = 78$. Beta-minus decay does not change the nucleon number but increases the proton number by 1. Therefore, for the final isotope to be $^{208}_{82}\text{Pb}$, the number of beta-minus decays required is $n = 82 - 78 = 4$.

QUESTIONS

4. The isotope uranium-235, $^{235}_{92}\text{U}$, decays to form a stable isotope of lead, $^{207}_{82}\text{Pb}$, by a series of alpha and beta-minus decays. In the series of decays, beta-minus occurs four times and alpha decay occurs n times. Determine the value of n.

5. The isotope fluorine-18, used as a tracer in PET scanning of the brain, is produced by irradiation of the stable isotope oxygen-18 with high-energy protons from a cyclotron. An incomplete equation of the process is shown below:

$$^{18}_{8}\text{O} + ^{...}_{...}\text{p} = ^{18}_{9}\text{F} + ^{...}_{...}\text{X}$$

Rewrite the equation but with the missing numbers added and identify the product particle X with the correct symbol.

6. The series of decays shown below forms a section of the actinium series from the unstable isotope polonium-215 to the stable isotope lead-207:

$$^{215}_{84}\text{Po} \rightarrow ^{215}_{85}\text{At} \rightarrow ^{211}_{83}\text{Bi} \rightarrow ^{207}_{81}\text{Tl} \rightarrow ^{207}_{82}\text{Pb}$$

a. Determine the neutron number for each of the isotopes shown in the section of the series shown.

b. i. The grid in Figure 8 is part of a neutron number versus proton number graph on which two of the isotopes in the above decay series have been plotted. On a copy of the grid, plot the other three isotopes from the part of the series shown above.

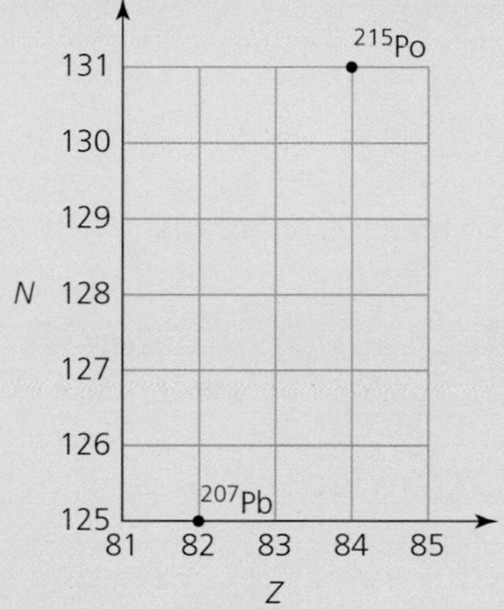

Figure 8 Section of an *N*–*Z* graph

ii. Draw arrows labelled α or β^- on your grid between the plotted isotopes to show the nature of each decay and to illustrate how a series of alpha or beta-minus decays results in a progression from a heavy unstable nucleus to a nucleus in the stability region.

KEY IDEAS

- Stable nuclei with proton number *Z* up to 20 have neutron number $N \approx Z$. Larger stable nuclei have $N/Z > 1$.
- Alpha emitters are usually heavy nuclei and they have the general decay equation

$$^{A}_{Z}X = {}^{A-4}_{Z-2}Y + {}^{4}_{2}He$$

- Beta-minus emitters tend to be over rich in neutrons and appear to the left of the stability region on a graph of *N* versus *Z*. Their general decay equation is

$$^{A}_{Z}X = {}^{A}_{Z+1}Y + {}^{0}_{-1}e + {}^{0}_{0}\bar{\nu}_e$$

- Beta-plus emitters tend to be over-rich in protons and appear to the right of the stability region on a graph of *N* versus *Z*. Their general decay equation is

$$^{A}_{Z}X = {}^{A}_{Z-1}Y + {}^{0}_{+1}e + {}^{0}_{0}\nu_e$$

- Less commonly, nuclei that are over-rich in protons decay by electron capture. These also appear to the right of the stability region on a graph of *N* versus *Z*. Their general decay equation is

$$^{A}_{Z}X + {}^{0}_{-1}e = {}^{A}_{Z-1}Y + {}^{0}_{0}\nu_e$$

10.2 NUCLEAR EXCITED STATES

The simple model of a nucleus made up of neutrons and protons, as represented in Figure 4, is useful when describing changes within the nucleus that occur during alpha and beta decay. However, there is a complication: gamma emission. The emission of a gamma photon sometimes occurs after a parent nucleus has undergone a decay by alpha or beta emission. The daughter nucleus must be formed with an excess of energy, which is emitted as a gamma photon. The gamma photons emitted from a particular nuclide always have the same energy. To account for the specific energies of the emitted gamma photons, the model of the nucleus has to be refined to include the existence of discrete **nuclear excited states**.

Cobalt-60 is a synthetic isotope made by bombarding the stable isotope cobalt-59 with neutrons in a nuclear reactor. Cobalt-60 decays by beta-minus emission to form nickel-60 with the emission of gamma photons of two possible energies, 1.17 MeV and 1.33 MeV. This indicates that there are at least two nuclear excited states at specific energies.

Nuclear excited states are represented diagrammatically by energy levels (Figure 9) in a similar way as atomic excited states. The lowest possible state in which a nucleus can exist is called the **ground state**. Taking the ground state to be 0.0 MeV, the first excited state of nickel-60 has energy 1.33 MeV and the second excited state is 2.50 MeV. The process of producing a cobalt-60 nucleus and the subsequent decay is represented by the following equations (where * denotes a nucleus in an excited state):

$$^{59}_{27}Co + {}^{1}_{0}n = {}^{60}_{27}Co$$

$$^{60}_{27}Co = {}^{60}_{28}Ni^{*} + {}^{0}_{-1}\beta + {}^{0}_{0}\bar{\nu}_e$$

$$^{60}_{28}Ni^{*} = {}^{60}_{28}Ni + \gamma$$

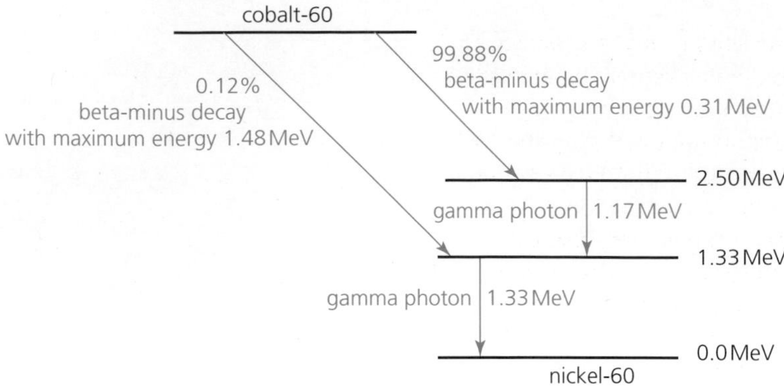

Figure 9 *Decay of cobalt-60 to nickel-60*

The naturally occurring isotope radium-226 is an alpha emitter and decays to form radon-222. However, in about 6% of the decays, the daughter nucleus $^{222}_{86}Rn$ can be created in an excited state (Figure 10). The subsequent return of the excited radon-222 nucleus to the ground state results in the emission of a gamma photon. As a consequence, a sample of radium-226 emits alpha particles of two distinct energies, 4.78 MeV and 4.60 MeV, along with gamma photons of energy 180 keV.

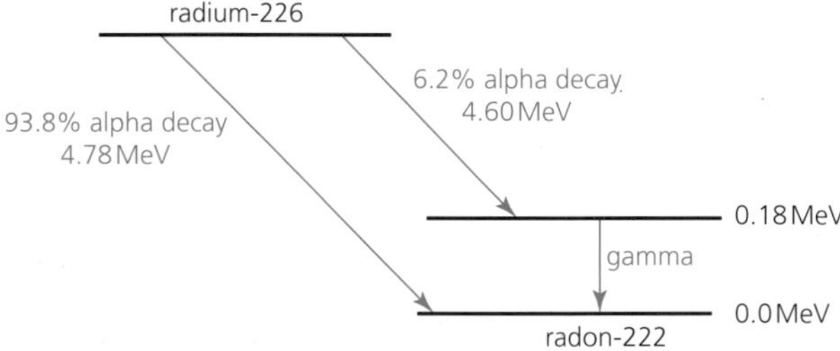

Figure 10 *Decay of radium-226*

QUESTIONS

7. Explain why alpha particles are emitted from a particular isotope with one or more distinct values of energy, whereas beta particles are emitted with a continuous range of energy up to a maximum.

8. A nucleus of magnesium-27 ($^{27}_{12}Mg$) decays by beta-minus emission to form aluminium-27 in an excited state. The aluminium-27 nucleus then emits one or two gamma photons on returning to its ground state, as shown in the energy level diagram of Figure 11.

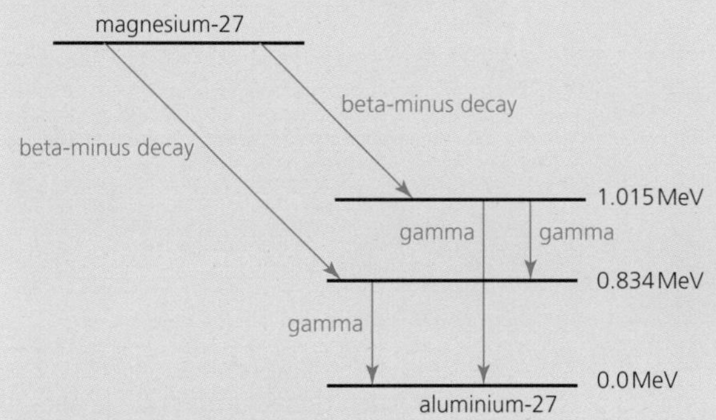

Figure 11 *Decay of magnesium-27 showing three possible nuclear transitions of the daughter nucleus, aluminium-27*

   a. Write equations to represent the decay.
   b. Determine the three possible energy values of the gamma photons that could be emitted when the excited aluminium-27 nucleus undergoes a transition to its ground state.

9. The isotope cobalt-57 decays to iron-57 by electron capture. Analysis of the emissions from a sample of cobalt-57 identifies three gamma photons with energies 14 keV, 122 keV and 136 keV.
   a. Write two equations to illustrate the decay processes.
   b. Determine the energies of the two nuclear excited states above the ground state for iron-57.

Metastable states

Most excited nuclei de-excite promptly, usually with a half-life of ≈ 10^{-12} s, but some remain in an excited state for much longer. Caesium-137 decays by beta-minus emission to form barium-137, with 94.6% of the decays producing the barium-137 nuclei in an excited state. However, the excited barium-137 nuclei are slow to return to their ground state and do so with a half-life of 153 s, emitting gamma photons of energy 0.662 MeV (Figure 12). The excited barium-137 nuclei are therefore described as being in a **metastable state**, which is labelled barium-137m. The term 'metastable state' is used for excited nuclei that return to their ground state with a half-life longer than 1 ns. The decay equations for the production of barium-137 are shown below:

$$^{137}_{55}\text{Cs} = {}^{137\text{m}}_{56}\text{Ba} + {}^{0}_{-1}\beta + {}^{0}_{0}\bar{\nu}_e$$

$$^{137\text{m}}_{56}\text{Ba} = {}^{137}_{56}\text{Ba} + \gamma$$

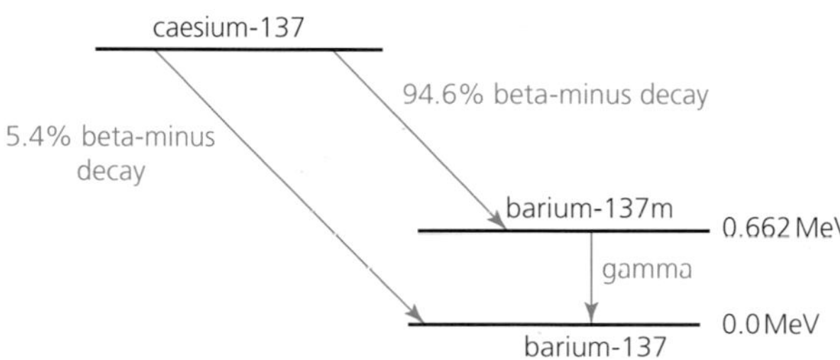

Figure 12 *Decay of caesium-137*

QUESTIONS

10. In the decay of caesium-137 (Figure 12), 94.6% of the beta particles are emitted with energies up to 0.512 MeV. Determine the maximum kinetic energy of the remaining 5.4%.

Metastable states are particularly useful in nuclear medicine, because, if the parent nuclei can be separated from the excited daughter nuclei, then radioactive material that emits only gamma rays can be produced. This radioactive material can then be introduced into the patient and used in conjunction with a gamma camera (Figure 13) without the patient being exposed to the beta particles emitted by the parent nuclei.

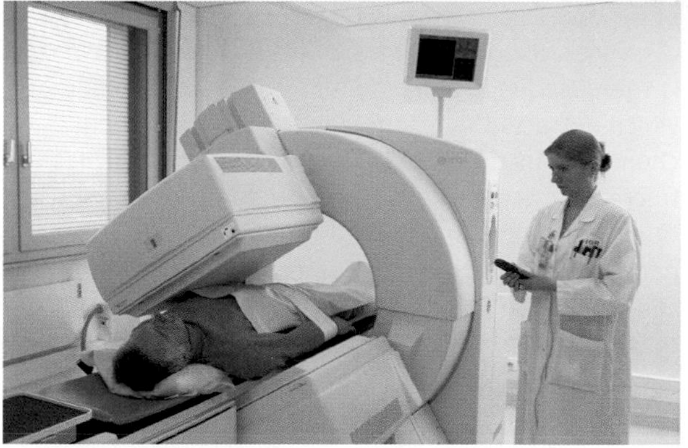

Figure 13 *A gamma camera*

The most commonly used medical radioisotope is technetium-99m. The parent nuclide is molybdenum-99 ($^{99}_{42}$Mo) and this undergoes beta-minus emission to form technetium-99m, which undergoes a transition to its ground state, emitting a gamma photon (Figure 14). In its ground state, technetium-99 is also radioactive but of low activity because of its very long half-life (of the order of 10^5 years), so it is not considered to be harmful to patients.

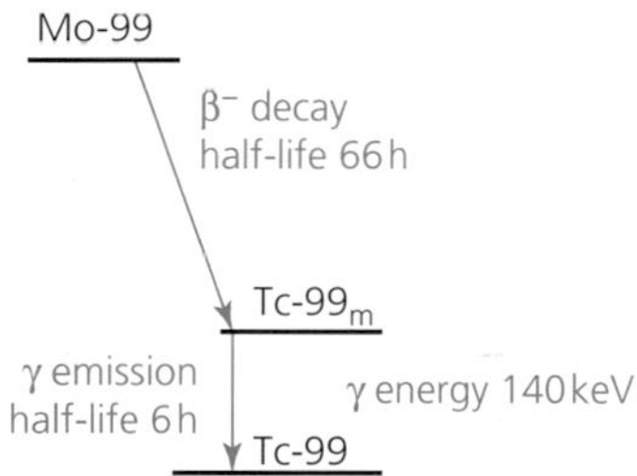

Figure 14 *Simplified energy level diagram showing the formation of technetium-99m*

Technetium-99m has a half-life of 6 hours, which is long enough to allow sufficient time for the tracer to reach the organ under investigation and enable the gamma camera to build an image, but short enough to keep the radiation exposure of the patient to a minimum and ensure that they are not still radioactive when they return home. The energy of gamma photons emitted by technetium-99m is suitable for use with a gamma camera but low enough (about 140 keV) to make them safe enough to use because of the substantially lower levels of ionisation compared with other gamma emitters.

The parent nuclei, molybdenum-99, are produced by fission of uranium-235 in nuclear reactors. The beta-minus-emitting molybdenum-99 has a half-life of 66 hours so can easily be transported over long distances from the reactor facility to hospitals, where it is then used in a 'molybdenum–technetium generator', which extracts the technetium-99m. This is then chemically incorporated into a variety of molecules to create tracers used to target different areas of the body. As the tracer travels through the body, it emits gamma radiation, which is detected by the gamma camera, which maps the functions of the body and diagnoses disorders.

QUESTIONS

11. Write decay equations to show the production and subsequent gamma decay of technetium-99m ($^{99m}_{43}$Te).
12. Technetium-99m emits gamma photons of energy 140 keV with a half-life of 6.0 hours. Calculate
    a. the wavelength of the gamma photons emitted
    b. the percentage of technetium-99m remaining in a sample after 24 hours.
13. Why is it important that a radioisotope used as a tracer emits only gamma radiation and not beta or alpha particles?
14. List the properties of technetium-99m that make it suitable as a medical tracer.

KEY IDEAS

- Gamma emission occurs after an alpha or a beta decay that has left the daughter nucleus in an excited state.
- A nuclear excited state lasting longer than 1 ns is called a metastable state.
- Isotopes in metastable states, such as technetium-99m, are used in medical diagnosis because, once separated from their parent isotopes, they are pure gamma emitters.

10.3 NUCLEAR RADIUS

What size is a nucleus? The simple model of the nucleus as a collection of closely packed protons and neutrons suggests that the volume V of a nucleus, of nucleon number A, would be approximately given by $V = A \times$ volume of one nucleon.

Assuming that a nucleus is a sphere of radius R, then the volume of a nucleus is $V = \frac{4}{3}\pi R^3$. Therefore

$$\frac{4}{3}\pi R^3 = V = A \times \text{volume of one nucleon}$$

which rearranges to give

$$R^3 = \frac{A}{\frac{4}{3}\pi} \times \text{volume of one nucleon}$$

Assuming that the volume of a nucleon is constant, this gives $R^3 \propto A$, and therefore nuclear radius $R \propto A^{1/3}$, which is usually written as

$$R = R_0 A^{1/3}$$

where R_0 is the constant of proportionality. To test the above theory requires data for the radius of nuclei. Based on his alpha particle scattering experiment, and using the mathematics of electrostatic repulsion between an alpha particle and a gold nucleus, Rutherford estimated an upper limit for the radius of a gold nucleus at $\approx 10^{-14}$ m (see section 9.1 of Chapter 9). A more accurate method, and one that allows measurements of nuclear radii of different elements, is to direct a beam of high-energy electrons at a thin solid sample of the element and measure the number of electrons scattered through various angles in a specified time interval (Figure 15).

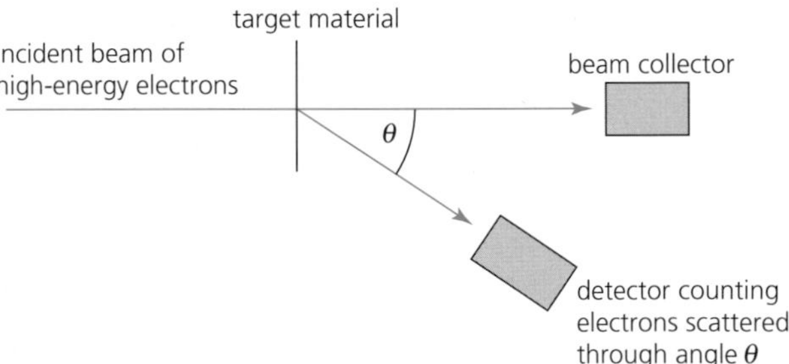

Figure 15 *Electron diffraction experiment*

Electrons are much more suitable as bombarding particles than alpha particles. Firstly electrons, being leptons, only interact with nuclei via the electromagnetic force as a result of their charge, which is well understood. Alpha particles, being hadrons, also interact with nuclei via the strong nuclear force, which is less well understood. Also, electrons, having a very much smaller mass than alpha particles, can be accelerated to very high speeds in a particle accelerator to give them a de Broglie wavelength similar in size to nuclear radii and therefore create diffraction patterns from which the size of nuclear radii can be determined (see section 8.5 in Chapter 8 of Year 1 Student Book).

The scattering of the electrons is due to their charge and the resulting attractive force between the electrons and the positively charged protons in the nucleus. However, the de Broglie wavelength for electrons accelerated through 400 MV is ≈ 10^{-15} m, which is approximately the radius of a nucleus. Therefore the wave behaviour of the electrons in the beam influences the scattering results – a diffraction pattern showing maxima and minima is superimposed. An example of the diffraction pattern produced by the scattering of electrons by oxygen nuclei is shown in Figure 16. The angle of the first minimum, θ_{min}, is about 38°, which gives a value for the radius of an oxygen nucleus of 2.7 fm.

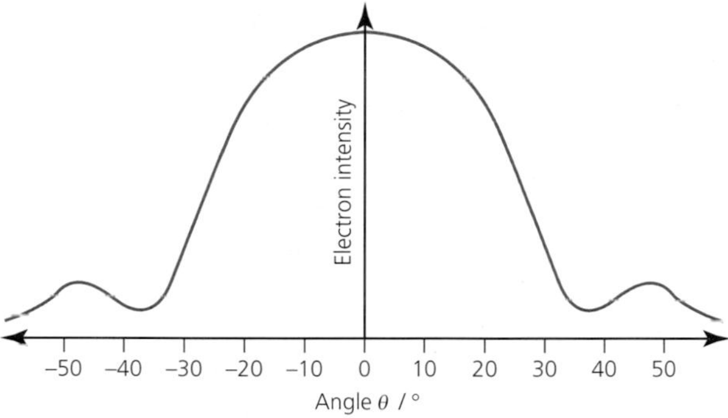

Figure 16 *Electron diffraction pattern due to an oxygen nucleus*

Electron diffraction results for the radius of nuclei show that a graph of nuclear radius R versus $A^{1/3}$ (Figure 17) is a straight line through the origin with a gradient of 1.05 fm, supporting the theory that $R \propto A^{1/3}$ and enabling nuclear radii to be calculated from $R = R_0 A^{1/3}$ where $R_0 = 1.05$ fm.

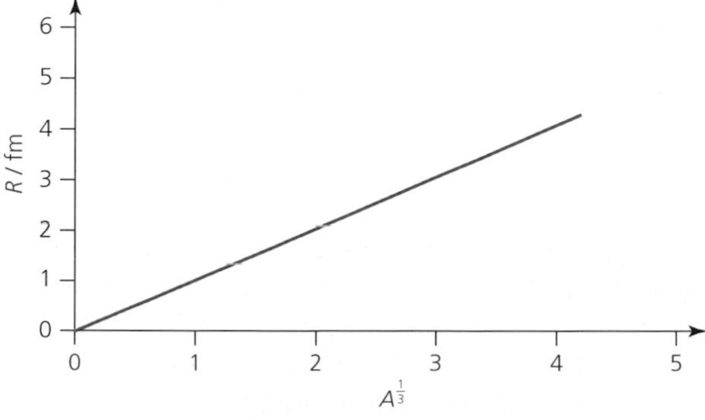

Figure 17 *Graph of R versus* $A^{1/3}$

Worked example

Determine the radius and the density of a copper-63 nucleus, assuming $R = R_0 A^{1/3}$ and taking $R_0 = 1.05$ fm.

Radius of a copper nucleus: $R = R_0 A^{1/3}$ $= 1.05 \times 10^{-15} \times 63^{1/3} = 4.2 \times 10^{-15}$ m.

$$\text{Mass of a copper atom} = \frac{\text{molar mass}}{\text{Avogadro constant}}.$$

$$= \frac{63}{6.02 \times 10^{23}} = 1.05 \times 10^{-22}\,\text{g} = 1.05 \times 10^{-25}\,\text{kg}$$

Assuming that the mass of a copper nucleus ≈ the mass of a copper atom gives the density ρ of the copper nucleus as

$$\rho = \frac{M}{V} = \frac{1.05 \times 10^{-25}}{\frac{4}{3}\pi \times (4.2 \times 10^{-15})^3} = 3.4 \times 10^{17}\,\text{kg m}^{-3}$$

Consider a nucleus of radius R and nucleon number A. The density ρ of the nucleus is given by

$$\rho = \frac{M}{V} = \frac{A \times 1\ \text{atomic mass unit}}{\frac{4}{3}\pi \times R^3}$$

$$= \frac{A \times 1.661 \times 10^{-27}}{\frac{4}{3}\pi \times (R_0 A^{1/3})^3} = \frac{1.661 \times 10^{-27}}{\frac{4}{3}\pi \times R_0^3}$$

$$= 3.4 \times 10^{17}\,\text{kg m}^{-3}$$

The density of a nucleus does not depend on the nucleon number.

The equation $R = R_0 A^{1/3}$, confirmed by experiment, leads to the fact that there is a constant density of nuclear material, so the density of a heavy nucleus is the same as the density of a light nucleus. The density of nuclei is huge compared with the densest naturally occurring element on Earth, osmium, which has a density of 2.26×10^4 kg m^{-3}.

QUESTIONS

15. Determine the radius and the density of a gold-197 nucleus, assuming $R = R_0 A^{1/3}$ and taking $R_0 = 1.05$ fm.

16. Calculate the mass of a teaspoon full (5 cm^3) of material that has the same density as atomic nuclei.

ASSIGNMENT 1: ANALYSING HIGH-ENERGY ELECTRON DIFFRACTION DATA FOR NUCLEAR RADII

(MS0.5, MS1.1, MS3.2, MS3.3, MS3.4, PS3.1, PS3.2, PS3.3)

The aim of this assignment is to analyse nuclear radius data produced by high-energy electron diffraction experiments to confirm the relation $R = R_0A^{1/3}$ and determine a value for R_0, the constant of proportionality.

Table A1 gives radius data produced in diffraction experiments involving high-energy beams of electrons fired at target nuclei.

Nucleus	Radius / fm
Carbon-12	2.40 ± 0.03
Oxygen-16	2.65 ± 0.03
Magnesium-24	3.04 ± 0.04
Silicon-28	3.20 ± 0.04
Sulfur-32	3.33 ± 0.04
Calcium-40	3.59 ± 0.05

Table A1

According to $R = R_0A^{1/3}$, a graph of nuclear radius R on the y-axis versus $A^{1/3}$ on the x-axis will generate a straight line. Calculation of the gradient of the best-fitting line enables a value for R_0 to be determined. The uncertainty in the value of R_0 can be found from the uncertainty in the gradient, which is given by

gradient uncertainty = best gradient − worst gradient

Uncertainties in the experimental nuclear radius values enable error bars to be added to points plotted. The error bars enable the steepest and least steep gradients to be determined, and the gradient that deviates most from the best-fitting line gradient can be taken as the 'worst gradient' in the above uncertainty equation.

Questions

A1 Plot a graph of R versus $A^{1/3}$ for the data in Table A1. The y-axis should start at 2.3 fm and the radius data should be plotted with error bars based on the uncertainties quoted. The $A^{1/3}$ axis should start at 2.2.

A2 **a.** Determine the gradient of the best-fitting line for the graph plotted in question **A1** to determine a value for R_0.

b. **i.** Determine the gradients of the steepest line and least steep line using error bars as a guide.

ii. Determine which line has the 'worst gradient' and calculate the uncertainty in the gradient of the best-fitting straight line to obtain an uncertainty in the value of R_0. Express R_0 with its uncertainty.

KEY IDEAS

- The radius of a nucleus can be estimated from the closest approach of alpha particles and determined more accurately from electron diffraction experiments.
- Nuclear radii are in the range 10^{-15} to 10^{-14} m.
- Nuclear radius R is related to nucleon number A by the equation $R = R_0A^{1/3}$.
- The equation $R = R_0A^{1/3}$ provides evidence for the constant density of nuclear material.

10.4 MASS AND ENERGY

Mass–energy equivalence

In 1905, Albert Einstein published his paper 'On the electrodynamics of moving bodies' in which he derived his now famous mass–energy equation:

$$E = mc^2$$

He concluded that the mass of a body is a measure of its energy content, and if energy is transferred to or from a body, its mass changes.

In the radioactive decay of a nucleus, energy is released in the form of kinetic energy of the decay products and in some cases as a gamma photon. According to Einstein, the mass before the decay should exceed the mass after the decay, and the **mass difference** should be equivalent to the total kinetic energy and/or electromagnetic energy of the products.

Consider the decay of radium-223 to radon-219 with the emission of an alpha particle, for which the decay equation is

$$^{223}_{88}\text{Ra} = {}^{219}_{86}\text{Rn} + {}^{4}_{2}\text{He} + \gamma$$

To determine the mass difference in a nuclear decay, the masses of the nuclei involved need to be very accurate, typically between six and eight significant figures.

Atomic and nuclear masses can be expressed in **atomic mass units** (u), where 1 u is defined as $\frac{1}{12}$th the mass of a carbon-12 atom. Data are available that give atomic and nuclear masses to an accuracy of at $\frac{1}{100000}$ u (0.000 01 u) or better.

The masses of the nuclei involved in the decay of radium-223 are

$^{223}_{88}\text{Ra}$: 222.970 26 u

$^{219}_{86}\text{Rn}$: 218.962 31 u

$^{4}_{2}\text{He}$: 4.001 50 u

The mass before the decay is simply the mass of the radium-223 nucleus. The total mass after the decay is the sum of the masses of the radon-219 nucleus and the alpha particle, which is 222.963 81 u.

The mass difference is therefore 222.970 26 − 222.963 81 = 0.006 45 u.

Since 1 atomic mass unit (1 u) = 1.661×10^{-27} kg, the mass difference = 1.071×10^{-29} kg, which according to Einstein is equal to an energy release given by

$$E = mc^2 = 1.071 \times 10^{-29} \times (3.00 \times 10^8)^2$$
$$= 9.639 \times 10^{-13}\,\text{J} = 6.02\,\text{MeV}$$

The energy of the emitted alpha particle has been measured at 5.78 MeV, which, allowing for recoil kinetic energy of the daughter nucleus (≈ 0.1 MeV) and the gamma photon energy (0.15 MeV) is in good agreement with Einstein.

Given that atomic and nuclear masses are usually expressed in atomic mass units, and particle energies are given in MeV, it is useful to convert directly between these two units:

1 u is equivalent to 931.5 MeV.

Einstein's equation $E = mc^2$ applies to *all energy changes* not just those involving subatomic particles. However, the mass differences occurring as a result of an energy transfer are only detectable in nuclear decays and nuclear reactions (see question 18).

QUESTIONS

17. Determine the decay energy released when magnesium-27 decays by beta-minus emission to form aluminium-27 given the following masses: magnesium-27, 26.977 76 u; aluminium-27, 26.974 40 u; electron, 0.000 55 u. Assume that the mass of the emitted electron antineutrino is negligible.

18. Determine the fractional mass difference when

a. a 800 kg car accelerates from rest to a speed of 10 m s^{-1}

b. an electron of rest mass 0.000 55 u is accelerated by a potential difference of 1 MV.

Nuclear binding energy

The **binding energy** of a nucleus is defined as the work that would need to be done to separate all of its nucleons.

Work is needed to separate the nucleons because the strong nuclear force that binds the nucleus together has to be overcome. Since doing work transfers energy, and according to Einstein transferring energy changes mass, the mass of the separated nucleons should be greater than the mass of the original nucleus (Figure 18).

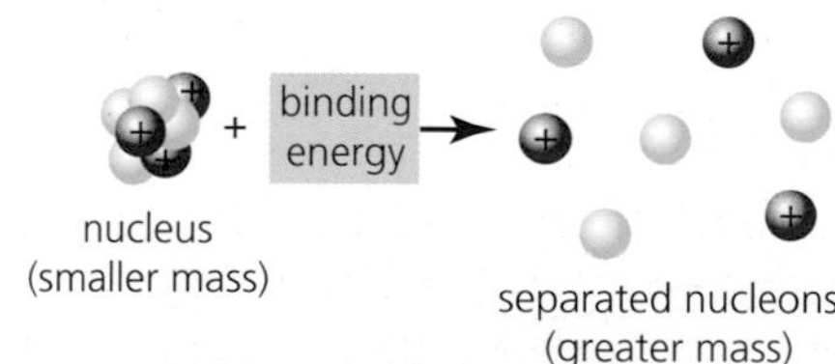

Figure 18 *Separating nucleons in a nucleus*

Consider a uranium-235 nucleus of mass of 234.993 42 u. The mass of a proton is 1.007 28 u and the mass of a neutron is 1.008 67 u so

$$\text{mass of constituent nucleons} = (92 \times 1.007\,28)$$
$$+ (143 \times 1.008\,67) = 236.909\,57\,\text{u}$$

which is 1.916 15 u *more* than the mass of the original uranium-235 nucleus. The energy equivalent of a mass of 1.916 15 u is $1.91615 \times 931.5 = 1785$ MeV, and therefore the binding energy of a uranium-235 nucleus is 1785 MeV.

It is inevitable that those nuclei with the greatest number of nucleons have the largest binding energy. Therefore, in order to judge which nuclei are the most tightly bound, the quantity **binding energy per nucleon** is defined as the energy required per nucleon to separate the nucleons in a nucleus. For example, since 1785 MeV is needed to separate all the nucleons in a uranium-235 nucleus, its binding energy per nucleon is $\frac{1785}{235} = 7.596$ MeV per nucleon.

A graph of binding energy per nucleon versus nucleon number is shown in Figure 19. The nucleus with the highest binding energy per nucleon is iron-56, and this therefore is the most tightly bound and therefore the most stable nucleus.

QUESTIONS

19. Determine the energy needed to completely separate all the nucleons in a radium-224 nucleus ($^{224}_{88}$Ra), given that its mass is 223.971 90 u, the proton mass is 1.007 28 u and the neutron mass is 1.008 67 u.

20. Determine the binding energy per nucleon of a helium-4 nucleus given the following masses: $^{4}_{2}$He, 4.001 50 u; proton, 1.007 28 u; neutron, 1.008 67 u.

Fusion

Consider the reverse of the process shown in Figure 18 – a nucleus being formed from a group of nucleons. Since the nucleus has less mass than its constituents, energy would have been transferred away from the nucleus on its formation. The more tightly bound the nucleus, the greater the reduction in mass and the larger the amount of energy transferred to the surroundings. Now suppose that, instead of being formed from its constituent nucleons, a nucleus is created from two lighter nuclei that fuse together. In order for this process to release energy, the nucleus formed would have to be more tightly bound than the two lighter nuclei that formed it. Given that the more tightly bound a nucleus, the higher its binding energy per nucleon, the condition that must be satisfied for a fusion process to release energy to the surroundings is that the two lighter nuclei must have lower values of binding energy per nucleon than the nucleus they form. A study of Figure 19 shows that the only nuclei that can satisfy this condition are nuclei with nucleon numbers to the left of the peak of the graph.

For example, the fusion of the nuclei of the two hydrogen isotopes, deuterium and tritium, which have binding energy per nucleon values of 1.11 MeV/nucleon and 2.83 MeV/nucleon, respectively, forms a nucleus of helium-4, which has a binding energy of 7.07 MeV/nucleon. The equation for this nuclear fusion reaction is

$$^{2}_{1}\text{H} + {}^{3}_{1}\text{H} = {}^{4}_{2}\text{He} + {}^{1}_{0}\text{n}$$

Since the helium-4 nucleus has a much higher binding energy per nucleon than either of the two nuclei that create it, the nuclear reaction will transfer energy to its surroundings and could potentially be a useful

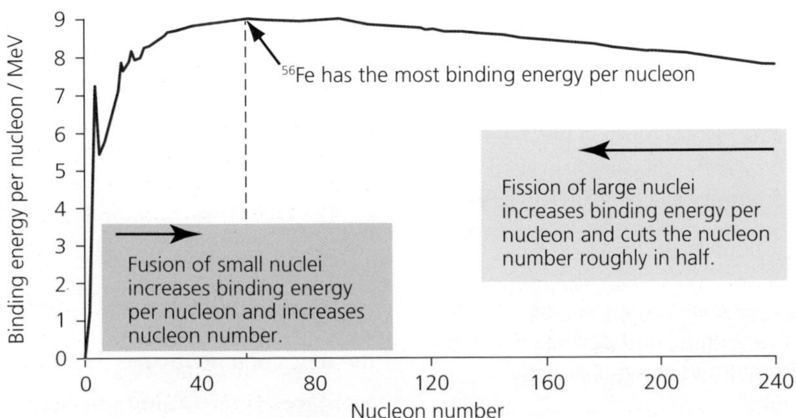

Figure 19 *Graph of binding energy per nucleon versus nucleon number*

source of energy. It is possible to determine the amount of energy released to the surroundings by calculating the change in binding energy:

change in binding energy =
$(4 \times 7.07) - (2 \times 1.11) - (3 \times 2.83) = 17.6\,\text{MeV}$

Alternatively, a calculation of the loss of mass can also determine the energy released. Given the masses of a deuterium nucleus (2.013 55 u), a tritium nucleus (3.015 50 u), a neutron (1.008 67 u) and a helium-4 nucleus (4.001 50 u), we have

mass loss = $2.01355 + 3.01550 - (4.00150 + 1.00867) = 0.01888\,\text{u}$

which is equivalent to $0.01888 \times 931.5 = 17.59\,\text{MeV}$.

The production of energy from nuclear fusion on Earth has struggled with the technical problems (see the introduction to this chapter) associated with overcoming the electrostatic repulsion between charged nuclei to get them close enough, $\approx 10^{-15}$ m, for the strong nuclear force to fuse them together. The heat and light radiated from stars are produced as a result of fusion reactions, with the most massive stars being able to fuse nuclei to produce iron. Fusion can take place inside stars because of the very high temperature, which means that the nuclei are moving fast enough to overcome the electrostatic repulsion between them.

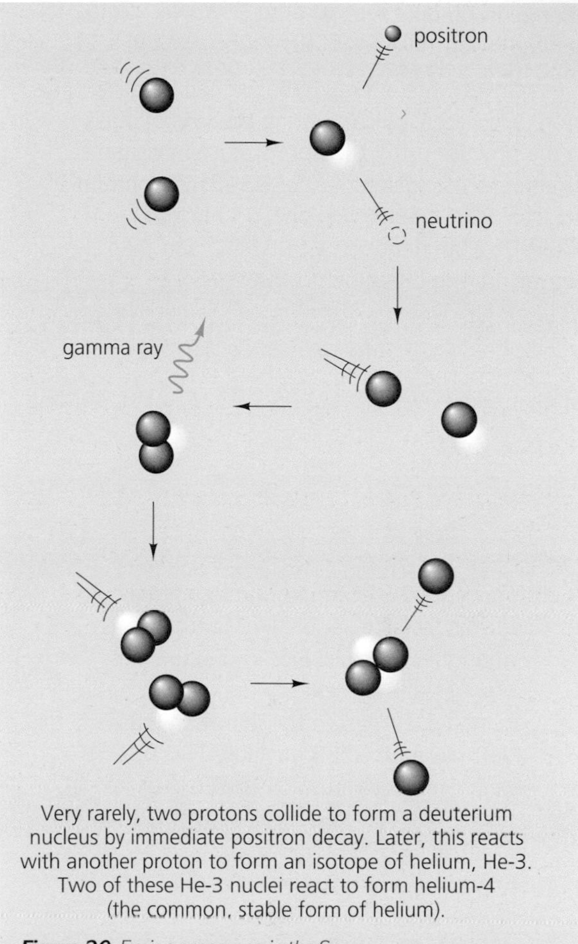

Very rarely, two protons collide to form a deuterium nucleus by immediate positron decay. Later, this reacts with another proton to form an isotope of helium, He-3. Two of these He-3 nuclei react to form helium-4 (the common, stable form of helium).

Figure 20 *Fusion processes in the Sun*

QUESTIONS

21. Determine the energy released when two deuterium nuclei fuse to form a helium-3 nucleus, given the following binding energy per nucleon data: deuterium nucleus, 1.11 MeV/nucleon; helium-3 nucleus, 2.57 MeV/nucleon.

22. The nuclear fusion process that fuels the Sun is shown in Figure 20. The overall equation for the process is

$$4\,{}^{1}_{1}\text{H} = {}^{4}_{2}\text{He} + 2\,{}^{0}_{+1}\text{e} + 2\,{}^{0}_{0}\nu_e$$

Calculate the loss of mass and the energy released into the surroundings in the formation of a helium-4 nucleus by this process, given the following data: proton mass, 1.007 28 u; positron mass, 0.000 55 u; helium-4 nucleus mass, 4.001 50 u.

Fission

A further study of the graph of binding energy per nucleon versus nucleon number (Figure 19) reveals a second process as a source of energy. Consider a very heavy nucleus to the far right of Figure 19 splitting into lighter nuclei but still with nucleon numbers greater than 56. This process would result in an overall increase in binding energy per nucleon and therefore a loss of mass, which would correspond to energy being transferred to the surroundings. For example, although a very rare event, a nucleus of uranium-235 can undergo spontaneous fission, producing two lighter nuclei such a strontium-90 and xenon-143. The equation for this fission event is

$$^{235}_{92}\text{U} = {}^{90}_{38}\text{Sr} + {}^{143}_{54}\text{Xe} + 2\,{}^{1}_{0}\text{n}$$

The energy released can be determined by considering the change in binding energy per nucleon values, which are: uranium-235, 7.59 MeV/nucleon; strontium-90, 8.70 MeV/nucleon; xenon-143, 8.20 MeV/nucleon. The energy released in the spontaneous fission of uranium-235 is given by

change in binding energy =
$$(90 \times 8.70) + (143 \times 8.20) - (235 \times 7.59) = 172\ \text{MeV}$$

Alternatively, the energy released can be determined more accurately by calculating the loss of mass in the process using the appropriate mass data (uranium-235 nucleus, 234.99342 u; strontium-90 nucleus, 89.88688 u; xenon-143 nucleus, 142.90525 u; neutron, 1.00867 u) and then converting the mass lost to MeV:

$$\text{mass loss} = 234.99342 - 89.88688 - 142.90525 - (2 \times 1.00867) = 0.18395\ \text{u}$$

which is equivalent to $0.18395 \times 931.5 = 171.35\ \text{MeV}$.

QUESTIONS

23. Fermium-256 can undergo spontaneous fission to form xenon-140, palladium-112 and four neutrons. Determine the loss of mass and the energy released in this fission event, given the following mass data: fermium-256 nucleus, 256.03688 u; xenon-140 nucleus, 139.89200 u; palladium-112 nucleus, 111.88206 u; and neutron, 1.00867 u.

KEY IDEAS

- The atomic mass unit (1 u) is defined as $\frac{1}{12}$th the mass of a carbon-12 atom.
- Einstein's equation $E = mc^2$ applies to all energy changes: transferring energy by doing work changes mass; a change in mass means there is an input or output of energy.
- Binding energy per nucleon is an indicator of nuclear stability.
- Nuclear changes that involve an increase in binding energy per nucleon result in energy being transferred to the surroundings.

10.5 INDUCED FISSION

The spontaneous fission of a nucleus is a very rare event, but it is possible to induce fission by bombarding heavy nuclei with neutrons, releasing typically 150–200 MeV per fission event. Each fission event produces two mid-sized nuclei called **fission fragments** along with a variable number of neutrons and gamma photons. An example is shown in Figure 21.

The key to obtaining large amounts of energy from fission is the creation of the free neutrons, which potentially could go on to cause more fission events in a **self-sustaining chain reaction** (Figure 22). In such a chain reaction, for every neutron absorbed and causing fission, at least one of the neutrons produced goes on to cause another fission event. The types of nuclei produced as fission fragments in the induced fission of uranium-235 nuclei vary, as does the number of neutrons produced, which averages at between two and three. The energy released per induced fission event appears in the form of the kinetic energy of the fission fragments and the neutrons, and can be found by calculating the loss of mass in the fission reaction.

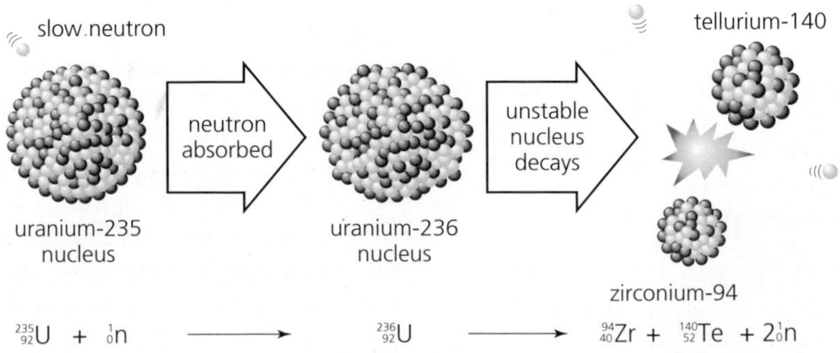

Figure 21 An induced fission event

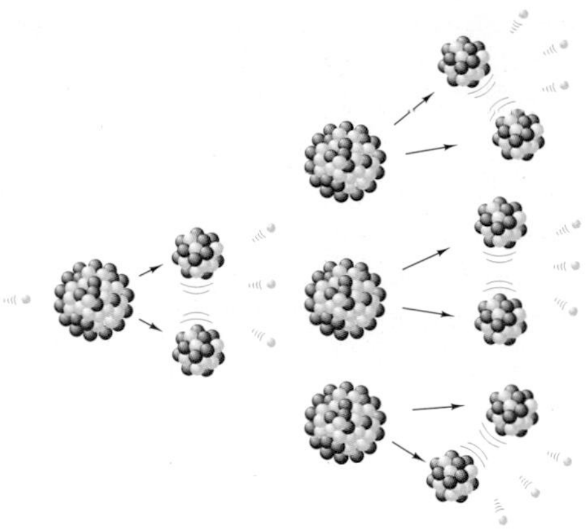

One fission may release 3 neutrons.
These can cause further fissions,
releasing 9 neutrons, then 27 neutrons, and so on.

Figure 22 A chain reaction

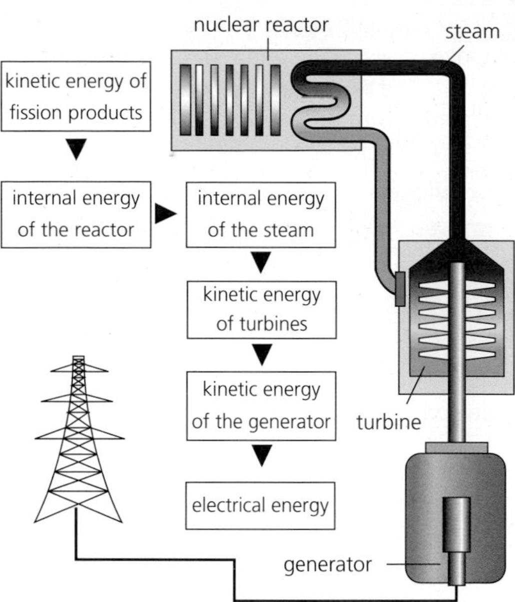

Figure 23 Energy flow at a nuclear power station

QUESTIONS

24. Determine the energy released in the fission of a uranium-235 nucleus forming the fission fragments, barium-141 and krypton-92, given the decay equation:

$$^{235}_{92}U + ^{1}_{0}n = ^{141}_{56}Ba + ^{92}_{36}Kr + 3^{1}_{0}n$$

and the following mass values: uranium-235 nucleus, 234.993 42 u; krypton-92 nucleus, 91.906 39 u; barium-141 nucleus, 140.883 67 u; neutron, 1.008 67 u.

Controlled fission in a nuclear reactor

Induced fission is the process occurring in a nuclear power station. The first self-sustaining chain reaction was produced at Chicago University in 1942, and the first nuclear power station started delivering electricity to the grid in Obninsk in Russia in 1954. The basic principle of operation is to establish a *controlled* self-sustaining chain reaction in nuclear fuel, such as uranium, and to harness the energy released. That energy is used to heat water to make steam at high pressure to drive a turbine, which in turn drives a generator to produce electricity that can be supplied to the national grid (Figure 23).

The fission events at a power station occur inside the nuclear reactor. The main technical issues to be considered in order that sufficient heat energy is generated in the reactor and then extracted to make high-pressure steam are as follows:

- The probability of fission occurring must be high to ensure that enough fission events take place every second in order to generate the required amount of heat energy.
- Enough fission neutrons go on to create more fission events, so that a self-sustaining chain can be established.
- The number of free neutrons in the reactor can be controlled.
- The heat energy generated in the reactor can be extracted efficiently.
- The people working at the power station are shielded from the radiation emitted from the reactor.

A neutron produced in a fission event has typically about 2 MeV of kinetic energy, which means that it is travelling at about $2 \times 10^7\,m\,s^{-1}$ and is usually referred to as a **fast neutron**. There is, however, a much higher probability that a nucleus of uranium-235 will absorb a neutron and then undergo fission if the neutron is travelling slowly, typically with its kinetic energy determined by the surrounding temperature. A neutron that is in thermal equilibrium with its surroundings is called a **thermal neutron** and has a kinetic energy $\ll$ 1 eV. The reactor must therefore contain a material made up of nuclei with which the fast neutrons can collide, resulting in their speed being reduced very significantly at each collision.

Consider a perfectly elastic head-on collision between two snooker balls, white and red, of the same mass (Figure 24). The white ball is struck with the cue so that it collides with the stationary red ball.

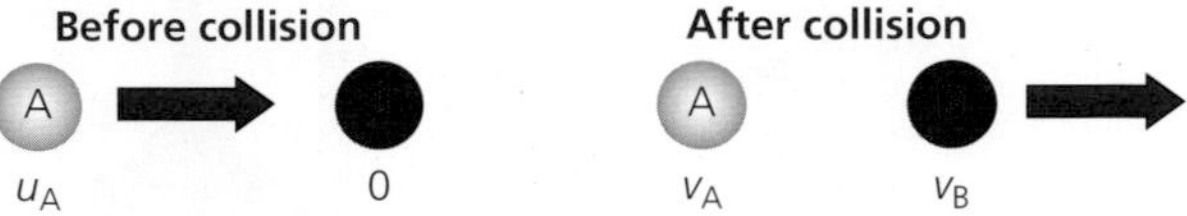

Figure 24 *An elastic head-on collision*

By conservation of momentum and by conservation of kinetic energy (*see sections 11.2 and 11.4 in Year 1 Student Chapter 11 of Book*) for this perfectly elastic collision:

$$mu_A = mv_A + mv_B$$

$$\frac{1}{2}mu_A^2 = \frac{1}{2}mv_A^2 + \frac{1}{2}mv_B^2$$

Combining the above two equations gives $v_A = 0$ (see question 26), which shows that the white ball, which was moving before the collision, is stationary after the collision, having transferred its kinetic energy to the red ball.

Now, going back to our nuclear collisions, if a neutron undergoes a head-on elastic collision with a stationary particle of the same mass, the neutron will lose its kinetic energy and become a thermal neutron. If the collision is not head-on, a series of collisions will be required to remove the neutron's kinetic energy. The material used to slow down the fast fission neutrons is known as the **moderator**. Since a fast neutron is most likely to lose a lot of its kinetic energy if it collides with something of the same mass, a commonly used moderator is water, since it contains protons, which have almost the same mass as neutrons.

To maintain a self-sustaining chain reaction, it is important that materials in the reactor core, other than the uranium-235, do not have a tendency to absorb neutrons. The material chosen as the moderator must therefore have a low probability of absorbing neutrons. The earliest reactors in operation in the UK were Magnox reactors, which used a carbon moderator in the form of graphite. Although about 12 times more massive than neutrons, carbon nuclei are still sufficiently small particles to be able to act as an adequate moderator and have the added advantage of having a very low probability of neutron absorption. Note that some of the neutrons' collisions with moderator nuclei can be inelastic and excite the moderator nuclei, resulting in the emission of gamma photons on de-excitation.

QUESTIONS

25. Although the moderator material is chosen for its low probability of absorbing neutrons, inevitably some neutrons will be absorbed, making the moderator radioactive.
What is the most likely type of radiation emitted by moderator nuclei that have absorbed neutrons?

Stretch and challenge

26. By applying conservation of momentum and conservation of kinetic energy to the elastic head-on collision shown in Figure 24, show that the velocity of the white snooker ball after collision is zero, given that the red and white snooker balls have the same mass.

The nuclear fuel used in a reactor is usually natural uranium (0.7% uranium-235 and 99.3% of uranium-238) or 'enriched' uranium (4% uranium-235 and 96% of uranium-238). It is the uranium-235 nuclei that undergo induced fission and are described as **fissile material**. A fuel rod is made up of a column of uranium oxide pellets clad and sealed into zirconium alloy tubes typically 4–5 m long. Zirconium is used as the cladding because it has a very low probability of absorbing neutrons. The fuel rods are assembled in bundles called fuel assemblies (Figure 25), which, in pressurised water reactors (see page 213), are about 20 cm across and have a square lattice arrangement capable of containing about 200 fuel rods. The fuel assemblies (typically about 150) are lowered into the reactor by a remote handling device (Figure 26).

Some of the rod positions in the assembly are left vacant to enable the insertion of movable **control rods**, neutron source rods and measuring instrumentation. The operation of a nuclear reactor requires a **critical chain reaction** in which, on average, *one* of the neutrons produced in a fission event goes on to produce another fission event. To maintain this critical condition, controls rods, made from a material with a high probability of absorbing neutrons – typically cadmium or boron – are raised from or lowered into the fuel assemblies (Figure 27). By controlling the neutron density, the control rods control the rate at which fission occurs and therefore the power output of the reactor.

Figure 25 A nuclear fuel assembly being checked. New fuel rods can be handled because uranium-238 and uranium-235 are only weak alpha emitters.

Figure 26 A nuclear fuel assembly being lowered into a reactor core

Additional control rods are positioned so they can be lowered very quickly into the core should an unsafe condition arise requiring the emergency shut-down of the reactor.

The neutron source rods provide the neutrons required for the initial start-up of the reactor and

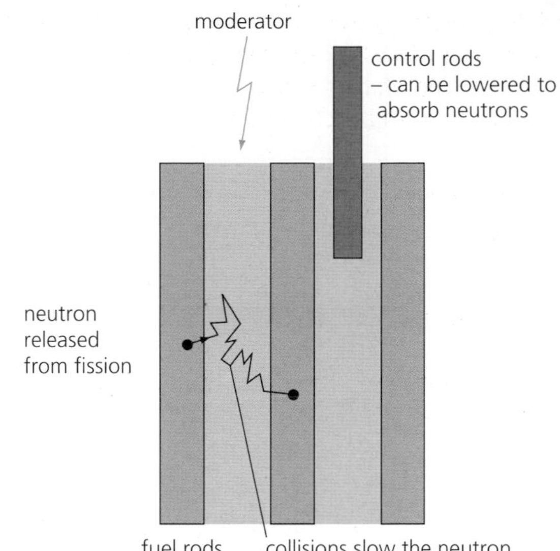

Figure 27 The roles of the moderator and control rods in a nuclear reactor

usually contain a material that generates neutrons by spontaneous fission, typically californium-252.

The kinetic energy of the fission fragments and neutrons generates a lot of heat in the reactor core. This heat energy needs to be transferred efficiently to convert water into high-pressure steam. The material that passes through the reactor core and absorbs the heat energy generated by fission is called the **coolant** and must have good heat transfer characteristics. These characteristics include having either a high specific heat capacity, such as liquid water, or the ability to be pumped very quickly around the system, such as carbon dioxide gas. Almost all the reactors in the UK use carbon dioxide as the coolant, whereas worldwide over 60% of reactors are pressurised water reactors (PWRs) in which water acts as both coolant and moderator, enabling a more compact design. In a PWR, the water flows around the primary cooling system (Figure 28), extracting heat from the reactor core and transferring heat to the secondary cooling system, where high-pressure steam is generated to drive the turbines, which, in turn, drive electrical generators. The water coolant remains liquid even though the core temperature is about 300 °C because it is kept under a pressure of about 150 times atmospheric pressure. The pressure vessel containing the core of the reactor is made of steel, as it has to withstand high temperatures.

Ordinary water, referred to as light water in this context, while having excellent moderating and heat transfer properties, is over 100 times more likely than carbon dioxide to absorb neutrons. Therefore, all light

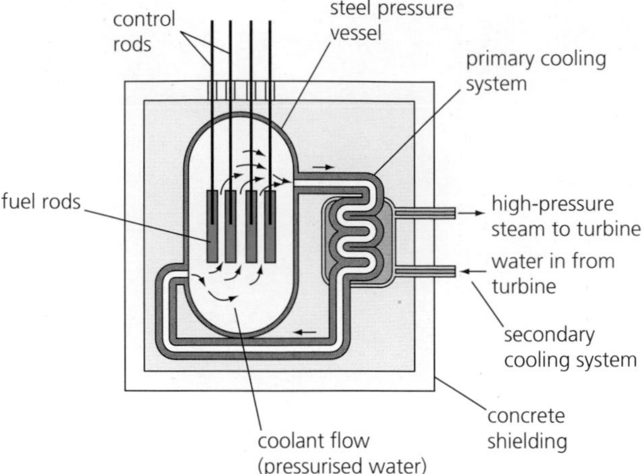

Figure 28 *Pressurised water reactor (PWR). In a PWR, water under high pressure acts as both the coolant and the moderator.*

water reactors require enriched uranium fuel rather than natural uranium.

The workforce at the power station are protected from radiation from the reactor by outer concrete **shielding**, which absorbs any escaping neutrons and gamma photons.

QUESTIONS

27. Describe and explain *one* advantage and *one* disadvantage of water over carbon dioxide as the coolant in a nuclear reactor.

28. Explain how the power output of a nuclear power station is affected by lowering the control rods into the reactor core.

29. How does enriched uranium differ from natural uranium?

30. Calculate the minimum rate at which the fuel rods in a reactor core must decrease in mass if the power output of the nuclear power station is 900 MW. Explain why the value obtained is a minimum.

31. a. What is meant by

- **i.** a fast neutron
- **ii.** a thermal neutron?

b. How does the speed of a neutron affect its ability to induce fission in a uranium-235 nucleus?

32. State what is meant by a *critical chain reaction*.

Critical mass of the fuel

Consider what could happen to a fission neutron produced by the fission of a uranium-235 nucleus within a fuel rod in the core of a nuclear reactor in operation:

1. The neutron could pass through the cladding of the fuel rod and undergo a series of collisions with the moderator atoms, causing it to slow down, before passing through the cladding of another fuel rod to cause a uranium-235 nucleus to undergo fission.
2. The neutron could be absorbed by a control rod.
3. The neutron could be absorbed by the fuel cladding, or by the material of the moderator or coolant.
4. The neutron could be absorbed by a uranium-238 nucleus (resulting in the production of a plutonium nucleus).
5. The neutron could be absorbed by a fission fragment nucleus or by a nucleus formed from the decay of a fission fragment.
6. The neutron could escape from the reactor core and be absorbed by the concrete shield.

There are therefore several possible fates of a neutron that would mean it would not contribute to the chain reaction. Fate 3 is reduced in likelihood by choice of material for the cladding, moderator and coolant. Fates 4 and 5 are more difficult to control. Fate 6 is important. The bigger the mass of fuel, the smaller is the surface area to volume ratio, which means a smaller percentage of the neutrons produced are likely to escape.

The **critical mass** of fuel is the minimum mass required to establish a self-sustaining chain reaction.

The value of the critical mass varies depending on the concentration of uranium-235 contained in the fuel and also on the geometry of the core and fuel rods, as this affects the surface area to volume ratio.

As time passes, the uranium-235 content of the rods in an assembly decreases and increasing amounts of fission fragments, their decay products and plutonium accumulate. After 4 to 5 years of operation, the fissile content of the fuel rods in the assembly is much reduced, and, given that fission fragments and their decay products can absorb neutrons, the efficiency of the fuel assembly becomes too low and it has to be removed from the reactor core during a shut-down period.

QUESTIONS

33. State at least *three* unwanted processes that could result in a reduction in the number neutrons available for fission in a nuclear reactor.

34. At a nuclear fuel production plant, safety procedures ensure that the unused uranium fuel is not stored in large quantities in one location. Explain why this is the case.

35. State what is meant by 'critical mass' and give *two* factors on which the critical mass depends.

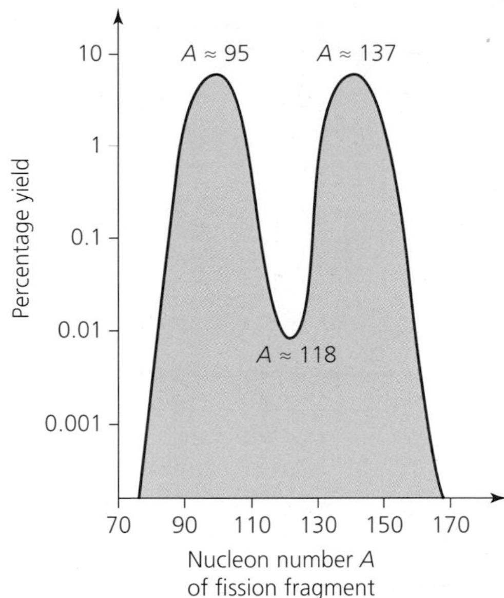

Figure 29 *Percentage yield of fission fragments versus nucleon number for the fission of uranium-235. Note the logarithmic scale on the percentage yield axis.*

Used (spent) fuel

The splitting of a uranium-235 nucleus is asymmetrical, with one fragment being significantly larger than the other. Figure 29 shows the fission fragment distribution for the fission of uranium-235. The graph shows that the majority of the fission fragments have nucleon numbers around 95 and 137. A very heavy nucleus, such as uranium-235, has a high neutron : proton ratio, and therefore, when undergoing fission, not only releases neutrons but also creates fragments that are over-rich in neutrons. Consequently, fission fragments are unstable and emit beta-minus and also gamma radiation, with some undergoing a series of beta-minus decays before forming a stable nucleus. The **spent fuel** is therefore very radioactive and generates a lot of heat as the fission fragments' decay energy is released. The spent fuel assemblies are removed from the reactor core by a remote handling device and transferred to cooling ponds (Figure 30), where the circulating water not only removes heat from the fuel rods but also provides an effective shield for the emitted beta and gamma radiation.

The fission fragments within the spent fuel rods have half-lives ranging from a few days to many thousands of years. Those with short half-lives are very radioactive, generating considerable amounts of heat from their decay, but cease to be radioactive after a short time. Those with very long half-lives will be radioactive for thousands of years, but their activity is low, with little heat generated. It is the fission fragments with half-lives of tens of years, for example caesium-137 and strontium-90, that pose the most risk, because they have considerable activity, generate a lot of heat and will be highly radioactive for hundreds of years.

Figure 30 *Spent fuel assembly being positioned in a cooling pond*

Flasks containing spent fuel rods are transported by train from nuclear power stations to a reprocessing plant. In the UK this is the Thorp reprocessing plant at Sellafield, Cumbria. The fuel rods are removed from the transport flasks under water and stored in ponds for further cooling before being **reprocessed**. During the reprocessing the spent fuel is dissolved in nitric acid, and the uranium and plutonium content chemically removed and stored for future use. The remainder, containing the fission fragments and their decay products, is highly radioactive, generating a lot of heat from their decay energy. It is defined by the nuclear power industry as **high-level waste** (**HLW**),

Category	Details	Example	Treatment/storage in the UK
High-level waste (HLW)	Has high levels of both radioactivity and heat, and therefore requires cooling as part of its safe storage	Reprocessed spent fuel	Stored as liquid in water-cooled tanks. Can also be evaporated to form a powder and contained within glass in a process known as vitrification. Requires thick concrete walls to shield operators from high levels of radiation. Currently stored above ground at Sellafield
Intermediate-level waste (ILW)	Waste that has the same or lower levels of radioactivity as HLW but does not generate enough heat to require cooling as part of its safe storage	Cladding that is removed from the outside of spent fuel rods	Some is stored in vaults and some is encapsulated in cement inside stainless steel drums. Requires thick concrete walls to shield operators from high levels of radiation
Low-level waste (LLW)	Waste that has much lower levels of radioactivity than HLW or ILW and does not generate heat	Contaminated clothing, including protective shoes and gloves	Compacted to a fraction of its original volume and stored in steel drums in concrete vaults (Figure 31)

Table 1 *Radioactive waste categories*

since its treatment and storage must be designed to deal with both its radioactivity and heat generation.

There are other types of nuclear waste created during the building, operation and decommissioning of nuclear power stations, described as **intermediate-level waste** (**ILW**) and **low-level waste** (**LLW**), which require different handling and storage. The definitions of the three types of waste along with the corresponding requirements for treatment and storage are shown in Table 1.

Figure 31 *Low-level waste storage containers*

Worldwide, more than 430 nuclear power stations are in operation in over 30 countries, and about 11% of the world's electricity is currently generated by nuclear power. The UK has 29 nuclear power reactors in permanent shut-down and 16 currently operational. The Nuclear Decommissioning Authority (NDA) is responsible for managing the clean-up of the nuclear sites across the UK and for delivering a deep geological disposal repository for high-level radioactive waste, although no suitable sites have as yet been identified. In France, where nearly 77% of electricity is generated from nuclear power, a waste repository site has been identified in the Marne area in eastern France, which, when constructed, will be 500 m deep in clay and occupy an area of 15 km^2.

Whether radioactive waste is stored at the Earth's surface or in deep underground vaults, the storage design has to ensure that the waste is prevented from entering the atmosphere, the water supply and the food chain for as long as the waste remains radioactive.

QUESTIONS

36. Some intermediate-level waste is as radioactive as high-level waste but is still treated differently. Explain why this is so.

37. Explain why spent fuel rods are much more hazardous than unused ones.

KEY IDEAS

- The fission of a uranium-235 nucleus can be induced by the absorption of a neutron. This is how the nuclear power industry generates electrical energy.
- A moderator is used to slow fission neutrons to thermal energies in order to increase the probability of a uranium-235 nucleus absorbing a neutron.
- Control rods are used to adjust the neutron density in a reactor core in order to maintain a controlled chain reaction.
- Heat energy is extracted from the reactor by the flow of the coolant, which transfers the energy to generate steam, which can drive a turbine/generator.
- Containment, shielding and careful control are crucial to the safe running of a nuclear power plant.
- Radioactive waste is classified as high, intermediate or low level.
- Reprocessed spent fuel constitutes high-level waste and contains highly radioactive fission fragments from the fission of uranium-235.
- High-level waste must be cooled under water.
- Long-term safe storage of high-level radioactive waste produced in fission reactors continues to be a major issue.

ASSIGNMENT 2: RESEARCHING THE NUCLEAR POWER DEBATE

Is nuclear power the solution to the world's energy needs and to the problem of climate change, or do the health concerns associated with radioactive waste and nuclear reactor accidents present too great a risk?

A1 Having increased your knowledge of nuclear power from your studies in this chapter, state your current attitude to nuclear power in the context of the above question.

A2 Figures A1 and A2 show data from the UK's Department for Energy and Climate Change illustrating the share of energy sources contributing to electricity generation in the UK in the second quarters of 2014 and 2015, respectively.

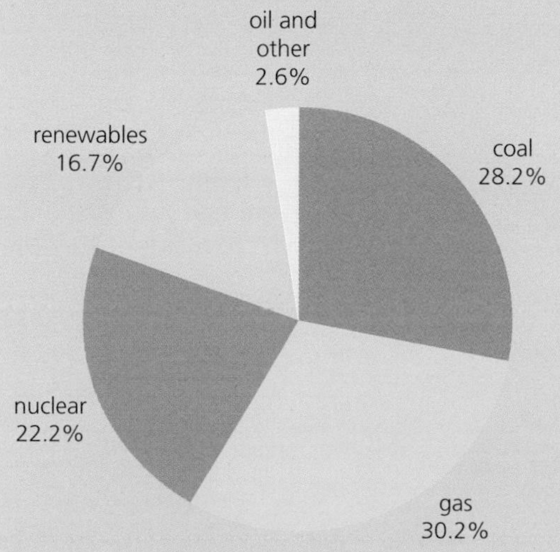

Figure A1 Electricity generation energy sources, UK, second quarter 2014

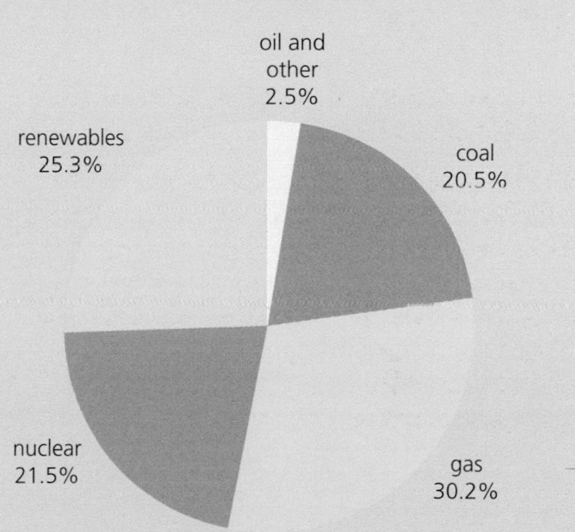

Figure A2 Electricity generation energy sources, UK, second quarter 2015

Study the two pie charts and identify specific data or changes from 2014 to 2015 that you think might:

a. be relevant to the problem of climate change

b. support the argument

i. for nuclear power

ii. against nuclear power.

A3 Three of the worst accidents associated with nuclear power in the last 60 years include:

- Three Mile Island in the USA, 1979
- Chernobyl in the Ukraine, 1986
- Fukushima in Japan, 2011.

Research one of the above accidents and write a brief report on its observed effect on the environment.

A4 **a.** Within the context of a class discussion, establish the main points that you and your classmates agree support the case for nuclear power and those that could be used to argue against nuclear power.

b. Imagine that you are a government minister required to make a decision regarding whether to push forward proposals for a programme of nuclear power plant construction in the UK. What further information would you be likely to request from your scientific advisers to enable you to make a well-informed decision?

c. Refer back to your answer to question **A1** and comment on whether your view on nuclear power has changed in the light of additional knowledge.

PRACTICE QUESTIONS

1. **a.** Describe the changes made inside a nuclear reactor to reduce its power output and explain the process involved.

b. State the main source of highly radioactive waste from a nuclear reactor.

c. In a nuclear reactor, neutrons are released with high energies. The first few collisions of a neutron with the moderator transfer sufficient energy to excite nuclei of the moderator.

i. Describe and explain the nature of the radiation that may be emitted from an excited nucleus of the moderator.

ii. The subsequent collisions of a neutron with the moderator are elastic. Describe what happens to the neutrons as a result of these subsequent collisions with the moderator.

AQA June 2013 Unit 5 Section 1 Q2

2. **a.** **i.** Explain what is meant by the *binding energy* of a nucleus.

ii. Determine the binding energy of an iron-56 nucleus ($^{56}_{26}Fe$), given that the mass of an iron-56 nucleus is 55.920 67 u.

iii. Sketch a graph to show the variation of binding energy per nucleon with nucleon number, giving approximate values on the axes, and plotting points corresponding to iron-56 and uranium-235. Then use your graph to explain why a heavy nucleus, such as uranium-235, releases energy when it undergoes fission to form two lighter nuclei.

b. The fusion of two protons to produce a deuterium nucleus ($^{2}_{1}H$) is the first stage of the fusion process that generates the energy radiated from a star. The equation for this nuclear fusion reaction is

$$^{1}_{1}H + ^{1}_{1}H = ^{2}_{1}H + ^{0}_{+1}e + \nu_e$$

i. Determine the mass difference, in atomic mass units, for this proton–proton nuclear reaction, given that the mass of a deuterium nucleus is 2.013 55 u.

ii. Determine the energy released in MeV for the proton–proton fusion reaction.

iii. Proton–proton fusion can take place in the core of a star because of the exceptionally high temperature. Explain why a high temperature is required for proton–proton fusion to take place.

3. The first artificially produced isotope, phosphorus $^{30}_{15}\text{P}$, was formed by bombarding an aluminium isotope, $^{27}_{13}\text{Al}$, with an α particle.

   a. Copy and complete the following nuclear equation by identifying the missing particle.

   $$^{27}_{13}\text{Al} + \alpha \rightarrow {}^{30}_{15}\text{P} + \ldots\ldots$$

   b. For the reaction to take place the α particle must come within a distance, d, from the centre of the aluminium nucleus. Calculate d if the nuclear reaction occurs when the α particle is given an initial kinetic energy of at least 2.18×10^{-12} J. The electrostatic potential energy between two point charges Q_1 and Q_2 is equal to $\frac{Q_1Q_2}{4\pi\varepsilon_0 r}$ where r is the separation of the charges and ε_0 is the permittivity of free space.

   AQA June 2011 Unit 5 Section 1 Q2

4. A series of decays involving the emission of two beta-minus particles and one alpha particle changes a radioactive nucleus of lead-212 into a stable nucleus of lead. What is the nucleon number of the stable nucleus of lead that is formed?

   A 207 **B** 208 **C** 209 **D** 210

5. A suitable material for the control rods in a nuclear reactor is

   A graphite **B** zirconium alloy
   C cadmium **D** steel

6. In the radioactive decay of a particular nucleus, the mass lost is 0.00254 u. What is the energy released in the decay, in joules?

   A 3.8×10^{-13} **B** 3.8×10^{-19}
   C 2.4×10^{-13} **D** 2.4×10^{-19}

ANSWERS TO IN-TEXT QUESTIONS

1 CIRCULAR MOTION

1. Since 10 minutes is the time for $\frac{1}{3}$ of a revolution, the corresponding angular displacement is equal to $\frac{1}{3} \times 2\pi = 2.1\,\text{rad}$.

2. Angular speed:

$$\omega = \frac{2\pi}{T} = \frac{2\pi}{12 \times 3600} = 1.454 \times 10^{-4}$$
$$= 1.45 \times 10^{-4}\ \text{rad s}^{-1}$$

Orbital speed:

$$v = r\omega = (6.37 \times 10^6 + 20200 \times 10^3) \times 1.454 \times 10^{-4}$$
$$= 3860\,\text{m s}^{-1}$$

3. B

4. A point on the equator completes a full circle of radius 6.37×10^6 m in 24 hours. Therefore the speed of this point, due to the Earth's rotation, is

$$v = \frac{2 \times \pi \times 6.37 \times 10^6}{24 \times 3600} = 463\,\text{m s}^{-1}$$

5. Radius of spin, r, at a latitude of 60 °N is given by $r = R \sin 30°$, where R is the radius of the Earth. This gives $r = 6.37 \times 10^6 \times \sin 30° = 3.185 \times 10^6$ m.

Linear speed, v, of a point on the Earth's surface at a latitude of 60 °N is given by

$$v = r\omega = 3.185 \times 10^6 \times \frac{2\pi}{24 \times 3600} = 232\,\text{m s}^{-1}$$

6. **a.** $F = \frac{mv^2}{r} = \frac{900 \times (30000 \div 3600)^2}{30}$
$= 2083 = 2.1$ kN

b. $5000 = \frac{mv^2}{r}$; hence maximum speed

$$v_{max} = \sqrt{\frac{5000 \times 30}{900}} = 13\,\text{m s}^{-1}$$

7. **a.** The gravitational force on the space station exerted by the Earth.

b. Centripetal acceleration

$$a = \frac{v^2}{r} = \frac{(27500 \times 10^3 \div 3600)^2}{6.37 \times 10^6 + 400 \times 10^3}$$
$$= 8.619 = 8.62\,\text{m s}^{-2}$$

Centripetal force $F = ma = 400 \times 10^3 \times 8.619$
$= 3.45 \times 10^6$ N

(Note that, since the centripetal force is provided by the force of gravity, the value for centripetal acceleration must be equal to the acceleration due to gravity at a height of 400 km above the Earth's surface.)

8. Centripetal acceleration

$$a = \frac{v^2}{r} = \frac{(2\pi \times 2.3 \times 10^{11} \div (687 \times 24 \times 3600))^2}{2.3 \times 10^{11}}$$
$$= 2.577 \times 10^{-3} = 2.6 \times 10^{-3}\,\text{m s}^{-1}$$

Therefore the Sun's gravitational field strength in the region of Mars' orbit is 2.6×10^{-3} N kg^{-1}.

9. The magnitude of vectors $\boldsymbol{v}_1$ and $\boldsymbol{v}_2$ is equal to speed v. By applying the small-angle approximation to the vector subtraction diagram of Figure 8, it can be seen that the velocity change Δv is equal to $v\Delta\theta$. Hence the centripetal acceleration can be rewritten $a = \frac{v\Delta\theta}{\Delta t}$. However, $\frac{\Delta\theta}{\Delta t} = \omega$, since angular speed ω is defined as angular displacement per unit time. This gives centripetal acceleration $a = v\omega$, which can be written as $a = \frac{v^2}{r}$ since $\omega = \frac{v}{r}$.

10. Rearranging the equation $\tan\theta = \frac{v^2}{rg}$ gives the linear speed as

$$v = \sqrt{rg\tan\theta} = \sqrt{3.3 \times 9.81 \times \tan 25°} = 3.9\ \text{m s}^{-1}$$

11. The banking angle is given by

$$\tan\theta = \frac{v^2}{rg} = \frac{(410 \times 10^3 \div 3600)^2}{5000 \times 9.81}$$

so banking angle $\theta = 15°$.

12. Assume the curved path to be part of a vertical circle. The centripetal force is equal to the car's weight minus the normal contact force of the road on the car. When the car loses contact with the ground, the contact force becomes zero. The centripetal force is therefore then equal to the weight, $mg = \frac{mv^2}{r}$, which gives the speed, v, when contact with the ground will be lost:

$$v = \sqrt{rg} = \sqrt{30 \times 9.81} = 17\,\text{m s}^{-1}$$

13. C. At the bottom of the circle, the tension is vertically upwards and the weight is vertically downwards. The tension must exceed the weight to provide a resultant centripetal force upwards. Hence the centripetal force $= T - W$.

2 OSCILLATION

1. As x is maximum when $t = 0$, use $x = A\cos(\omega t)$, with $A = 6\,\text{mm}$ and $\omega = 2\pi f = 10\,000\pi$.

Suitable values of t need to be chosen to determine values of x for plotting. (Values of $t = 1, 2, 3$, etc. are too large and represent points that are thousands of oscillations apart.)

The oscillation has a time period of $T = \frac{1}{f} = \frac{1}{5000} = 0.0002\,\text{s} = 0.2\,\text{ms}$ and it has maximum displacement at 0, 0.2, 0.4, 0.6 ms, etc. It will have a minimum (negative) displacement at 0.1, 0.3, 0.5 ms, etc., and zero displacement in between these points, at 0.05, 0.15, 0.25, 0.35 ms, etc.

See Figure 1.

2. As $x = 0$ when $t = 0$, use $x = A\sin(\omega t)$, with $A = 6\,\text{mm} = 6 \times 10^{-3}\,\text{m}$, $\omega = 2\pi \times 5000\,\text{s}^{-1}$ and $t = 0.24\,\text{ms} = 0.24 \times 10^{-3}\,\text{s}$. So

$$\begin{aligned} x = A\sin(\omega t) &= 6 \times 10^{-3}\sin(2\pi \times 5000 \times 0.24 \times 10^{-3}) \\ &= 6 \times 10^{-3}\sin(7.54\,\text{rad}) = 6 \times 10^{-3} \times 0.95 \\ &= 5.7 \times 10^{-3}\,\text{m} \end{aligned}$$

3. See Figure 2.

4. Since SHM acceleration $a = -\omega^2 x$, the maximum acceleration is $a_{max} = \omega^2 A = \left(\frac{2\pi}{T}\right)^2 A$. Therefore

$$a_{max} = \left(\frac{2\pi}{0.017}\right)^2 \times 0.05 = 6800\,\text{m s}^{-2}$$

Velocity is

$$v = \pm 2\pi f\sqrt{A^2 - x^2} = \pm\frac{2\pi}{T}\sqrt{0.05^2 - 0.02^2} = \pm 17\,\text{m s}^{-1}$$

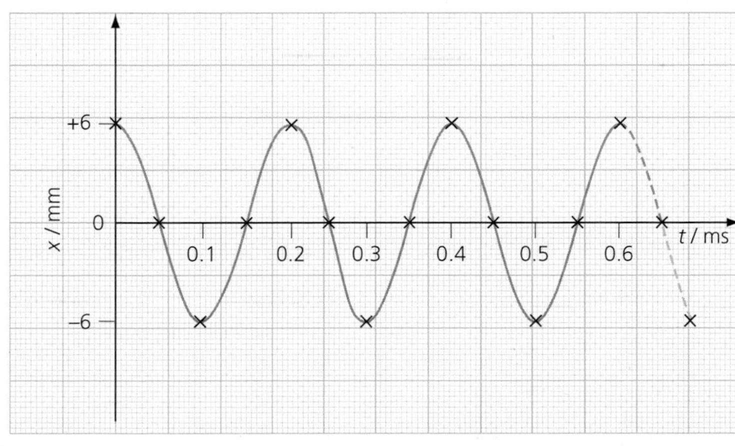

Figure 1 *Answer to question 1*

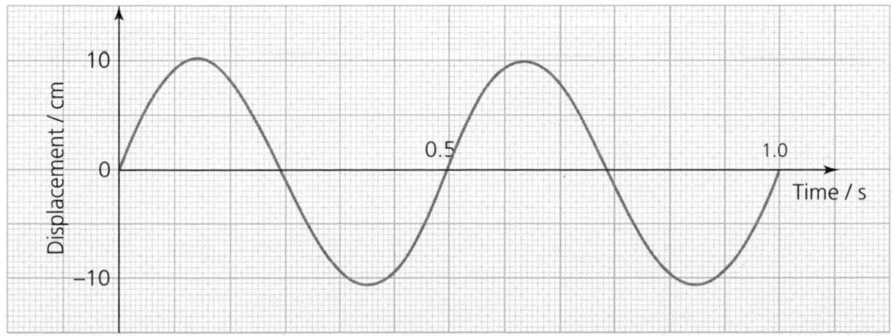

Figure 2 *Answer to question 3*

5. Acceleration
$a = -\omega^2 x = -(2\pi f)^2 x = -(2\pi \times 2)^2 \times 0.2 = -32\,\text{m s}^{-2}$

6. a. The gradient of the line is equal to $\frac{-800}{20} = -40\,\text{s}^{-2}$ and therefore the equation relating a and x is $a = -40x$.

b. Since, for SHM, $a = -\omega^2 x$, the gradient of the graph is equal to $-\omega^2$. However, $\omega = 2\pi f$, so $\omega^2 = 4\pi^2 f^2$, which shows that, if the frequency is doubled, the magnitude of the gradient increases by a factor of 4.

7. Time period $T = 2\pi\sqrt{\frac{m}{k}} = 2\pi\sqrt{\frac{0.25}{30}} = 0.57\,\text{s}$

8. a. $T = 2\pi\sqrt{\frac{m}{k}} = 2\pi\sqrt{\frac{0.5}{20}} = 1.0\,\text{s}$

b. $T = 2\pi\sqrt{\frac{m}{k}} = 2\pi\sqrt{\frac{0.5}{80}} = 0.5\,\text{s}$

9. Phase difference in terms of fraction of a cycle is $\frac{0.4}{1.2} = \frac{1}{3}$ cycle. Since one cycle is 2π radians, the phase difference between the two oscillating systems is $\frac{2}{3}\pi$ rad.

10. Time period $T = 2\pi\sqrt{\frac{m}{k}}$. Squaring both sides and rearranging gives

$$k = \frac{4\pi^2 m}{T^2} = \frac{4\pi^2 \times 0.2}{1.46^2} = 3.7\,\text{N m}^{-1}$$

11. B. ($T = 2\pi\sqrt{\frac{m}{k}}$ so doubling m and halving k increases $\frac{m}{k}$ by a factor of 4, and hence $\sqrt{\frac{m}{k}}$ by a factor of 2.)

12. Rearranging the equation for time period $T = 2\pi\sqrt{\frac{l}{g}}$ to make l the subject:

$$l = \frac{T^2 g}{4\pi^2} = \frac{9.81}{4\pi^2} = 0.25\,\text{m}$$

13. B. ($T = 2\pi\sqrt{\frac{l}{g}}$ so doubling the length would increase T by a factor of $\sqrt{2}$.)

14. a. The additional gravitational potential energy is converted to kinetic energy as the mass passes through the equilibrium position. As the mass decelerates as it approaches the other extreme position, the kinetic energy is converted to gravitational potential energy.

b. When $t = 0$ the mass is at its extreme position and its potential energy is at a maximum. The graph of potential energy versus time varies as a cosine-squared graph, as shown in Figure 3.

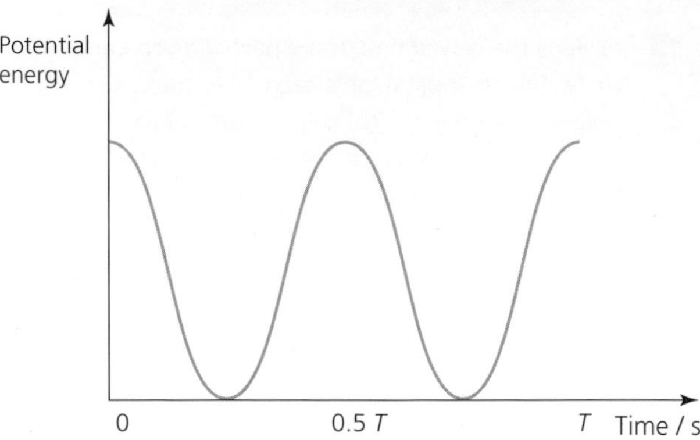

Figure 3 *Answer to question 14(b)*

c. i. The additional gravitational potential energy is $\Delta E_p = mg\Delta h = 0.5 \times 9.81 \times 0.05 = 0.245\,\text{J}$.

ii. As the mass passes through the equilibrium position, all the additional potential energy is converted to kinetic energy, so the maximum kinetic energy is 0.245 J.

iii. The maximum kinetic energy is $\frac{1}{2}mv^2 = 0.245$, which rearranges to give the maximum speed:

$$v^2 = \frac{2 \times 0.245}{0.5} = 0.98$$

$$v = 0.99\,\text{m s}^{-1}$$

15. a. Elastic energy at one extreme changes to kinetic energy at the equilibrium position and then back to elastic energy at the other extreme.

b. Maximum speed $v_{max} = A\omega = 1 \times 10^{-3} \times 2 \times \pi \times 256$
$= 1.6\,\text{m s}^{-1}$

16. Quarter of a cycle, $\frac{\pi}{2}$ radians.

17. The resonant frequency decreases as damping increases.

3 THERMAL PHYSICS

1. Since the temperature of the brass block is increasing, the vibrational kinetic energy of the atoms must have increased. The block will have expanded very slightly, so the potential energy of the atoms will also have increased.

2. **a.** Energy needed to heat the pan is

$Q = mc\,\Delta\theta = 2.0 \times 461 \times 82 = 75600\,\text{J}$

Energy needed to heat the milk is

$Q = mc\,\Delta\theta = \rho V c\,\Delta\theta$

$= 1033 \times 5.7 \times 10^{-4} \times 3930 \times 82 = 189750\,\text{J}$

Total energy supplied is

$75600 + 189750 = 265350\,\text{J}$

Power supplied

$= \dfrac{265350}{120} = 2.21 \times 10^3\,\text{W} = 2.2\text{ kW}$

b. A flame or a conventional electric ring would heat the surrounding air as well as the pan. An induction hob does not itself generate heat. A conventional electric ring would remain hot after the cooking was complete. With an induction hob, there is no heating effect once the pan has been removed.

3. Time taken t = (energy transfer)/power, so

$$t = \frac{\text{energy supplied to heat kettle} + \text{energy supplied to heat water}}{\text{power}}$$

$$= \frac{(0.5 \times 502 \times 82) + (1.0 \times 4190 \times 82)}{2200} = 166\,\text{s}$$

4. **a.** As the water warms after the brass is placed in it, some of its energy will be lost through evaporation. Some will also be transferred through the sides and base of the beaker, unless the insulation is perfect.

b. The unwanted energy transfers from the water mean that the final temperature θ_2 is too low. This will cause a systematic error in the final result.

c. Ensure that the material of the beaker is as good an insulator as possible and include an insulated lid.

5. Equating the energy lost by the steel to the energy gained by the water:

$mc_{\text{steel}} \times (100 - \theta_2) = 1000 \times V \times 4190 \times (\theta_2 - \theta$

Rearranging gives

$$(100 - \theta_2) = \frac{1000 \times 100 \times 10^{-6} \times 4190 \times (\theta_2 - 18.2)}{50 \times 10^{-3} \times 502}$$

Therefore

$$\begin{aligned} 100 - \theta_2 &= 16.69 \times (\theta_2 - 18.2) \\ &= 16.69\theta_2 - 303.8 \\ 403.8 &= 17.69\theta_2 \end{aligned}$$

which gives the final temperature as $\theta_2 = 22.8\,°\text{C}$.

6. The temperature difference is

$$\Delta\theta = \frac{gh}{c} = \frac{9.81 \times 51}{4190} = 0.12\,°\text{C}$$

7. **a.** From the equation in the text, the specific heat capacity of lead is

$$c = \frac{Nmgh}{m\Delta\theta} = \frac{Ngh}{\Delta\theta} = \frac{50 \times 9.81 \times 1.0}{2.6} = 190\,\text{J kg}^{-1}\,°\text{C}^{-}$$

b. Since the work that is actually done on the lead shot will in reality be less than $Nmgh$, the measured temperature rise $\Delta\theta$ is smaller than it would otherwise be, making the value for c an overestimate.

8. Internal energy transferred to surroundings per second is

$$Q = mc\,\Delta\theta = \frac{80}{60} \times 3900 \times (94 - 81) = 68000\,\text{W}$$

9. Power supplied = (energy transferred)/time, so

$$\begin{aligned} \text{power} &= \frac{mc\,\Delta\theta}{t} \\ &= \frac{4 \times 10^{-5} \times 1000 \times 4190 \times (39 - 14)}{60} \\ &= 6983\,\text{W} = 7.0\,\text{kW} \end{aligned}$$

10. Power output of the radiator is

P = energy loss per second from the radiator water

= mass flowing per second $\times c \times \Delta\theta$

Substituting data gives power

$P = 0.013 \times 4190 \times 20 = 1089\,\text{W} = 1.1\,\text{kW}$

11. The total energy needed is equal to the energy needed to raise the temperature of the copper to its melting point plus the energy needed to change the state of the copper. So

$$\begin{aligned} \text{total energy} &= mc\,\Delta\theta + ml \\ &= [1.0 \times 385 \times (1083 - 20)] + [1.0 \times 2.07 \times 10^5] \\ &= 6.2 \times 10^5\,\text{J} \end{aligned}$$

12. **a.** Use m = 20 g to denote the mass of ice and M = 200 g to denote the mass of water. The energy needed to raise the temperature of the ice from $-10\,°\text{C}$ to $0\,°\text{C}$ is

$Q_1 = mc_{\text{ice}}\,\Delta\theta = 0.020 \times 2000 \times 10 = 400\text{ J}$

The energy needed to melt the ice is

$Q_2 = ml = 0.020 \times 3.34 \times 10^5 = 6680\,\text{J}$

This energy all comes from the water, so the temperature drop of the water is

$$\Delta\theta = \frac{Q}{Mc_{\text{water}}} = \frac{Q_1 + Q_2}{Mc_{\text{water}}} = \frac{400 + 6680}{0.2 \times 4190} = 8.4°\text{C}$$

b. No, because the water is now mixed with the melted ice. There is $M = 200$ g of water at 11.6 °C and $m = 20$ g of water at 0 °C. Heat will pass between them until an equilibrium final temperature θ_f is reached:

$$Q = mc_{\text{water}}\theta_f = Mc_{\text{water}}(11.6 - \theta_f)$$

giving

$$\theta_f = \frac{11.6M}{m + M} = \frac{11.6 \times 200}{220} = 10.5°\text{C}$$

13. **a.** The gradient represents the rate of temperature change.

b. During the first change of state, the molecules that previously were in the solid state and restricted to vibrational motion become able to rotate and move around.

c. During the second change of state, the total random kinetic energy is unchanged but the random potential energy decreases.

14. In 1 minute, $Q = 500 \times 60$ s. Rearranging $Q = ml$ gives the mass of sweat per minute:

$$m = \frac{Q}{l} = \frac{500 \times 60}{2.3 \times 10^6} = 0.013\,\text{kg}\,\text{min}^{-1}$$

15. The energy needed per second to heat 200 kg of water from 15 °C to 100 °C and convert it to steam at 100 °C is

$$mc\Delta\theta + ml = (200 \times 4190 \times 85) + (200 \times 2.3 \times 10^6) = 5.3 \times 10^8\,\text{W}$$

16. Apply Boyle's law $p_1V_1 = p_2V_2$. Substituting data gives

$$1.0 \times 10^5 \times 15 = 1.4 \times 10^5 \times V_2$$

which gives the new volume $V_2 = 11\,\text{cm}^3$.

17. Applying Charles's law and substituting data:

$$\frac{V_1}{T_1} = \frac{V_2}{T_2}$$

$$\frac{3.5 \times 10^{-3}}{100 + 273.15} = \frac{V_2}{250 + 273.15}$$

$$\frac{3.5 \times 10^{-3}}{373.15} = \frac{V_2}{523.15}$$

gives the new volume $V_2 = 4.9 \times 10^{-3}\,\text{m}^3$.

18. Since the volume is constant, we can apply the pressure law. Substituting data:

$$\frac{p_1}{T_1} = \frac{p_2}{T_2}$$

$$\frac{1.1 \times 10^5}{300} = \frac{p_2}{400}$$

gives $p_2 = 1.5 \times 10^5$ Pa.

19. **a.** Assuming that the air behaves as an ideal gas, the number of moles, n, is

$$n = \frac{pV}{RT} = \frac{2.0 \times 10^7 \times 2.2 \times 10^{-2}}{8.31 \times (10 + 273)} = 187.1 = 187\,\text{mol}$$

The mass of air required can be found by multiplying the number of moles by the molar mass:

$$\text{mass of air required} = 187.1 \times 0.029 = 5.426 = 5.4\,\text{kg}$$

b. The number of molecules N is

$$N = nN_A = 187 \times 6.02 \times 1023 = 1.13 \times 1026$$

20. The ideal gas equation $pV = nRT$ can be rearranged to

$$n = \frac{pV}{RT}$$

Number of moles released is

$$n_1 - n_2 = \frac{p_1V}{RT} - \frac{p_2V}{RT} = (p_1 - p_2) \times \frac{V}{RT} = (90 \times 10^3) \times \frac{7.2 \times 10^{-2}}{8.31 \times 290} = 2.7\,\text{mol}$$

21. Using the rearranged equation gives

$$V_2 = \frac{p_1V_1T_2}{T_1p_2} = \frac{101 \times 10^3 \times 1 \times (273 + 4)}{(273 + 20) \times 79 \times 10^3} = 1.2\,\text{m}^3$$

22. The total pressure at a depth of 45 m is 550 kPa, 450 kPa due to the water and 100 kPa due to atmospheric pressure.

The initial conditions of the gas are: $p_1 = 550$ kPa, $V_1 = 2.0 \times 10^{-5}\,\text{m}^3$, $T_1 = 278$ K

The final conditions of the gas are: $p_2 = 100$ kPa, V_2 is unknown, $T_2 = 283$ K

The number of moles in the bubble is constant, so $\frac{p_1V_1}{T_1} = \frac{p_2V_2}{T_2}$ can be applied.

Rearranging gives

$$V_2 = \frac{p_1V_1T_2}{T_1p_2} = \frac{550 \times 10^3 \times 2 \times 10^{-5} \times 283}{278 \times 100 \times 10^3} = 1.1 \times 10^{-4}\,\text{m}^3$$

23. The 1.5 litre of water increases to $1.5 \times 1600 = 2400$ litre of steam, so the volume increase is 2398.5 litre, which is equivalent to 2.4 m^3. Work done in expanding against the atmosphere is given by $W = p\Delta V = 1.0 \times 10^5 \times 2.4 = 2.4 \times 10^5$ J.

24. a. At any point in time, a larger particle would be getting hit by many more molecules, so the effect of a slight imbalance in the numbers hitting any particular bit of the particle will be insignificant, so Brownian motion will not be observed.

b. The air molecules are moving faster at higher temperature and so have more impact on the smoke particles.

c. The lack of direct evidence for the existence atoms, owing to their very small size, meant that for a long time the atomic theory could not be verified conclusively.

25. a. The average speed of the molecules is unchanged because the temperature is constant.

b. There are more collisions of molecules with the cylinder walls per second because the molecules are contained in a smaller space and travel shorter distances between collisions.

c. The pressure increases because the force on the cylinder walls has increased because of the greater number of collisions with the wall every second.

26. a. Since the temperature has increased, the average speed of the molecules has increased.

b. The average force on the piston increases because faster molecules experience a greater change of momentum on colliding then bouncing off the piston.

c. If the force exerted on the piston by the molecules increases but the pressure is unchanged, the molecules must be hitting the piston less frequently.

27. a. The gas temperature has been raised, so the average speed of the molecules has increased.

b. A faster molecule exerts a bigger force on the container walls because it experiences a larger momentum change on hitting and bouncing off the container walls.

c. The molecules are on average travelling faster, so even though the distance travelled between collisions is unchanged, the time between collisions with the container walls is reduced, making collisions more frequent.

d. Both the increased force on the container walls and the increased frequency of collisions cause the gas pressure to increase.

28. a. $(c_{rms})^2 = 1.639 \times 10^5\,\text{m}^2\,\text{s}^{-2}$, which gives $c_{rms} = 405\,\text{m}\,\text{s}^{-1}$

b. Average speed $= 391\,\text{m}\,\text{s}^{-1}$

c. There is a 3.5% difference between the two values.

29. Rearranging $pV = \frac{1}{3}Nm(c_{rms})^2$ gives

$$c_{rms} = \sqrt{\frac{3pV}{Nm}} = \sqrt{\frac{3 \times 101 \times 10^3 \times 1}{1.84}} = 405\,\text{m}\,\text{s}^{-1}$$

30. First write down and vertically align the equations for the individual atoms. The vertical dots represent all the atoms between molecule 2 and molecule N:

$$\begin{aligned} c_1^2 &= u_1^2 + v_1^2 + w_1^2 \\ c_2^2 &= u_2^2 + v_2^2 + w_2^2 \\ \vdots &= \vdots \\ c_N^2 &= u_N^2 + v_N^2 + w_N^2 \end{aligned}$$

Now add up all the terms on the left hand side of the equation and similarly the right hand side:

$$\begin{aligned} c_1^2 + c_2^2 + \ldots + c_N^2 &= (u_1^2 + u_2^2 + \ldots + u_N^2) \\ &+ (v_1^2 + v_2^2 + \ldots + v_N^2) + (w_1^2 + w_2^2 + \ldots + w_N^2) \end{aligned}$$

Recalling from the main text that

$$(c_{rms})^2 = \frac{c_1^2 + c_2^2 + \cdots + c_N^2}{N}$$

and similarly for the u, v and w components, the left hand side terms add up to $N(c_{rms})^2$ and the right hand side terms add up to $N(u_{rms})^2 + N(v_{rms})^2 + N(w_{rms})^2$. Therefore the equation becomes

$$N(c_{rms})^2 = N(u_{rms})^2 + N(v_{rms})^2 + N(w_{rms})^2$$

Dividing throughout by N gives

$$(c_{rms})^2 = (u_{rms})^2 + (v_{rms})^2 + (w_{rms})^2$$

showing that Pythagoras's theorem can be applied to mean square speeds.

31. a. The ideal gas equation $pV = NkT$ can be rearranged to $N = \frac{pV}{kT}$, which gives the number of atoms as

$$N = \frac{210 \times 10^3 \times 0.51 \times 10^{-3}}{1.38 \times 10^{-23} \times 290} = 2.676 \times 10^{22}$$
$$= 2.7 \times 10^{22}\,\text{atoms}$$

b. Total kinetic energy is

$$N \times \frac{3}{2}kT = 2.676 \times 10^{22} \times 1.5 \times 1.38 \times 10^{-23} \times 290$$
$$= 161\,\text{J}$$

4 GRAVITY

1. Gravitational force is

$$F = \frac{Gm_1m_2}{r^2} = \frac{6.67\times10^{-11}\times1.31\times10^{22}\times1.52\times10^{21}}{(19640\times10^3)^2}$$
$$= 3.44\times10^{18}\,\text{N}$$

2. Gravitational force is

$$F = \frac{Gm_1m_2}{r^2} = \frac{6.67\times10^{-11}\times5.97\times10^{24}\times1123}{(6.37\times10^6+685\times10^3)^2}$$
$$= 8980\,\text{N}$$

3. Typical values: mass of each person 60 kg; separation of their centres of mass 1 m. So

$$F = \frac{Gm_1m_2}{r^2} = \frac{6.67\times10^{-11}\times60\times60}{1^2} = 2.40\times10^{-7}\,\text{N}$$
$$= 2\times10^{-7}\,\text{N to 1 s.f.}$$

4. Gravitational field strength is

$$g = \frac{GM}{r^2} = \frac{6.67\times10^{-11}\times5.97\times10^{24}}{(6.37\times10^6+685\times10^3)^2} = 8.00\,\text{N kg}^{-1}$$

5. The gravitational field strength at the Earth's surface is 9.81 N kg^{-1}. Half this value is 4.905 N kg^{-1}. The equation $g = \frac{GM}{r^2}$ rearranges to give

$$r = \sqrt{\frac{GM}{g}} = \sqrt{\frac{6.67\times10^{-11}\times5.97\times10^{24}}{4.905}} = 9.01\times10^6\,\text{m}$$

Therefore, the height above the Earth's surface is $9.01\times10^6 - 6.37\times10^6 = 2.64\times10^6\,\text{m}$.

Alternative method, using ratios:
Since $g \propto \frac{1}{r^2}$ we can write $\frac{g_h}{g_R} = \frac{R^2}{r^2} = \frac{1}{2}$, with R representing the Earth's radius and r the radius at which gravitational field strength is half the surface value. Therefore $r^2 = 2R^2$ and radius $r = \sqrt{2}\times R = 9.01\times10^6\,\text{m}$ giving height $h = r - R = 2.64\times10^6\,\text{m}$ as before.

6. Equating the formula for the gravitational field strengths of the Sun and the Earth gives

$$\frac{r_1}{r_2} = \sqrt{\frac{M_E}{M_S}} = \sqrt{\frac{5.972\times10^{24}}{1.989\times10^{30}}} = 0.001733$$

where r_2 is the distance of the Sun from the neutral point and r_1 is the distance of the Earth to the neutral point. Therefore $r_1 = 0.001\,733r_2$. Since $r_1 + r_2 = 1.4960\times10^{11}$, we can write $0.001733r_2 + r_2 = 1.4960\times10^{11}$, which gives

$$1.001733r_2 = 1.4960\times10^{11}$$

Therefore $r_2 = 1.4934\times10^{11}$

and so the distance of the neutral point from the Earth is equal to

$$1.4960\times10^{11} - 1.4934\times10^{11} = 2.6\times10^8\,\text{m}$$

7. Since the gravitational potential at a point beyond the influence of the asteroid's field is zero, the probe needs to move through a gravitational potential difference of

$$\Delta V = 40\times10^3\,\text{J kg}^{-1}$$

Minimum output energy = work done in escaping the field

$$\Delta W = m\Delta V = 510\times40\times10^3 = 20\,\text{MJ}$$

8. A (since $V \propto \frac{1}{r}$)

9. Output energy = work done, so
$$\Delta W = m\Delta V = 800\times(34.5-33.9) = 480\,\text{MJ}$$

10. The equipotential surfaces are equidistant planes, parallel to the Earth's surface, as shown in Figure 4.

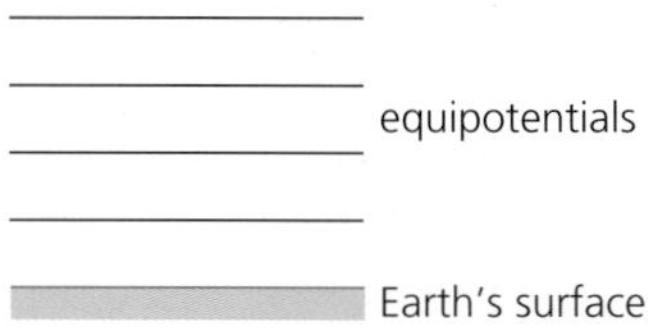

equipotentials

Earth's surface

Figure 4

11. The magnitude of the potential gradient at a point in the Earth's field, $\frac{\Delta V}{\Delta r}$, equals the gravitational field strength, g. Therefore, the gravitational force on the satellite is $mg = 900\times7.9 = 7100\,\text{N}$.

12. The area of each large grid square is equal to $2\times\frac{1}{2}R = 2\times\frac{1}{2}\times6.37\times10^6 = 6.37\times10^6\,\text{J}$. The approximate number of large squares between distance R and distance $4R$ is 8, which is equivalent to 50 MJ kg^{-1}. So the gravitational potential changes by 50 MJ kg^{-1} and therefore the work done to move a mass of 1 kg from the Earth's surface to a height of three Earth radii above the Earth's surface is approximately 50 MJ.

13. The gradient of the graph gives the gravitational field strength. Using a ruler to estimate the gradient at a distance equal to 2.5R from the Earth's centre gives a value of 1.6 N kg^{-1}.

14.
$$W = \int_r^\infty \frac{GM}{r^2}\,dr = -GM\left[\frac{1}{r}\right]_r^\infty$$
$$= -GM\left[\frac{1}{\infty}\right] - \left(-GM\left[\frac{1}{r}\right]\right) = \frac{GM}{r}$$

15. Angular speed of the Earth in its orbit is
$$\omega = \frac{2\pi}{T} = \frac{2\pi}{365\times24\times3600} = 1.99\times10^{-7}\,\text{rad s}^{-1}$$

Thus mass of the Sun is

$$m_1 = \frac{r^3\omega^2}{G} = \frac{(1.5\times10^{11})^3\times(1.99\times10^{-7})^2}{6.67\times10^{-11}} = 2.0\times10^{30}\,\text{kg}$$

16. Angular speed of the solar system is

$$\omega = \frac{2\pi}{T} = \frac{2\pi}{240\times10^6\times365\times24\times3600}$$
$$= 8.3\times10^{-16}\,\text{rad s}^{-1}$$

and thus the mass within its orbit is

$$m_1 = \frac{r^3\omega^2}{G} = \frac{(27000\times9.46\times10^{15})^3\times(8.3\times10^{-16})^2}{6.67\times10^{-11}}$$
$$= 1.7\times10^{41}\,\text{kg}$$

17. Orbital speed is

$$v = \sqrt{\frac{GM}{r}} = \sqrt{\frac{6.67\times10^{-11}\times5.97\times10^{24}}{6.37\times10^6+547\times10^3}}$$
$$= 7.59\times10^3\,\text{m s}^{-1}$$

18. Angular speed of a GPS satellite is

$$\omega = \frac{2\pi}{T} = \frac{2\pi}{12\times3600} = 1.454\times10^{-4}\,\text{rad s}^{-1}$$

Equating Newton's law and centripetal force formula, and rearranging as in the text gives

$$\frac{GM}{\omega^2} = r^3$$

Thus

$$r = \sqrt[3]{\frac{GM}{\omega^2}} = \sqrt[3]{\frac{6.67\times10^{-11}\times5.97\times10^{24}}{(1.454\times10^{-4})^2}}$$
$$= 2.66\times10^7\,\text{m}$$

which gives an altitude of

$$h = 2.66\times10^7 - 6.37\times10^6 = 2.02\times10^7\,\text{m}$$

19. a. Equating Newton's law to the formula for the required centripetal force for the satellite's orbit gives

$$\frac{GMm}{r^2} = mr\omega^2$$

Substituting $\omega = \frac{2\pi}{T}$ into the above equation gives

$$\frac{GM}{r^2} = r\left(\frac{2\pi}{T}\right)^2$$

which rearranges to gives

$$T^2 = \frac{4\pi^2r^3}{GM} = \frac{4\pi^2\times(6.37\times10^6+810\times10^3)^3}{6.67\times10^{-11}\times5.97\times10^{24}}$$
$$= 3.67\times10^7\,\text{s}^2$$

so $T = 6.06\times10^3\,\text{s}$

b. Number of orbits in one day $= \frac{24\times3600}{6.06\times10^3} = 14$

c. Orbital speed is

$$\frac{\text{circumference of orbit}}{T} = \frac{2\pi r}{T}$$
$$= \frac{2\pi(6.37\times10^6+810\times10^3)}{6.06\times10^3}$$
$$= 7.44\times10^3\,\text{m s}^{-1}$$

20. a. Kinetic energy of *SMAP* is $E_k^{SMAP} = \frac{GMm}{2r}$, so

$$E_k^{SMAP} = \frac{6.67\times10^{-11}\times5.97\times10^{24}\times1123}{2\times(6.37\times10^6+685\times10^3)} = 3.17\times10^{10}\,\text{J}$$

b. Kinetic energy of Earth in orbit around the Sun $E_k^{Earth} = \frac{GMm}{2r}$, so

$$E_k^{Earth} = \frac{6.67\times10^{-11}\times1.99\times10^{30}\times5.97\times10^{24}}{2\times1.50\times10^{11}}$$
$$= 2.64\times10^{33}\,\text{J}$$

21. Estimate of star's mass ≈ solar mass ≈ 10^{30} kg.
Estimate of planet's mass ≈ Jupiter's mass ≈ 10^{27} kg.
Estimate of orbital radius
≈ 10 × Earth's orbital radius ≈ $10\times1.5\times10^{11}$ m.
Kinetic energy of planet

$$\approx \frac{GMm}{2r} \approx \frac{7\times10^{-11}\times10^{30}\times10^{27}}{2\times10\times1.5\times10^{11}} = 2.3\times10^{34} \approx 10^{34}\,\text{J}$$

22. Escape velocity is

$$v = \sqrt{\frac{2GM}{R}} = \sqrt{\frac{2\times6.67\times10^{-11}\times7.35\times10^{22}}{1.74\times10^6}}$$
$$= 2.37\times10^3\,\text{m s}^{-1}$$

5 ELECTRIC FIELDS

1. a. It is negative, the same as the charge on the electroscope. We can tell this because electrons from the electroscope's disc are repelled towards the gold leaf, making it diverge further.

b. The gold leaf would fall, because electrons on the leaf and plate would be attracted up towards the disc, so reducing the repulsive effect between plate and leaf.

2. When the negatively charged rod is brought near to the sphere, electrons on the sphere move through the metal away from the rod. When the sphere is momentarily earthed, the electrons at the surface of the sphere flow to earth. When the rod is removed, the charge distribution on the sphere becomes even again, and there is a lack of electrons, so the sphere has a net positive charge.

3. Force between spheres is

$$F = \frac{Q_1Q_2}{4\pi\varepsilon_0 r^2} = \frac{(9\times10^{-9})^2}{4\pi\times8.85\times10^{-12}\times0.12^2}$$
$$= 5.058\times10^{-5}\,\text{N}$$

Balance reading

$$= \frac{F}{g} = \frac{5.058\times10^{-5}}{9.81} = 5.16\times10^{-6}\,\text{kg} = 0.005\,\text{g}$$

4. D (New charges are $+3Q$ and $+Q$ so force is repulsive and three times bigger)

5. Since $F = \frac{Q_1Q_2}{4\pi\varepsilon_0 r^2}$, doubling each charge increases the force 4 times, then halving the separation increases the force 4 times, making the force 16 times bigger. So the new force equals $16F$.

6. a. Electrostatic force between two protons is

$$F_e = \frac{Q_1Q_2}{4\pi\varepsilon_0 r^2} = \frac{(1.6\times10^{-19})^2}{4\pi\times8.85\times10^{-12}\times(2\times10^{-15})^2}$$
$$= 57\,\text{N}$$

This is repulsive.

b. Gravitational force between two protons is

$$F_g = \frac{Gm_1m_2}{r^2} = \frac{6.67\times10^{-11}\times(1.67\times10^{-27})^2}{(2\times10^{-15})^2}$$
$$= 4.7\times10^{-35}\,\text{N}$$

This is attractive.

The gravitational force between the protons is negligible compared with the electrostatic force.

7.
$$F = \frac{Q_1Q_2}{4\pi\varepsilon_0 r^2} = \frac{2\times79\times(1.6\times10^{-19})^2}{4\pi\times8.85\times10^{-12}\times(50\times10^{-15})^2}$$
$$= 14.55\,\text{N} = 15\,\text{N (to 2 s.f.)}$$

8. Electric field strength is

$$E = \frac{Q}{4\pi\varepsilon_0 r^2} = \frac{79\times1.6\times10^{-19}}{4\pi\times8.85\times10^{-12}\times(50\times10^{-15})^2}$$
$$= 4.5\times10^{19}\,\text{NC}^{-1}$$

9. a. Electric field strength is

$$E = \frac{Q}{4\pi\varepsilon_0 r^2} = \frac{20\times10^{-9}}{4\pi\times8.85\times10^{-12}\times0.2^2}$$
$$= 4.5\times10^3\,\text{NC}^{-1}$$

b. Force on the ion is

$$F = EQ = 4.4\times10^3\times1.6\times10^{-19} = 7.2\times10^{-16}\,\text{N}$$

towards the sphere.

10. Electric field strength at P due to either one of the charges is

$$E = \frac{Q}{4\pi\varepsilon_0 r^2} = \frac{5\times10^{-9}}{4\pi\times8.85\times10^{-12}\times0.04^2}$$
$$= 2.81\times10^4\,\text{NC}^{-1}$$

Field strength is a vector, so the field strength at P due to both charges is given by vector addition. By Pythagoras's theorem, the resultant is

$$\sqrt{(2.81\times10^4)^2 + (2.81\times10^4)^2} = 4.0\times10^4\,\text{NC}^{-1}$$

acting at angle of 45° as in Figure 5.

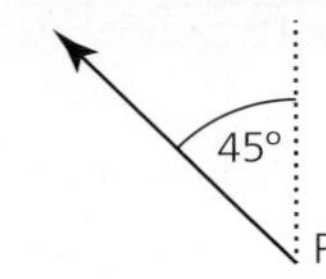

Figure 5

11.

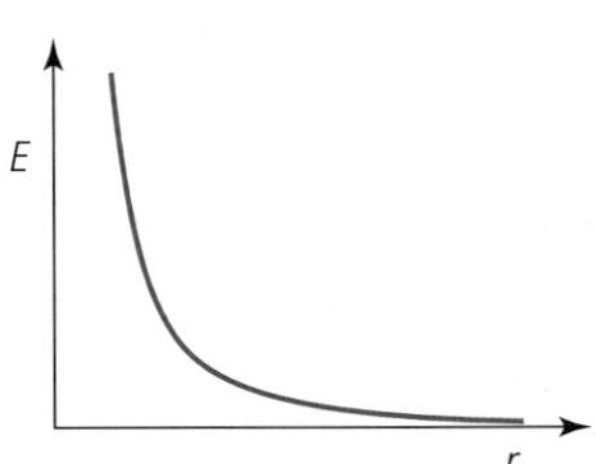

Figure 6

See Figure 6. This is an inverse square graph.

12. a. Acceleration is

$$a = \frac{F}{m} = \frac{EQ}{m} = \frac{Ve}{md} = \frac{50\times1.6\times10^{-19}}{9.11\times10^{-31}\times0.15}$$
$$= 5.9\times10^{13}\,\text{ms}^{-2}$$

b. Work done by electric field in accelerating the electron through a potential difference V is

$W = VQ = Ve = 50\times1.6\times10^{-19} = 8.0\times10^{-18}$ J. Since work done = energy transferred, the kinetic energy gained by the electron = 8.0×10^{-18} J.

13. a. From Newton's second law, $F = ma$, acceleration is

$$a = \frac{F}{m} = \frac{EQ}{m} = \frac{25\times10^3\times1.6\times10^{-19}}{9.11\times10^{-31}}$$
$$= 4.4\times10^{15}\,\text{ms}^{-2}$$

b. Work done

$W = VQ = 2000\times1.6\times10^{-19} = 3.2\times10^{-16}$ J. Since work done = energy transferred, the kinetic energy gained = 3.2×10^{-16} J.

14. a. i. $E = \frac{V}{d} = \frac{50}{0.05} = 1000\,\text{Vm}^{-1}$

ii. Time $t = \frac{\text{length of plate}}{\text{horizontal speed}} = \frac{0.08}{4.0\times10^6} = 2.0\times10^{-8}$ s

b. i. Vertical acceleration is

$$a = \frac{F}{m} = \frac{EQ}{m} = \frac{1000\times1.6\times10^{-19}}{9.11\times10^{-31}} = 1.756\times10^{14}$$
$$= 1.8\times10^{14}\,\text{ms}^{-2}$$

ii. Vertical velocity is

$$v = at = 1.756\times10^{14}\times2.0\times10^{-8}$$
$$= 3.513\times10^6 = 3.5\times10^6\,\text{ms}^{-1}$$

iii. Vertical displacement is

$$s = \frac{1}{2}at^2 = 0.5 \times 1.756 \times 10^{14} \times \left(2.0 \times 10^{-8}\right)^2$$
$$= 3.512 \times 10^{-2} = 0.035\text{m}$$

iv. Velocity on leaving the field is

$$\sqrt{(4 \times 10^6)^2 + (3.513 \times 10^6)^2} = 5.323 \times 10^6$$
$$= 5.3 \times 10^6\,\text{ms}^{-1}$$

at an angle θ to the horizontal given by

$$\theta = \tan^{-1}\left(\frac{\text{vertical velocity}}{\text{horizontal velocity}}\right)$$
$$= \tan^{-1}\left(\frac{3.513 \times 10^6}{4.0 \times 10^6}\right) = 41°$$

15. Kinetic energy of each electron = work done on electron by the electric fields, so

$$E_k = 200 \times V \times e = 200 \times 25\,000 \times 1.6 \times 10^{-19}$$
$$= 8.0 \times 10^{-13}\,\text{J}$$

The equation $E_k = (\gamma - 1)m_0c^2$ can be rearranged to give

$$(\gamma - 1) = \frac{E_k}{m_0c^2} = \frac{8.0 \times 10^{-13}}{9.11 \times 10^{-31} \times 9 \times 10^{16}} = 9.76$$

Then, using the equation for γ, we have

$$\gamma = 10.76 = \frac{1}{\sqrt{1 - \frac{v^2}{c^2}}}$$

which finally gives the electron speed as $v = 2.99 \times 10^8\,\text{ms}^{-1}$.

16. Potential at P due to upper charge is

$$V = \frac{Q}{4\pi\varepsilon_0 r} = \frac{10 \times 10^{-9}}{4\pi \times 8.85 \times 10^{-12} \times 0.1} = 900\text{V}$$

Potential at P due to lower charge is also 900V.

Potential at P due to middle charge is

$$V = \frac{Q}{4\pi\varepsilon_0 r} = \frac{10 \times 10^{-9}}{4\pi \times 8.85 \times 10^{-12} \times 0.08} = 1124\text{V}$$

Total potential at P is

$$V_{\text{total}} = 900 + 900 + 1124 = 2900\text{V}$$

17. Suppose the point where the potential is zero is at a distance x from the –5.0 nC charge. Then

$$\frac{-5 \times 10^{-9}}{4\pi\varepsilon_0 x} + \frac{8 \times 10^{-9}}{4\pi\varepsilon_0(0.04 - x)} = 0$$

Hence,

$$\frac{8}{(0.04 - x)} = \frac{5}{x}$$

which rearranges to

$$8x = 0.2 - 5x$$

which gives $x = 15.4$mm.

18. a. i. $V_{AB} = 0$

ii. $V_{BC} = 150 - 100 = 50\text{V}$

iii. $V_{CD} = 100 - 50 = 50\text{V}$

iv. $V_{AD} = 150 - 50 = 100\text{V}$

b. i. $\Delta W = Q\Delta V = 0$

ii. $\Delta W = Q\Delta V = 1 \times 10^{-9} \times 50 = 5 \times 10^{-8}\,\text{J}$

iii. $\Delta W = 5 \times 10^{-8}\,\text{J}$

iv. $\Delta W = Q\Delta V = 1 \times 10^{-9} \times 100 = 1 \times 10^{-7}\,\text{J}$

c. Since the charge returns to the same equipotential that it started from, the net work done is zero.

19. A (There is no field inside the sphere at $r/2$. Outside the sphere, doubling the distance from the centre of the sphere reduces the field strength according to the inverse square law.)

20. Any difference in potential that arose due to an uneven charge distribution would very quickly revert to zero because the free electrons in the conducting material would flow to cancel the imbalance.

21. Figure 7 shows the graph for electric potential. A point charge may be positive or negative. The electric potential due to a point positive charge is always positive, since external work needs to be done to move a positive test charge from infinity towards the charge. The electric potential due to a point negative charge is always negative, since no external work has to be done in moving a positive test charge from infinity towards the charge; the work is done by the field itself, resulting in a loss of electric potential energy. The gravitational potential due to a point mass (Figure 8) can only be negative since gravitational forces are always attractive.

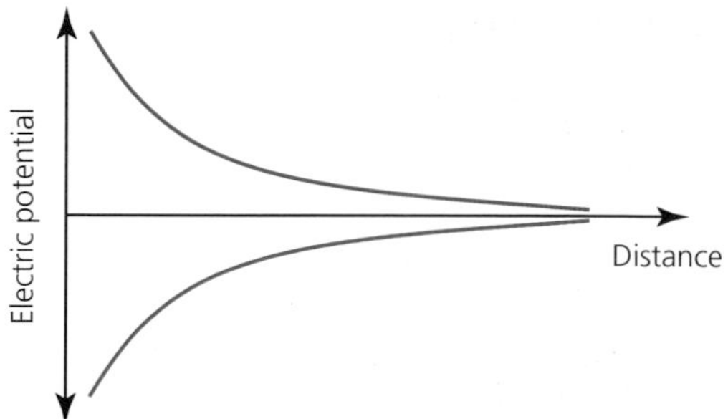

Figure 7

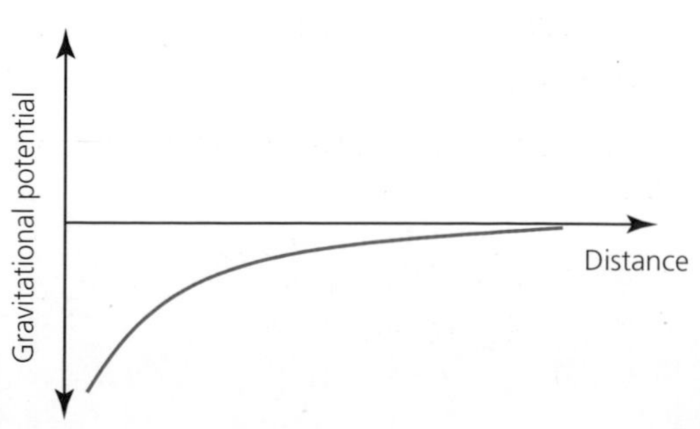

Figure 8

6 CAPACITANCE

1. $Q = CV = 1000 \times 10^{-6} \times 35 = 0.035\,\text{C}$
2. Voltage $V = \frac{Q}{C} = \frac{500 \times 10^{-6}}{100 \times 10^{-6}} = 5\,\text{V}$
3. The capacitance would increase by a factor of 8.
4. The lengths of foil that are rolled up could have a much bigger plate area; the distance between the plates is much smaller; the space is filled with a dielectric. These three factors make the capacitance $C = \frac{A\varepsilon_0\varepsilon_r}{d}$ greater.
5. Capacitance is

$$C = \frac{A\varepsilon_0\varepsilon_r}{d} = \frac{0.15 \times 0.20 \times 8.85 \times 10^{-12} \times 1.00}{10 \times 10^{-3}}$$
$$= 2.7 \times 10^{-11}\,\text{F} = 27\,\text{pF}$$

6. With an air gap, capacitance is

$$C = \frac{A\varepsilon_0\varepsilon_r}{d} = \frac{0.04 \times 8.85 \times 10^{-12} \times 1.00}{5 \times 10^{-3}}$$
$$= 7.08 \times 10^{-11}\,\text{F} = 71\,\text{pF (2 s.f.)}$$

With polyester resin dielectric, capacitance is
$3.5 \times 7.08 \times 10^{-11} = 2.48 \times 10^{-10}\,\text{F} = 250\,\text{pF}$

With polycarbonate dielectric, capacitance is
$3 \times 7.08 \times 10^{-11} = 2.12 \times 10^{-10}\,\text{F} = 210\,\text{pF}$

With mica dielectric, capacitance is
$7 \times 7.08 \times 10^{-11} = 4.96 \times 10^{-10}\,\text{F} = 500\,\text{pF}$

With polythene dielectric, capacitance is
$2.4 \times 7.08 \times 10^{-11} = 1.70 \times 10^{-10}\,\text{F} = 170\,\text{pF}$

7. **a.** The polar molecule rotates and aligns with the field, so that the negative part of the molecule faces the positive plate and the positive part of the molecule faces the negative plate.

   b. The overall electric field across the capacitor is reduced, because the dielectric's internal field is in the opposite direction.

8. Maximum charge $Q = CV = 4700 \times 10^{-6} \times 16$
$= 7.5 \times 10^{-2}\,\text{C}$

9. **a.** Energy stored

$$E = \frac{1}{2}CV^2 = \frac{1}{2} \times 1000 \times 10^{-6} \times 30^2 = 0.45\,\text{J}$$

   b. Maximum pd across capacitor = 25 V. So maximum energy stored is

$$E = \frac{1}{2}CV^2 = \frac{1}{2} \times 1000 \times 10^{-6} \times 25^2 = 0.31\,\text{J}$$

10. **a. i.** Charge that flows is
$Q = CV = 1000 \times 10^{-6} \times 6 = 6.0 \times 10^{-3}\,\text{C}$

   ii. Energy stored is

$$E = \frac{1}{2}CV^2 = \frac{1}{2} \times 1000 \times 10^{-6} \times 6^2 = 1.8 \times 10^{-2}\,\text{J}$$

   iii. Energy supplied by battery is
$E = QV = 6.0 \times 10^{-3} \times 6 = 3.6 \times 10^{-2}\,\text{J}$

   b. The energy stored on the capacitor is $\frac{1}{2}QV$, which is half the energy supplied by the battery. This is because the pd across the capacitor increases steadily from zero to V while charge Q flows onto the plates. Half the energy supplied by the battery is transferred in the resistance of the circuit.

11. A (Power output

$$= \frac{\frac{1}{2}CV^2}{\text{time}} = \frac{0.5 \times 50 \times 10^{-6} \times (20 \times 10^3)^2}{5 \times 10^{-3}} = 2 \times 10^6\,\text{W}$$

12. Energy stored is equal to the area between the line and the pd axis:

$$E = \frac{1}{2}QV = \frac{1}{2} \times 200 \times 10^{-6} \times 20 = 0.002\,\text{J}$$

13. **a.** Energy stored

$$E = \frac{1}{2}CV^2 = 0.5 \times 100 \times 2.7^2 = 365\,\text{J}$$

   b. Because they can store a lot of electrical energy when charged, which can give a serious electric shock. (They should be stored uncharged with the terminals strapped together for safety.)

14. **a.** Potential difference between the underside of the ground and the surface of the Earth is
$V = Ed = 300\,000 \times 2000 = 600\,\text{MV}$.

Energy stored

$$E = \frac{1}{2}QV = \frac{1}{2} \times 40 \times 600 \times 10^6 = 12\,\text{GJ}$$

   b. Power output

$$= \frac{\text{energy}}{\text{time}} = \frac{12 \times 10^9}{30 \times 10^{-6}} = 4 \times 10^{14}\,\text{W}$$

15. **a.** The maximum current flows at the instant the circuit is complete and before any charge has flowed onto the capacitor. At this instant, the pd across the resistor is equal to the cell's emf.

Maximum current

$$I_0 = \frac{\text{emf of cell}}{\text{circuit resistance}} = \frac{E}{R} = \frac{9}{18 \times 10^3} = 5.0 \times 10^{-4}\,\text{A}$$

   b. The current halves in the time interval called the half-life, $T_{1/2}$.

$T_{1/2} = 0.69\,RC = 0.69 \times 18 \times 10^3 \times 1000 \times 10^{-6}$
$= 12\,\text{s}$

c. Time constant $RC = 18 \times 10^3 \times 1000 \times 10^{-6} = 18\,\text{s}$

d. Time to be fully charged is $5RC = 5 \times 18 = 90\,\text{s}$

16. **a.** The maximum pd across the capacitor occurs when the capacitor is fully charged. At this point the maximum pd $V_0 = \mathcal{E}$, the emf of the cell. So $V_0 = 4.5\,\text{V}$.

b. The charge on a capacitor during charging is

$$Q = Q_0(1 - e^{-t/RC}) = CV_0(1 - e^{-t/RC})$$
$$= 1000 \times 10^{-6} \times 4.5$$
$$\times (1 - e^{-10/(47\times10^3\times1000\times10^{-6})})$$
$$= 1000 \times 10^{-6} \times 4.5 \times (1 - e^{-10/47})$$

which gives $Q = 8.6 \times 10^{-4}\,\text{C}$.

c. The time for the charge stored to reach 63% of its maximum value is equal to the time constant, which is equal to $RC = 47 \times 10^3 \times 1000 \times 10^{-6} = 47\,\text{s}$.

17. The charge stored is equal to the area under the I–t curve. The total number of small squares between the line and the time axis up to 60 s is approximately 770. Each small square has area of 1 μC, which gives a total charge of 770 μC.

18. Total charge Q flowing onto the capacitor = constant current × time, so

$$Q = 20 \times 10^{-6} \times 70 = 1.4 \times 10^{-3}\,\text{C}$$

Capacitance $C = \frac{Q}{V} = \frac{1.4 \times 10^{-3}}{3} = 4.7 \times 10^{-4}\,\text{F} = 470\,\mu\text{F}$

19. The gradient of the line on a ln I versus time graph is equal to $-\frac{1}{RC}$. Since the gradient of the line is equal to $-0.04\,\text{s}^{-1}$, the time constant

$RC = \frac{1}{0.04} = 25\,\text{s}$.

20. Current is rate of change of charge. The current at 20 s is equal to the gradient of the tangent at that point, which is $\frac{6 \times 10^{-3}}{60} = 1 \times 10^{-4}\,\text{A} = 100\,\mu\text{A}$

21. Gradient $= -\frac{1}{\text{time constant}} = -\frac{3}{100}\,\text{s}^{-1}$

which gives the time constant as 33 s.

22. Current $= \frac{\text{pd across } R}{R}$. However, during the discharge of a capacitor, the pd across the resistor is equal to the pd across the capacitor. Therefore

$$I = \frac{\text{pd across capacitor}}{R} = \frac{Q}{CR}$$

and so

$$Q = CRI = 500 \times 10^{-6} \times 60 \times 10^3 \times 60 \times 10^{-6}$$
$$= 1.8 \times 10^{-3}\,\text{C}$$

23. Integrating:

$$\int_0^t \left(-\frac{dt}{CR}\right) = \int_{Q_0}^{Q} \frac{dQ}{Q}$$

which gives

$$-\frac{t}{CR} = \log_e\left(\frac{Q}{Q_0}\right)$$

Taking inverse natural logs gives

$$e^{-t/CR} = \frac{Q}{Q_0}$$

which rearranges to give

$$Q = Q_0\, e^{-t/RC}$$

7 MAGNETIC FIELDS

1. **a.** When the right hand end of the solenoid is viewed, the current is flowing in an anticlockwise direction, so this end acts like an N pole.

b. Magnet field lines are directed out of an N pole and into an S pole. Therefore, within the iron, the field is from left to right.

2. Applying Fleming's left hand rule shows that the force on the wire is downwards (towards the bottom of the page).

3. The force on the wire can be determined from the change in the balance reading:

$$F = mg - \frac{298.100 - 284.591}{1000} \times 9.81 = 0.1325\,\text{N}$$

Magnetic flux density

$$B = \frac{F}{Il} = \frac{0.1325}{3.75 \times 0.115} = 0.307\,\text{T} = 307\,\text{mT}$$

4. **a.** The force on the aluminium rod is 'out of the page'.

b. $F = BIl = 330 \times 10^{-3} \times 5.2 \times 0.05 = 8.6 \times 10^{-2}\,\text{N}$

5. Equating magnetic force and centripetal force gives

$$BQv = \frac{mv^2}{r}$$

which rearranges to give

$$v = \frac{BQr}{m} = \frac{3 \times 1.6 \times 10^{-19} \times 0.11}{1.67 \times 10^{-27}} = 3.2 \times 10^7\,\text{ms}^{-1}$$

6. Equating magnetic force and centripetal force gives

$$BQv = \frac{mv^2}{r}$$

which rearranges to give

$$r = \frac{mv}{BQ}$$

Since the particle's charge Q and the magnetic flux density B are both constant, the radius of curvature $r \propto mv$, where mv is the particle's momentum.

7. B ($BQv = \frac{mv^2}{r}$ so $r = \frac{mv}{BQ}$. Increasing m and v would increase r but increasing B and Q would decrease r.)

8. A ($F = BQv$ and, for the singly charged particle, the charge is half but the speed is 10 times bigger.)

9. **a.** The centripetal force required for circular motion is provided by the magnetic field:

$$BQv = \frac{mv^2}{r}$$

which rearranges to

$$v = \frac{BQr}{m} = \frac{1.1 \times 1.6 \times 10^{-19} \times 0.2}{1.67 \times 10^{-27}} = 2.108 \times 10^7 \text{m s}^{-1}$$

Proton energy

$$= \frac{1}{2}mv^2 = \frac{1}{2} \times 1.67 \times 10^{-27} \times (2.108 \times 10^7)^2$$

$$= 3.71 \times 10^{-13} \text{J} = 2.3 \text{MeV}$$

b. Frequency of the applied pd is given by

$$f = \frac{BQ}{2m\pi} = \frac{1.1 \times 1.6 \times 10^{-19}}{2\pi \times 1.67 \times 10^{-27}} = 1.677 \times 10^7 = 17 \text{MHz}$$

10. **a.** Number of times N the proton passes between the dees before exiting is

$$N = \frac{5 \times 10^6}{1000} = 5000$$

b. The proton completes $N/2 = 2500$ circles in the cyclotron before exiting, so the time t for which the proton is spiralling is

$$t = \frac{2500}{8 \times 10^6} = 3.1 \times 10^{-4} \text{s}$$

11. **a.** Magnetic force $F = BQv = 2.8 \times 1.6 \times 10^{-19} \times 1.6 \times 10^7 = 7.168 \times 10^{-12} = 7.2 \times 10^{-12}$ N

b. Magnetic force $F = \frac{mv^2}{r} = 7.168 \times 10^{-12}$ N, which rearranges to give

Radius

$$r = \frac{mv^2}{7.168 \times 10^{-12}} = \frac{6.4 \times 10^{-27} \times (1.6 \times 10^7)^2}{7.168 \times 10^{-12}} = 0.23 \text{m}$$

12. Total proton energy E = kinetic energy + rest energy = 1938 MeV. Since $E = mc^2$, then $1938 \times 10^6 \times 1.6 \times 10^{-19} = m \times (3 \times 10^8)^2$, which gives $m = 3.44 \times 10^{-27}$ kg. So

increase in inertial mass $= (3.44 - 1.67) \times 10^{-27} = 1.77 \times 10^{-27}$ kg

13. Magnetic flux through the coil is $\Phi = BA$ $= 300 \times 10^{-3} \times \pi \times (0.75 \times 10^{-2})^2 = 5.3 \times 10^{-5}$ Wb

14. Magnetic flux = area of the coil × component of magnetic flux density normal to coil

$= A \times B \cos sq = \pi(0.01)^2 \times 150 \times 10^{-3} \times \cos 40°$

$= 3.6 \times 10^{-5}$ Wb

15. Magnetic flux linkage $= N \times BA \cos\theta = 20 \times \pi(5 \times 10^{-3})^2 \times 150 \times 10^{-3} \times \cos 16° = 2.3 \times 10^{-4}$ Wb turns

16. **a.** Magnetic flux $\Phi = BA \cos\theta$

$= 120 \times 10^{-3} \times 0.05 \times 0.03 \times \cos 30° = 1.559 \times 10^{-4}$

$= 1.6 \times 10^{-4}$ Wb

b. Magnetic flux linkage $= N\Phi = 100 \times 1.559 \times 10^{-4}$

$= 1.6 \times 10^{-2}$ Wb turns

17. **a.** Flux linkage $= N\Phi = NBA = 5000 \times 1.1 \times \pi \times (5 \times 10^{-3})^2 = 0.432$ Wb $= 0.43$ Wb turns

b. Flux linkage $= 0.432 \times \cos 30° = 0.37$ Wb turns

8 ELECTROMAGNETIC INDUCTION AND ALTERNATING CURRENT

1. End A becomes positive, from Fleming's right hand rule.

2. Current flows from the centre of the disc to the rim (through the meter and back to the centre), from Fleming's right hand rule.

3. **a.** Induced emf is

$$\mathcal{E} = N\frac{\Delta\Phi}{\Delta t} = \frac{\Delta\Phi}{\Delta t} = \frac{BA}{\Delta t} = \frac{0.3 \times 0.045 \times 0.02}{0.25}$$

$$= 1.08 \times 10^{-3} \text{V} = 1.1 \times 10^{-3} \text{V}$$

b. Induced current is

$$I = \frac{V}{R} = \frac{1.08 \times 10^{-3}}{0.5} = 2.2 \times 10^{-3} \text{A}$$

4. **a.** Vertical component of the Earth's magnetic field is

$B_v = 1.7 \times 10^{-4} \cos 30° = 1.472 \times 10^{-4}$

$= 1.5 \times 10^{-4}$ T

b. In one second, the aircraft travels 200 m. The area swept out by the wings is $A = 200 \times 12 = 2.4 \times 10^3 \text{m}^2$.

c. The induced emf between the wing tips is

$$\mathcal{E} = \frac{\Delta\Phi}{\Delta t} = \frac{B_v \times A}{1} = 1.472 \times 10^{-4} \times 2.4 \times 10^3$$

$$= 0.35 \text{V}$$

5. Induced emf is

$$\mathcal{E} = N\frac{\Delta\Phi}{\Delta t} = N\frac{BA}{\Delta t} = 20 \times \frac{40 \times 10^{-3} \times \pi \times 0.02^2}{0.2}$$

$$= 5.03 \times 10^{-3} \text{V} = 5.0 \text{mV}$$

6. Induced emf is

$$\mathcal{E} = N\frac{\Delta\Phi}{\Delta t} = N\frac{BA}{\Delta t} = 5000 \times \frac{1.2 \times 10^{-3} \times \pi \times (5 \times 10^{-3})^2}{4}$$

$$= 1.2 \times 10^{-4} \text{V}$$

7. At any point, as the magnetised neodymium pellet falls through the copper tube, that portion of tube cuts through the field lines of the magnet, causing currents to be induced in the copper. These induced currents, called eddy currents, create magnetic fields that exert a force on the magnet, opposing its fall (by Lenz's law), so that it falls much more slowly than the unmagnetised neodymium pellet.

8. a. Flux linkage $N\Phi = BAN \cos \theta$.

So maximum flux linkage is

$$BAN = 80 \times 10^{-3} \times 0.06 \times 0.035 \times 100$$
$$= 1.68 \times 10^{-2} \text{ Wb turns} = 1.7 \times 10^{-2} \text{ Wb turns}$$

b. The induced emf is $\mathcal{E} = BAN\,\omega \sin \omega t$.

So maximum induced emf is

$$BAN\,\omega = BAN \times 2\pi f = 1.68 \times 10^{-2} \times 2 \times \pi \times 25$$
$$= 2.6 \text{ V}$$

9. a. The peak voltage = maximum emf, so

$$V_p = \mathcal{E} = BAN\, 2\pi f = 0.2 \times 0.1 \times 0.05 \times 100 \times 2\pi \times 25$$
$$= 16 \text{ V}$$

b. See Figure 9.

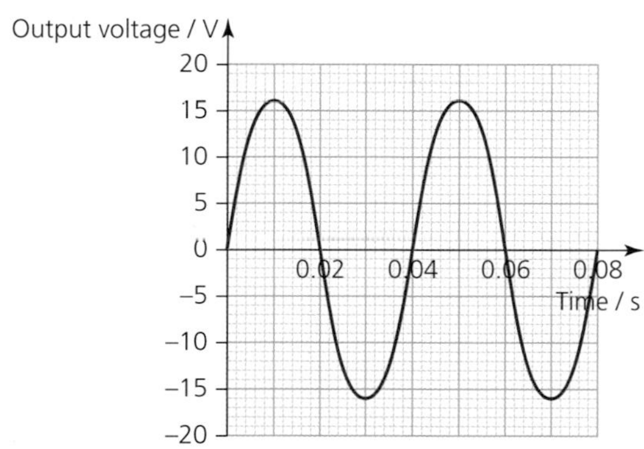

Figure 9

10. C (Peak voltage = maximum emf $\mathcal{E} = BA \times N \times 2\pi f = V_0$. New peak voltage is $4BA \times 2N \times 2\pi \times f / 2 = 4 \times BAN2\pi f = 4V_0$)

11. The coil of the motor is rotating in the magnetic field, and therefore, according to Faraday's law, an emf is induced in the coil. According to Lenz's law, this emf would try to drive a current in a direction to oppose the change causing it. Therefore, the induced emf will act in opposition to the power supply's emf, reducing the size of the current in the coil of the motor.

12. a. Peak voltage across loud speaker is

$$V_0 = 20 \times \frac{60}{100 + 60} = 7.5\text{V}$$

rms voltage $V_{rms} = \frac{V_0}{\sqrt{2}} = \frac{7.5}{\sqrt{2}} = 5.3\text{V}$

b. The rms current $I_{rms} = \frac{V_{rms}}{R} = \frac{5.3}{60} = 0.088\text{A}$

c. Average power output $= V_{rms} \times I_{rms} = 5.3 \times 0.088$
$= 0.47 \text{ W}$

13. The rms voltage $V_{rms} = \frac{V_0}{\sqrt{2}}$

peak voltage $V_0 = \sqrt{2} \times V_{rms} = \sqrt{2} \times 120 = 170\text{V}$

peak-to-peak voltage $= 2V_0 = 340\text{V}$

14. a. The cell's emf is found by multiplying the number of divisions from the central horizontal axis to the horizontal line of the trace by the *y*-gain setting. (This assumes that the *y*-gain control of the oscilloscope has been set to its calibrated position.)

b. The *y*-gain control would have to be set to a more sensitive setting so that the line of the trace is higher up the screen. Also, instead of positioning the line along the central axis when the oscilloscope input is initially set to GD, the line could have been positioned nearer to the bottom of the screen, allowing an even more sensitive setting of the *y*-gain to be selected so that the trace takes up more of the screen.

15. a. Peak-to-peak voltage $= 6.0 \times 0.5 = 3.0 \text{ V}$

Percentage uncertainty in the number of divisions is $\frac{0.2}{6} \times 100 = \pm 3.3\%$, which gives an uncertainty in the peak-to-peak voltage of $\frac{3.3}{100} \times 3.0 = \pm 0.1\text{V}$.

b. The rms voltage $V_{rms} = \frac{V_0}{\sqrt{2}} = \frac{1.5}{\sqrt{2}} = 1.06\text{V}$, with the same percentage uncertainty as the peak-to-peak value. Therefore the uncertainty in the rms voltage $= \pm\frac{3.3}{100} \times 1.06 = \pm 0.03\text{V}$.

c. Time for one cycle is $T = 5.8 \times 50 \times 10^{-6} = 2.90 \times 10^{-4}$ s

which gives a frequency

$$f = \frac{1}{T} = \frac{1}{2.90 \times 10^{-4}} = 3448 = 3450\text{Hz}$$

Percentage uncertainty in the number of divisions is $\frac{0.2}{5.8} \times 100 = \pm 3.4\%$, and therefore the uncertainty in the frequency is $\frac{3.4}{100} \times 3448 = 117.3 = \pm 120\text{Hz}$.

16. **a.** One cycle takes up 4.9 divisions at 1 ms/div. The time for one cycle is therefore 4.9 ms.

Frequency $f = \frac{1}{T} = \frac{1}{4.9 \times 10^{-3}} = 204\,\text{Hz}$

b. Reduce the intensity of the trace so that the horizontal sections are much thinner, so that the peak-to-peak voltage can be measured more precisely.

17. The vertical scale needs to cover at least 5.2 V and there are 8 vertical divisions, so select the y-gain setting to be 1 V/div, giving the largest trace that fits vertically on the screen. Time for one cycle = $\frac{1}{20} = 0.05\,\text{s}$, so the time for two cycles is 0.1 s. For 10 horizontal divisions to be equivalent to 0.1 s, each division has to be 10 ms. There are 10 horizontal divisions, so select time base setting to be 10 ms/div.

18. **a.** The transformer equation $\frac{N_s}{N_p} = \frac{V_s}{V_p}$ can be rearranged to give

$$N_s = \frac{V_s}{V_p} \times N_p = \frac{12}{230} \times 480 = 25 \text{ turns}$$

b. Efficiency $\frac{I_s V_s}{I_p V_p} = 0.8$, which rearranges to give

$$I_s = 0.8 \times \frac{I_p V_p}{V_s} = 0.8 \times \frac{0.2 \times 230}{12} = 3.1\,\text{A}$$

19. According to Faraday's law, an emf is only induced in a coil when there is a change in the magnetic flux through the coil. An alternating current flowing in the primary of a transformer creates a changing magnetic flux through the secondary and hence an emf is induced in the secondary. If the current in the primary was dc, the core would still be magnetised but the flux density would be constant and no emf would be induced in the secondary.

20. The induced current in the ring would be greater and so the magnetic force on the ring would be greater and the ring would be pushed to a greater height before reaching equilibrium.

21. The resistance in the aluminium ring will be significantly reduced as a result of being placed in the very cold liquid nitrogen. Therefore, a much larger induced current would flow in the ring. The magnetic force pushing the ring upwards would be much greater, and may even be sufficient for the ring to be fired upwards high enough to hit the ceiling.

22. Current in the cable $I = \frac{P}{V} = \frac{2 \times 10^6}{33000} = 60.6\,\text{A}$

Power loss in cables $P = I^2R = 60.6^2 \times 20 = 73\,\text{kW}$

23. **a.** Cable current $I = \frac{P}{V} = \frac{4000}{240} = 16.7\,\text{A}$

Power loss in the cables $P = I^2R = 16.7^2 \times 1.2 = 335\,\text{W}$

b. Voltage drop across the cables when the generator is running at full power $= IR = 16.7 \times 1.2 = 20\,\text{V}$, so the voltage then available at the farmhouse is 220 V. This is the minimum value.

9 RADIOACTIVITY

1. **a.** The scattering angle is affected by the speed of the alpha particles and needs to be a variable that is kept constant.

b. **i.** The beam had to be narrow and parallel so that the region on the foil that the beam hit was small and the original beam direction was clearly defined, so that the angle of scattering between the scattered direction and the original beam direction could be accurately measured.

ii. The gold foil had to be very thin so that it was likely that an alpha particle was scattered only once as it passed through the foil.

c. Since the alpha particle is a hadron, the alpha particle and the gold nucleus could also interact via the strong nuclear force if they are less than 3 fm apart.

d. An alpha particle is detected when it strikes the zinc sulfide screen on the end of the microscope, causing a flash of light to be emitted.

2. $\frac{\text{atomic radius}}{\text{nuclear radius}} = \frac{10^{-10}}{10^{-14}} = 10^4$

3. An alpha particle could be back-scattered only if it collided with a dense massive object with the same charge. The observed scattering of alpha particles through large angles supported Rutherford's idea that all the positive charge and most of the mass of the atom was concentrated in a central nucleus. Thomson's plum pudding model could only predict small angles of scattering.

4. **a.** A larger impact parameter would mean that the alpha particle would pass the nucleus at a greater distance, and therefore the electrostatic repulsive force on the alpha particle would be smaller, resulting in a smaller scattering angle.

b. A higher-speed alpha particle would experience the electrostatic repulsive force for a shorter time and therefore be scattered by a smaller angle.

5. **a.** **i.** Number of protons is 26 and the number of neutrons is 30 (i.e. 56 – 26).

ii. Number of protons is 26 and the number of neutrons is 33 (i.e. 59 – 26).

b. Number of atoms in 1 g

$= \frac{1}{56} \times 6.02 \times 10^{23} = 1.1 \times 10^{22}$.

6. An alpha source is more dangerous if ingested, as it ionises more intensely, so its energy is concentrated in a smaller space, causing significant cell damage.

7. The beta particle will travel the greater distance in air because it has a smaller mass and smaller charge than an alpha particle and therefore creates fewer ion pairs per centimetre of its path.

8. **a.** Because beta particles have a range in air (depending on their energy) of 2–3 m and can penetrate the skin easily, causing damage to tissues by ionisation.

b. Because the lead is sufficient to absorb the energy of the beta radiation, so while in the box the source is not harmful.

9. Into the page (from Fleming's left hand rule)

10. Since $I = \frac{k}{x^2}$, then $C = \frac{\text{constant}}{x^2}$, since $C \propto I$.

Therefore $C \times x^2 = \text{constant}$:

$12.9 \times 30^2 = 8.7 \times x^2$, which gives $x = 36.5\,\text{cm}$.

11. **a.** Since the GM tube detects 5% of the gammas entering the tube, the actual number of gammas entering the tube in one second $= 20 \times 15.4 = 308$. The area of the GM tube pointing towards the source is $\pi r^2 = \pi \times 0.012 = 3.14 \times 10^{-4}\,\text{m}^2$.

The surface area of a sphere of radius 25 cm $= 4\pi r^2 = 4\pi \times 0.25^2 = 0.785\,\text{m}^2$. Therefore, the number of gamma photons passing through the surface of a sphere of radius 25 cm in one second is

$$308 \times \frac{0.785}{3.14 \times 10^{-4}} = 7.70 \times 10^5$$

Since the gamma photons are unaffected by the air, it can be assumed that 7.70×10^5 gamma photons are emitted from the source every second.

b. The intensity of electromagnetic radiation is defined as the radiation energy passing normally through an area of 1 m^2. Therefore, the intensity I at a distance of 2.0 m is given by

$$I = \frac{\text{photons emitted per second} \times \text{photon energy}}{\text{area of a sphere of radius 2m}}$$
$$= \frac{7.70 \times 10^5 \times 100 \times 10^3 \times 1.6 \times 10^{-19}}{4\pi \times 2^2}$$
$$= 2.45 \times 10^{-10}\,\text{W m}^{-2}$$

12. Because a significant proportion of its energy is absorbed as it ionises the air molecules.

13. The sudden drop when the GM tube is moved a few centimetres from the sample suggests that alpha radiation is emitted from the sample, since alpha only has a short range in air. The count rate is well above background when a 3 mm thick aluminium absorber is used, which indicates the emission of gamma radiation from the sample. However, no beta radiation is being emitted, as indicated by the unchanged count rate when the 3 mm thick aluminium absorber was used.

14. The source should be kept in its lead-lined box and removed and manipulated only by using long tongs. The duration of the experiment should be kept to as short as possible, and the source returned to its lead-lined storage box as soon as the experiment is completed.

15. A radiographer or a worker in the nuclear power industry.

16. The Earth's atmosphere will absorb some of the cosmic rays. The Earth's magnetic field will deflect cosmic rays consisting of charged particles away from the Earth.

17. Number of atoms present is

$$N = \frac{1}{210} \times 6.02 \times 10^{23} = 2.87 \times 10^{21}$$

Activity $A = \lambda N = 5.8 \times 10^{-8} \times 2.87 \times 10^{21}$
$= 1.7 \times 10^{14}\,\text{Bq}$

18. For the corrected count rate to fall from 200 to 100 takes 1.2 hours. To fall from 100 to 50 takes 1.2 hours and to fall from 50 to 25 takes 1.2 hours. Therefore, the half-life is 1.2 hours.

19. Since 25 years corresponds to 5 half-lives, the fraction remaining after 25 years will be $\frac{1}{2^5} = \frac{1}{32}$.
Therefore, after 25 years, the activity will have fallen to $\frac{1}{32} \times 1.2 \times 10^{16} = 3.75 \times 10^{14}\,\text{Bq}$.

20. Initial activity

$$A_0 = \lambda N_0 = 1.0 \times 10^{-6} \times \frac{1}{131} \times 6.02 \times 10^{23}$$
$$= 4.595 \times 10^{15}\,\text{Bq}$$

Activity after 24 hours

$$A = A_0 e^{-\lambda t} = 4.595 \times 10^{15} \times e^{-1.0\times10^{-6}\times24\times3600}$$
$$= 4.0 \times 10^{15}\,\text{Bq}$$

21. **a.** Decay constant

$$= \frac{\ln 2}{T_{1/2}} = \frac{\ln 2}{25 \times 60} = 4.621 \times 10^{-4} = 4.6 \times 10^{-4}\,\text{s}^{-1}$$

b. Number of atoms remaining after 15 minutes is

$$N = N_0 e^{-\lambda t} = 2.0 \times 10^{16} \times e^{-4.621\times10^{-4}\times15\times60}$$

which gives $N = 1.32 \times 10^{16}$ atoms. The number of atoms that have decayed is therefore equal to $2.0 \times 10^{16} - 1.32 \times 10^{16}$ $= 6.8 \times 10^{15}$.

c. Energy released = number of decays $\times 5 \times 10^6 \times 1.6 \times 10^{-19} = 6.8 \times 10^{15} \times 5 \times 10^6 \times 1.6 \times 10^{-19}$ $= 5.4 \times 10^3$ J

22. Integrating:

$$\int_{N_0}^{N} \frac{dN}{N} = \int_0^t -\lambda dt$$

which gives

$$\ln N - \ln N_0 = -\lambda t$$

which can be written as

$$\ln \frac{N}{N_0} = -\lambda t$$

Taking inverse natural logarithms gives

$$\frac{N}{N_0} = e^{-\lambda t}$$

which rearranges to give $N = N_0 e^{-\lambda t}$.

23. The gradient of the graph is equal to –0.10 years^{-1}.

Half-life $T_{1/2} = \frac{\ln 2}{\lambda} = \frac{\ln 2}{0.10} = 7.0$ years

24. The equation $N = N_0 e^{-\lambda t}$ can be rearranged to give

$$\frac{N}{N_0} = e^{-\lambda t}$$

The ratio $\frac{N}{N_0} = 0.392$, which taking logarithms and substituting gives

$$\ln \frac{N}{N_0} = -\lambda t$$
$$\ln 0.392 = -3.84 \times 10^{-12} \times t$$

and therefore time $t = 2.439 \times 10^{11}$ s = 7730 years

10 NUCLEAR ENERGY

1. **a.** $^{226}_{88}\text{Ra} \rightarrow ^{222}_{86}\text{Rn} + ^{4}_{2}\text{He}$

b. $^{214}_{82}\text{Pb} \rightarrow ^{214}_{83}\text{Bi} + ^{0}_{-1}\text{e} + ^{0}_{0}\bar{\nu}_e$

c. $^{210}_{84}\text{Po} \rightarrow ^{206}_{82}\text{Pb} + ^{4}_{2}\text{He}$

2. Kinetic energy $E = \frac{1}{2}mv^2$. The decay is governed by the conservation of momentum. Therefore $E \times m = \frac{1}{2}mv^2 m = \frac{1}{2}p^2$, where p is momentum. Assuming that the parent nucleus was stationary, then by conservation of momentum $0 = p_\alpha + p_D$, where p_α is the momentum of the alpha and p_D is the momentum of the daughter.

Since the magnitudes of p_α and p_D are equal, then $E \times m$ is the same for the alpha particle and daughter. Therefore $E_\alpha m_\alpha = E_D m_D$ and rearranging gives

$$\frac{E_\alpha}{E_D} = \frac{m_D}{m_\alpha}$$

3. Beta-plus decay: $^{18}_{9}\text{F} \rightarrow ^{18}_{8}\text{O} + ^{0}_{+1}\text{e} + ^{0}_{0}\nu_e$

Electron capture: $^{18}_{9}\text{F} + ^{0}_{-1}\text{e} \rightarrow ^{18}_{8}\text{O} + ^{0}_{0}\nu_e$

4. A beta-minus decay does not change the nucleon number, whereas alpha decay changes the nucleon number by 4. Since the series of decays changes the nucleon number by 235 – 207 = 28, there must have been seven alpha decays, therefore $n = 7$.

5. The complete equation is $^{18}_{8}\text{O} + ^{1}_{1}\text{p} = ^{18}_{9}\text{F} + ^{1}_{0}\text{n}$, showing that the product particle is a neutron.

6. **a.** Neutron numbers: $^{215}_{84}\text{Po}$ has 131 neutrons; $^{215}_{85}\text{At}$ has 130 neutrons; $^{211}_{83}\text{Bi}$ has 128 neutrons; $^{207}_{81}\text{Tl}$ has 126 neutrons; $^{207}_{82}\text{Pb}$ has 125 neutrons.

b. **i.** and **ii.** See Figure 10.

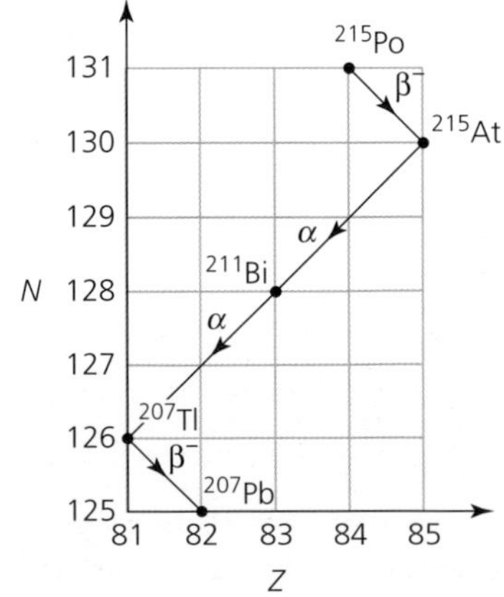

Figure 10

7. In alpha emission, the only products are the daughter nuclide and the alpha particle. The decay energy is shared between these in a fixed ratio (because of momentum and kinetic energy conservation). The alpha particle thus has a distinct energy. It may have a lower distinct energy if a gamma photon is also emitted. In beta emission, as well as the daughter nuclide and the electron/positron, an antineutrino/neutrino is also emitted. The relative proportions of energy of the emitted particles can vary continuously.

8. a. $^{27}_{12}\text{Mg} = ^{27}_{13}\text{Al}^* + ^{0}_{-1}\beta + ^{0}_{0}\bar{\nu}_e$

$^{27}_{13}\text{Al}^* = ^{27}_{13}\text{Al} + \gamma$

b. Gamma photon energies:

$1.015 - 0.834 = 0.181\,\text{MeV}$

$1.015 - 0 = 1.015\,\text{MeV}$

$0.834 - 0 = 0.834\,\text{MeV}$

9. a. $^{57}_{27}\text{Co} + ^{0}_{-1}\text{e} = ^{57}_{26}\text{Fe}^* + ^{0}_{0}\nu_e$

$^{57}_{26}\text{Fe}^* = ^{57}_{26}\text{Fe} + \gamma$

b. Taking the ground state to be 0 keV, the first excited state is 14 keV and the second excited state is 136 keV.

10. Since the energy of the emitted gamma photon is 0.662 MeV, the maximum kinetic energy of the remaining 5.4% of beta particles = 0.512 MeV + energy of gamma photon = 0.512 + 0.662 = 1.174 MeV.

11. $^{99}_{42}\text{Mo} = ^{99m}_{43}\text{Te} + ^{0}_{-1}\beta + ^{0}_{0}\bar{\nu}_e$

$^{99m}_{43}\text{Te} = ^{99}_{43}\text{Te} + \gamma$

12. a. Photon energy $E = \frac{hc}{\lambda}$, which can be rearranged to give $\lambda = \frac{hc}{E}$. Therefore the wavelength of the gamma photons is

$$\lambda = \frac{6.63\times10^{-34}\times3\times10^{8}}{140\times10^{3}\times1.6\times10^{-19}} = 8.9\times10^{-12}\,\text{m}$$

b. 24 hours is equal to 4 half-lives. The fraction remaining is $\frac{1}{2^4} = 0.0625$. Therefore, the percentage of technetium-99m remaining is 6.25%.

13. Because the tracer is introduced to the inside of the body, and beta and alpha particles are much more damaging to tissue – whereas gamma radiation largely penetrates the body and emerges to be detected by the gamma camera.

14. The six-hour half-life is long enough to allow sufficient time for the tracer to reach the required organ and for the gamma camera to get a suitable image, but short enough to minimise the patient's exposure to radiation. The gamma photons produced have energies that make them suitable for detection by a gamma camera, but low enough to keep ionisation to a minimum to prevent harm to the patient. It can be incorporated into a range of different biologically active compounds and so target different organs of the body.

15. Radius of a gold nucleus: $R = R_0A^{1/3}$ $= 1.05\times10^{-15}\times197^{1/3} = 6.1\times10^{-15}\,\text{m}$.

Mass of a gold atom $= 197\times1.661\times10^{-27}$ $= 3.27\times10^{-25}\,\text{kg}$.

Assuming that the mass of a gold nucleus ≈ the mass of a gold atom gives the density ρ of the gold nucleus as

$$\rho = \frac{M}{V} = \frac{3.27\times10^{-25}}{\frac{4}{3}\pi\times(6.1\times10^{-15})^3} = 3.4\times10^{17}\,\text{kg}\,\text{m}^{-3}$$

16. Density $\rho = \frac{M}{V}$ rearranges to gives $M = \rho V$ $= 3.4\times10^{17}\times5\times10^{-6} = 1.7\times10^{12}\,\text{kg}$

17. Mass difference = 26.977 76 – (26.974 40 + 0.000 55) = 0.002 810 u.

Therefore the decay energy is 0.002 810 × 931.5 = 2.618 MeV.

18. a. Kinetic energy

$$E_k = \frac{1}{2}mv^2 = 0.5\times800\times10^2 = 40000\,\text{J}$$
$$= \frac{40000}{(3\times10^8)^2}$$
$$= 4.4\times10^{-13}\,\text{kg}$$

which is a fraction 5.6×10^{-16} of the car's stationary mass of 800 kg.

b. The electron gains 1 MeV of energy, which is equivalent to 0.001 07 u, which is a fractional increase in mass of 1.9 of the electron's rest mass of 0.000 55 u.

19. Mass difference = (88 × 1.007 28) + (136 × 1.008 67) – 223.971 90 = 1.847 86 u

Therefore the energy needed to separate the nucleons = 1.847 86 × 931.5 = 1721 MeV

20. Mass difference = (2 × 1.007 28) + (2 × 1.008 67) – 4.001 50 = 0.030 40 u

Binding energy per nucleon = 0.030 40 × 931.5 ÷ 4 = 7.079 MeV per nucleon.

21. Change in binding energy
$= (3 \times 2.57) - (2 \times 1.11) - (2 \times 1.11) = 3.27\,\text{MeV}$

22. Mass loss
$= (4 \times 1.00728) - 4.00150 - (2 \times 0.00055)$
$= 0.0265\,\text{u}$

Energy released to the surroundings
$= 0.0265 \times 931.5 = 24.7\,\text{MeV}$

23. Mass loss $= 256.03688 - 139.89200 - 111.88206 - (4 \times 1.00867) = 0.22814\,\text{u}$ which is equivalent to $0.22814 \times 931.5 = 212.5\,\text{MeV}$ of energy released.

24. Mass loss $= 234.99342 + 1.00867 - 140.88367 - 91.90639 - (3 \times 1.00867) = 0.18602\,\text{u}$ which is equivalent to $0.18602 \times 931.5 = 173.3\,\text{MeV}$

25. The excess neutrons would mean that the moderator nuclei would emit beta-minus (and gamma) radiation.

26. By conservation of kinetic energy:

$$\frac{1}{2}mu_A^2 = \frac{1}{2}mv_A^2 + \frac{1}{2}mv_B^2$$

which becomes

$$u_A^2 = v_A^2 + v_B^2 \qquad \text{(i)}$$

By conservation of momentum:

$$mu_A = mv_A + mv_B$$

which becomes

$$u_A = v_A + v_B \qquad \text{(ii)}$$

Substituting equation (ii) into equation (i) gives

$$(v_A + v_B)^2 = v_A^2 + v_B^2$$

Multiplying out the brackets gives

$$v_A^2 + v_B^2 + 2v_Av_B = v_A^2 + v_B^2$$

which cancels to give

$$2v_A v_B = 0$$

so either v_A or v_B is zero. We can conclude that $v_A = 0$ since v_B can only be zero if no collision had occurred.

27. Advantage: Water can also be used as the moderator, allowing for a more compact moderator design. Water is a good moderator because the hydrogen nucleus in the water molecule is a proton, which has a similar mass to a neutron. Therefore, when a fast neutron collides with a hydrogen nucleus, it loses its kinetic energy in large chunks, quickly becoming a thermal neutron.

Also: Water has a larger specific heat capacity than carbon dioxide and so can more easily transfer energy away from the reactor core.

Disadvantage: Water has a much higher probability of absorbing neutrons than does carbon dioxide, and so, to maintain the level of neutrons required for a critical reaction, an enriched fuel is needed.

28. When the control rods are lowered into the reactor core, neutrons are absorbed at a greater rate, and so fewer neutrons are available to cause fission. A reduction in the number of fissions per second reduces the power output of the power station.

29. The ratio of uranium-235 (the fissile material) to uranium-238 is greater in enriched fuel.

30. Einstein's equation $E = mc^2$ can be rearranged to give the mass loss: $m = \frac{E}{c^2}$. The rate of mass loss is therefore given by

$$\frac{\text{power output}}{c^2} = \frac{900 \times 10^6}{(3 \times 10^8)^2} = 1 \times 10^{-8}\,\text{kg}\,\text{s}^{-1}$$

However, the power station is not 100% efficient, so in practice the rate of loss of mass would have to be greater.

31. **a. i.** A fast neutron is one that has just been created by a fission event and has a large kinetic energy, typically 2 MeV.

ii. A thermal neutron is in thermal equilibrium with its surroundings and has very little kinetic energy, typically much less than 1 eV.

b. The probability of a neutron being absorbed by a uranium-235 nucleus is much greater if the neutron is moving slowly.

32. In a critical chain reaction, on average one of the neutrons produced in a fission event goes on to produce another fission event.

33. Neutrons could be absorbed by: cladding, fission fragments, moderator, coolant, uranium-238; some neutrons escape from the reactor core.

34. Since spontaneous fission can occur in uranium-235, a chain reaction could become established if the mass of fuel stored in one place exceeded the critical mass.

35. The critical mass is the minimum mass of nuclear fuel required to establish a self-sustaining chain reaction. It depends on the concentration of uranium-235 in the fuel and on the shape of the fuel rod arrangement.

36. High-level waste also generates considerable amounts of heat, which must be removed as part of its treatment.

37. Unused fuel rods contain isotopes of uranium, which are alpha emitters and have long half-lives and therefore low levels of radioactivity. Spent fuel rods contain fission fragments, which emit beta-minus and gamma radiation and have high levels of radioactivity because some of the fission fragments have short half-lives.

GLOSSARY

Absolute zero The lowest possible temperature at which particles have a minimum kinetic energy; particles are stationary except for quantum-mechanical motion. Absolute zero is 0 K or −273.15° C.

Activity In the context of radioactivity, the number of decays per second is defined as the activity, A, and is measured in becquerel (Bq) where 1 Bq is equal to one decay per second.

Amplitude The amplitude of an oscillation or vibration is the maximum displacement, measured from the position of equilibrium.

Angular displacement The angular displacement, θ, is the angle about a fixed axis described by a rotating or orbiting object in degrees or radians from a fixed point.

Angular velocity The angular velocity, or angular speed, ω, is the angular displacement per unit time and is measured in rad s^{-1}.

Atomic mass unit (u) A small unit of mass used in nuclear physics. 1 u is defined as one-twelfth of the mass of a carbon-12 atom (1 u = 1.66043×10^{-27}kg.)

Average molecular kinetic energy The average molecular kinetic energy of an ideal gas is the average value of the translational kinetic energy of each molecule (vibrational and rotational kinetic energy is negligible in an ideal gas). It is given by the expression: $\frac{1}{2}m(c_{rms})^2$.

Avogadro constant The Avogadro constant, N_A is the number of particles in one mole of a substance and therefore equal to the number of carbon atoms in exactly 12 grams of carbon-12. The numerical value of the Avogadro constant is 6.02×10^{23} mol^{-1}.

Avogadro's law A law stating that equal volumes of gases, at the same temperature and pressure, contain the same number of molecules.

Background radiation Background radiation is a radiation dose rate at a specified location which is generated by natural and artificial ionising radiation sources existing in the environment.

Becquerel (Bq) The becquerel, Bq, is a measure of activity of a radioactive sample, where 1 Bq is equal to one decay per second.

Binding energy The energy needed to pull an atomic nucleus apart, i.e. to separate the individual neutrons and protons.

Binding energy per nucleon The binding energy of the nucleus divided by the nucleon number.

Boltzmann constant The Boltzmann constant, k, is the molar gas constant, R, divided by the Avogadro constant, N_A. $k = \frac{R}{N_A}$. k has a value of 1.38×10^{-23} JK^{-1}.

Boyle's law States that for a fixed mass of an ideal gas at a constant temperature, the pressure of the gas is inversely proportional to its volume: pV = constant.

Brownian motion The observable random movements of particles such as smoke particles, caused by the high speed thermal motion of liquid or gas molecules.

Capacitance The capacitance, C, of a capacitor is defined as the charge stored per unit potential difference across the plates: $C = \frac{Q}{V}$.

Carbon dating A method of dating once-living artefacts containing carbon by comparing the ratio of carbon-12 to carbon-14 atoms in a sample. The half-life of carbon-14 is 5730 years, making it suitable for dating artefacts between 200 and 60 000 years old.

Centripetal acceleration An acceleration towards the centre of a circular path: $a = \frac{v^2}{r} = \omega^2 r$.

Centripetal force A force causing an object to move in a circular path; it acts towards the centre of a circle: $F = \frac{mv^2}{r} = m\omega^2 r$.

Charles' law States that for a fixed mass of an ideal gas at constant pressure, the volume of the gas is proportional to its absolute temperature: $\frac{V}{T}$ = constant.

Coolant A fluid used to cool a nuclear reactor core, most commonly water but sometimes carbon dioxide or molten sodium.

Control rods Rods which can be raised or lowered into a nuclear reactor core to control the rate of nuclear reaction by absorbing neutrons. They typically contain cadmium or boron.

Coulomb repulsion The Coulomb repulsion is the repulsive force between two similarly charged objects such as an alpha particle and a gold nucleus. The Coulomb repulsion is governed by Coulomb's law.

Coulomb's law Coulomb's law states the force between two point charges Q_1 and Q_2 separated by a distance r in a vacuum is directly proportional to the product of the two charges and inversely proportional to the square of their separation: $F = \frac{kQ_1Q_2}{r^2}$.

Critically damped A system is said to be critically damped when the resistive forces are just enough to prevent oscillation and the object returns to equilibrium in the minimum possible time.

Critical chain reaction A nuclear chain reaction where exactly one neutron from each fission event is allowed to cause another fission event.

Cyclotron A circular particle accelerator in which the accelerated charged particles follow an outward spiralling path.

Damping Damping is the removal of energy from an oscillating system. The extent of the damping determines how long it takes for an oscillation to die away.

Daughter nucleus The nucleus formed from a radioactive decay event such as alpha decay, beta decay or fission.

Decay energy The energy output from a radioactive decay which appears as kinetic energy of the daughter nucleus or nuclei and emitted particles.

Decay series A chained series of radioactive decays ending with a stable nuclide. The chain may be branched or unbranched and is often shown on a graph of neutron number against proton number.

Decay constant The probability of an individual nucleus of a particular radioisotope decaying per second is called the decay constant, λ, which has the unit s^{-1}.

Dielectric A material which is an electrical insulator than can be polarised by an applied electric field.

Dielectric constant (a.k.a relative permittivity) Is the factor by which the electric field between two charges is decreased by the presence of the dielectric, relative to a vacuum.

Dissipation The transfer of energy from a source to its surroundings such that it can no longer be used for doing useful work.

Displacement The position of an object relative to an origin. Displacement, x, is a vector quantity and so the position is defined by a distance and a direction. In the context of simple harmonic motion, the displacement is the position relative to the equilibrium position.

Dosimeter badge A dosimeter badge is worn by someone for monitoring the cumulative amount of radiation their body has absorbed. The badge measures the dose in sievert, which takes account of the relative biological effects of the ionising radiation.

Driving frequency The driving frequency in hertz, Hz, is the frequency of 'pushes' causing a forced oscillation.

Eddy currents An alternating magnetic flux through the an iron core induces emfs in the core, which drive eddy currents. These generate heat in the core resulting in energy wastage.

Efficiency The energy efficiency of a system is defined as the ratio of the useful energy output to the energy input or the ratio of the useful power output to the power input. For a transformer, the efficiency is given by the equation: efficiency $= \frac{I_s V_s}{I_p V_p}$.

Electric field An electric field is a region around a charged particle or object within which a force would be exerted on other charged particles or objects.

Electric field strength The electric field strength, E, at a point in an electric field is defined as the force per unit charge on a positive test charge placed at that point. Electric field strength is a vector quantity and its unit is N C^{-1}. The defining formula for E is $E = \frac{F}{Q}$ where F is the force on test charge Q.

Electric potential At a point is the work done per unit positive charge in bringing a small positive test charge from infinity to the point.

Electric potential difference Electric potential difference, ΔV, is defined as the work done per unit charge in moving the charge between two points: $\Delta V = \frac{W}{Q}$.

Electrolytic capacitor A capacitor made from aluminium electrodes separated by paper soaked in aluminium borate. An electrolytic capacitor is capable of storing much higher charge than a standard capacitor.

Electromagnetic induction Electromagnetic induction is the production of an electromotive force across a conductor exposed to a varying magnetic field.

Electron capture During electron capture, an electron in an atom's inner shell is drawn into the nucleus where it combines with a proton, forming a neutron and a neutrino. The neutrino is ejected from the atom's nucleus.

Empirical Based on observation or experimental investigation.

Equipotential surface A surface where the potential is the same everywhere. For example, in an electric field no work is done in moving a charge along an equipotential surface.

Exponential decay A change occurring in which the rate of decrease of a quantity is directly proportional to the size of the quantity at that instant in time.

Faraday's law Faraday's law states that when the magnetic flux linking a circuit changes, an electromotive force is induced in the circuit proportional to the rate of change of the flux linkage. The equation linking the emf, E to the flux linkage is: $E = N\frac{\Delta\Phi}{\Delta t}$.

Fast neutron An energetic neutron produced by the fission of an unstable nucleus.

Field A region of space where a force is exerted on an object by virtue of the object's mass, charge or magnetic properties.

Fissile material A radioactive isotope which is capable of sustaining a nuclear fission chain reaction.

First law of thermodynamics A law stating the conservationof energy in heat engines: $Q = \Delta V + W$.

Fission fragments The atomic fragments left after a large atomic nucleus undergoes nuclear fission.

Fleming's left-hand rule Fleming's left-hand rule predicts the direction of the force on a current-carrying conductor in a magnetic field. For example, it predicts the direction of rotation of an electric motor.

Fleming's right-hand rule Fleming's right-hand rule predicts the direction in which an induced current flows. For example, predicting the direction of the induced current when a wire is moved through the magnetic field between two magnets.

Flux linkage For a conducting coil with several turns, flux linkage, $N\Phi$, is the product of the magnetic flux and the number of turns on the coil. The SI unit of flux linkage is the weber turn (Wb turn).

Forced oscillation An oscillation imposed upon a body or system by an external energy source. The body or system is forced to oscillate at the frequency of the external source rather than at its natural frequency.

Free oscillation A free oscillation is one in which a body or system oscillates at its natural frequency. For example, allowing a simple pendulum to swing without external interference.

Frequency The number of oscillations, vibration or waves, in one second. The unit of frequency is the hertz, Hz. Frequency is related to the time period by the expression $f = \frac{1}{T}$.

Frequency In the context of circular motion, the number of revolutions per unit time is known as the frequency, f, the SI unit of which is the hertz (Hz), where 1 Hz is one revolution per second. In the context of waves and oscillations, the frequency, f, is the number of waves or oscillations, also measured in hertz (Hz). The frequency of a rotation, oscillation or vibration is related to the time period by the equation: $f = \frac{1}{T}$.

Geiger–Müller (GM) tube A Geiger–Muller (GM) tube is a device which registers a voltage pulse each time an ionising particle, e.g. an alpha or beta particle, ionises a gas atom or molecule inside the tube. The electrical pulses may be counted by using a digital counter.

Generator In the context of electromagnetic induction: A generator consists of a rotating coil within a magnetic field and is used to generate an alternating current by electromagnetic induction.

Geostationary orbit An orbit in which a satellite moves in the same direction as the Earth's rotation, above the equator with an orbital period of 24 hours.

Gravitational constant The gravitational constant, G, is the constant of proportionality in the equation describing Newton's law of universal gravitation. H

Gravitational field strength The gravitational field strength g, at a point, is defined as the gravitational force per unit mass at that point. The unit of gravitational field strength is N kg^{-1}.

Gravitational potential The gravitational potential, V, at a point is the work done (or gain in potential energy) per unit mass to move a mass from infinity to that point. The unit is J kg^{-1}.

Gravitational potential difference Gravitational potential difference, ΔV, is the change in energy, ΔW, of a body of mass m caused by a change in position within a gravitational field: $\Delta W = m\,\Delta V$.

Gravitational field A region of space where a force is exerted on an object by virtue of the object's mass

Ground state An atom is said to be in its ground state when its electrons all occupy the lowest possible allowed energy levels.

Half-life The time taken for half the nuclei in a sample of a radioisotopes to decay; *or* the time taken for the activity of a radioactive source to drop by half.

Half-life (capacitor) The time for the current to halve as a capacitor charges; also, the time for the current, charge and voltage to halve as a capacitor discharges.

High-level waste (HLW) Radioactive waste which has high levels of radioactivity and heat generation.

Ideal transformer A transformer in which the power output is equal to the power input (it is 100% efficient). In reality, the best transformers attain 99.5% efficiency with 98% being more common.

Ideal gas A gas that obeys Boyle's law under all conditions: a gas whose molecules are infinitely small and exert no force on each other, except during collisions.

Ideal gas equation An equation of state linking the bulk properties (pressure, p, volume, V, and absolute temperature, T) of n moles of an ideal gas: $pV = nRT$.

Induced charge A charge created on an object by the presence of a nearby object that is already charged.

Intermediate-level waste (ILW) Radioactive waste which has high levels of radioactivity but does not generate heat.

Internal energy The energy of a substance due to the sum of the kinetic energy and potential energy of all its constituent particles.

Inverse square law Law describing how radiation, gamma radiation for example, spreads out in three dimensions from a point source; the intensity of radiation is inversely proportional to the square of the distance from the source. The strength of and electric field of a point charge and the gravitational field strength of a planet or star also obey an inverse square law.

Isotherm A curve on a pressure versus volume graph joining points at the same temperature.

Isotopes Atoms of an element can exist in different forms, called isotopes, which have the same number of protons and electrons but different numbers of neutrons.

Kinetic theory A theory developed in the 1860s and 1870s by James Maxwell, Ludwig Boltzmann and others in which liquids and gases were considered to be made up of small particles (atoms or molecules) which were in constant random motion. The theory attempted to explain the gas properties of pressure and temperature in terms of the movement of these particles.

Lenz's law The direction of an induced current opposes the change of magnetic flux that produces it.

Low-level waste (LLW) Radioactive waste which has low levels of radioactivity and does not generate heat.

Magnetic flux Magnetic flux, Φ, is the number of magnetic field lines passing through a closed surface, such as an area. It is calculated from the product of the magnetic flux density (perpendicular to the area) and the area. The SI unit of magnetic flux is the weber (Wb).

Magnetic flux density The magnetic flux density is a quantity which represents the strength of a magnetic field and is assigned the symbol, B. The tesla, T, is the unit of magnetic flux density. A flux density of 1 T causes a force of 1 N to be exerted on every 1 m length of a wire carrying a current of 1 A in a direction perpendicular to the field.

Mass difference The difference between the mass of a nucleus and the total mass of its constituent nucleons; sometimes called the mass defect.

Mean square speed The sum of the squares of all the molecules' speeds of a gas divided by the number of molecules:

$$(c_{rms})^2 \frac{c_1^2 + c_2^2 + c_3^2 + \cdots c_N^2}{N}.$$

Mechanical oscillation A mechanical oscillation is a periodic conversion of energy such as from potential energy to kinetic energy to potential energy. For a body is to oscillate it must be acted on by a restoring force which is directed towards the equilibrium position. The simplest type of oscillation is called simple harmonic motion.

Metastable state An excited state of a nucleus with a relatively long half-life.

Moderator A material such as graphite which is used in a nuclear reactor to slow down neutrons without absorbing them.

Molar gas constant (*R*) The constant of proportionality in the ideal gas equation: $pV = nRT$.

Molar mass The mass of one mole of an element or compound.

Molecular kinetic theory model A theory which considers gas molecules to be moving around, colliding with each other and with the walls of their container, exerting pressure and moving faster if the temperature of the gas is raised. To make the mathematics of the kinetic theory model more straightforward the gas is considered to be an ideal gas.

Monoenergetic Radioactive particles with a single distinct energy produced by a nuclear reaction.

Natural frequency The frequency at which a vibrating object undergoes free vibrations; for example, the frequency at which a tuning fork will oscillate when struck.

Neutron The neutron is a subatomic particle, symbol n, with no net electric charge and a mass slightly larger than that of a proton. It is composed of three quarks: down, down and up. It is found in the nucleus of atoms and so is a nucleon.

Newton's law of universal gravitation The law states that any two point masses attract each with a force F that is directly proportional to the product of their masses m_1 m_2 and inversely proportional to the square of their separation r: $F = \frac{Gm_1m_2}{r^2}$.

Nuclear excited states Higher energy states of a nucleus produced following alpha or beta decay. Nuclear excited states reduce their energy by gamma emission.

Nucleon number The total number of protons and neutrons in a nucleus; also referred to as the mass number; symbol A.

Nuclide A nuclide is a type of nucleus with a specific number of protons, a specific number of neutrons and a specific energy state. There are 339 naturally occurring nuclides of which 254 have never been observed to decay and are therefore described as stable.

Orbital speed In circular motion, orbital speed, v, is the distance the object covered per unit time. The speed can be calculated by dividing the circumference of the circle by the time period. Orbital speed is also called linear speed.

Oscilloscope A cathode ray oscilloscope is used for measuring dc and ac voltage, measuring time intervals and frequencies and also displaying alternating waveforms.

Parent nucleus The nucleus which undergoes a radioactive decay event to produce one or more daughter nuclei.

Parabola A parabola is a curve with the general equation, $y = ax^2 + bx + c$. The trajectory of a moving charged particle entering a uniform electric field initially at right angles is a parabola, as is the path of a projectile under the influence of gravity.

Pascal The SI unit of pressure derived from the force in newton divided by the area in m^2.

Peak voltage This is the maximum value for an alternating emf and so is the highest amplitude value of the emf.

Peak-to-peak voltage This is the difference between the maximum and minimum values for an alternating emf and so is twice the amplitude of the emf.

Periodic motion A repeating pattern of motion, such as rotation or oscillation.

Permittivity of free space The charge per unit area on oppositely charged parallel plates in a vacuum when the electric field strength between the plates is 1 volt per metre.

Phase differences The difference in position in the cycle of two oscillating systems or waves, expressed in degrees or radians.

Polar molecules Molecules such as water which have a slightly positive part and a slightly negative part.

Polar orbit An orbit in which a satellite passes over both the North and South poles of the Earth.

Polarisation The creation of an electric field by the movement of charge. The alignment of transverse waves so the vibrations are confined to one direction.

Polarised A transverse wave is polarised when the vibration of the wave is confined to one direction.

Potassium–argon dating A method of dating rocks containing potassium by comparing the ratio of potassium-40 atoms to argon-40 atoms in a sample. Because of the long half-life of potassium-40 (1250 million years), potassium–argon dating can be used to date rocks from 100 000 years to 4.5 million years in age.

Probability The mathematical likelihood of an event happening.

Proton number The number of protons in a nucleus, also called the atomic number; has the symbol *Z*.

Proton A positively charged hadron composed of three quarks: up, up and down; it is believed to be the only stable hadron.

Radial field A field in which the field lines are straight and either converge or diverge from a single point.

Radian The angle subtended when the arc length is equal to the radius; 1 rad = 57.3°.

Radioisotope A form of a nucleus (see Isotopes) that is radioactive.

Relative permittivity The factor by which a dielectric raises the capacitance of a parallel plate capacitor.

Reprocessing In the context of nuclear fuel, the physical and chemical separation of useful unused nuclear fuel from highly radioactive and often useless fission and decay products.

Resonance Large amplitude vibrations caused when the driving frequency matches the natural frequency of a system.

Resonant frequency The natural frequency at which an object or system oscillates.

Restoring force A force that acts so as to return an object to its equilibrium position. It is the restoring force within a system which causes oscillation in simple harmonic motion.

Root mean square (rms) The value of an ac current or potential difference that is equal to the dc value that would lead to the same power being dissipated in a resistor; the rms. values of *I* and *V* are linked to the peak values; $I_{rms} = \frac{I_0}{\sqrt{2}}$ and $V_{rms} = \frac{V_0}{\sqrt{2}}$.

Root mean square speed A measure of the speed of particles in a gas which is most convenient for problem solving within the kinetic theory of gases: c_{rms}.

Rutherford scattering The elastic scattering of charged particles such as alpha particles by similarly charged particles such as nuclei.

Scintillation A flash of light in a transparent material caused by the passage of an energetic particle or photon such as an alpha or beta particle or a gamma ray.

Self-sustaining chain reaction A nuclear fission reaction in which at least one of the neutrons produced goes on to cause another fission event.

Shielding In the context of fields, the blocking of a magnetic or electric field. In the context of nuclear energy, the blocking of harmful radiation by thick steel and concrete.

Simple harmonic motion The periodic motion in which the restoring force, *F*, is proportional to the displacement, *x*, from the equilibrium position and acts in the opposite direction: $F = -kx$.

Small-angle approximation For an small angle, the sine or tangent of the angle is approximately equal to the value of the angle in radians. This approximation holds true for angles up to about 10°.

Specific heat capacity The energy required to raise the temperature of a 1 kg mass of a substance by 1 K; unit J kg^{-1} K^{-1}.

Specific latent heat of fusion The energy needed for 1 kg of a solid to change to a liquid, with no increase in temperature; unit J kg^{-1}.

Specific latent heat of vaporisation The energy needed for 1 kg of a solid to change to a liquid with no increase in temperature; unit J kg^{-1}.

Spent fuel Fuel from a nuclear reactor which is no longer useful and is removed from the reactor for reprocessing.

Static electricity An electic charge within a material or on its surface which remains stationary until it is discharged.

Stationary waves A non-progressive wave formed when two progressive waves of the same frequency travel in opposite directions and superpose, e.g. on a fixed stretched string; characterised by nodes and antinodes.

Step-up transformer A transformer which converts low voltage ac to a higher voltage.

Step-down transformer A transformer which converts high voltage ac to a lower voltage.

Synchronous orbit An orbit in which an orbiting body (such as a satellite) has an orbital period equal to the rotational period of the body (such as a planet) it orbits, in the same direction of rotation as that body.

Synchrotron A particle accelerator in which particles are accelerated as they travel in a circular path of constant radius.

Tesla A unit of magnetic flux density; 1 T is the magnetic flux density when 1 m of wire carrying 1 A of current at right angles to a magnetic field experiences a force of 1 N.

Thermal enery transfer The transfer of energy occurring as a result of a temperature difference.

Thermal neutron A slow moving neutron which can be captured by a fissile nucleus, causing nuclear fission.

Time base The control on an oscilloscope that determines the speed at which the electron beam moves horizontally across the screen.

Time constant The time taken for the charge on (or voltage across or current flowing off) a discharging capacitor to drop to approximately 0.37 of its original value; it is equal to *CR*, where *C* is the capacitance and *R* is the total resistance of the circuit. The time constant has the unit seconds.

Time period The time taken for an object to rotate through one complete circle, usually given the symbol, *T*. Time period is often abbreviated to period and is related to frequency by the equation: $f = \frac{1}{T}$.

Tracers (radioactive tracer) A radioactive isotope used for medical imaging and which is tracked and detected by an external radiation detector.

Transformer A device that uses electromagnetic induction to change the voltage (and current) of an ac signal.

Transformer equation $\frac{V_s}{V_p} = \frac{N_s}{N_p}$

DATA SECTION

FUNDAMENTAL CONSTANTS AND VALUES

Quantity	Symbol	Value	Unit
speed of light in vacuo	c	3.00×10^{8}	$m\ s^{-1}$
permeability of free space	μ_0	$4\pi \times 10^{-7}$	$H\ m^{-1}$
permittivity of free space	ε_0	8.85×10^{-12}	$F\ m^{-1}$
charge of electron (magnitude)	e	1.60×10^{-19}	C
the Planck constant	h	6.63×10^{-34}	J s
gravitational constant	G	6.67×10^{-11}	$N\ m^2\ kg^{-2}$
the Avogadro constant	N_A	6.02×10^{23}	mol^{-1}
molar gas constant	R	8.31	$J\ K^{-1}\ mol^{-1}$
the Boltzmann constant	k	1.38×10^{-23}	$J\ K^{-1}$
the Stefan constant	σ	5.67×10^{-8}	$W\ m^{-2}\ K^{-4}$
the Wien constant	α	2.90×10^{-3}	m K
electron rest mass	m_e	9.11×10^{-31}	kg
electron charge/mass ratio	e/m_e	1.76×10^{11}	$C\ kg^{-1}$
proton rest mass	m_p	$1.67(3) \times 10^{-27}$	kg
proton charge/mass ratio	e/m_p	9.58×10^{7}	$C\ kg^{-1}$
neutron rest mass	m_n	$1.67(5) \times 10^{-27}$	kg
gravitational field strength	g	9.81	$N\ kg^{-1}$
acceleration due to gravity	g	9.81	$m\ s^{-2}$
atomic mass unit	u	1.661×10^{-27}	kg

SI UNIT PREFIXES AND SYMBOLS

Multiplication factor		Prefix	Symbol
1000 000 000 000	10^{12}	tera	T
1000 000 000	10^{9}	giga	G
1000 000	10^{6}	mega	M
1000	10^{3}	kilo	k
	10^{2}	centi	c
0.001	10^{-3}	milli	m
0.000 001	10^{-6}	micro	μ
0.000 000 001	10^{-9}	nano	n
0.000 000 000 001	10^{-12}	pico	p
0.000 000 000 000 001	10^{-15}	femto	f

ALGEBRAIC EQUATION

quadratic equation $x = \frac{-b \pm \sqrt{b^2 - 4ac}}{2a}$

GEOMETRICAL EQUATIONS

circumference of circle $= 2\pi r$

area of circle $= \pi r^2$

surface area of cylinder $= 2\pi rh$

volume of cylinder $= \pi r^2 h$

surface area of sphere $= 4\pi r^2$

volume of sphere $= \frac{4}{3}\pi r^3$

arc length $= r\theta$ (θ in radians)

$\sin^2\theta + \cos^2\theta = 1$

for small angles, $\sin\theta \approx \tan\theta \approx \theta$ in radians

equation of a straight line $y = mx + c$ (m = gradient; c = y intercept)

FUNDAMENTAL PARTICLES

Class	Name	Symbol	Rest energy / MeV
photon	photon	γ	0
lepton	(electron) neutrino	ν_e	0
	(muon) neutrino	ν_μ	0
	electron	$e^\pm$	0.510999
	muon	$\mu^\pm$	105.659
meson	π meson (pion)	$\pi^\pm$	139.576
		π^0	134.972
	K meson (kaon)	$K^\pm$	493.821
		K^0	497.762
baryon	proton	p	938.257
	neutron	n	939.551

PROPERTIES OF QUARKS

Type	Charge	Baryon number	Strangeness
u	$+\frac{2}{3}e$	$+\frac{1}{3}$	0
d	$-\frac{1}{3}e$	$+\frac{1}{3}$	0
s	$-\frac{1}{3}e$	$+\frac{1}{3}$	−1

Antiquarks $\bar{u}$, $\bar{d}$, $\bar{s}$ have opposite signs.

PROPERTIES OF LEPTONS

					Lepton number
particles	e^-	ν_e	μ^-	ν_μ	+1
antiparticles	e^+	$\bar{\nu}_e$	μ^+	$\bar{\nu}_\mu$	−1

PHOTONS AND ENERGY LEVELS

photon energy $E = hf = hc/\lambda$

in photoelectric effect $hf = \phi + E_{k\,(max)}$

energy levels $hf = E_1 - E_2$

de Broglie wavelength $\lambda = h/p = h/mv$

WAVES

wave speed $c = f\lambda$

$f = 1/T$ (where T = period)

for first harmonic on a string $f = (1/2l)\sqrt{(T/\mu)}$

interference fringe spacing $w = \lambda D/s$

diffraction grating formula $d \sin \theta = n\lambda$

refractive index of substance $n = c/c_s$

law of refraction $n_1 \sin \theta_1 = n_2 \sin \theta_2$

for critical angle $\sin \theta_c = n_2/n_1$ for $n_1 > n_2$

MECHANICS

$v = \Delta s/\Delta t$

$a = \Delta v/\Delta t$

$v = u + at$

$s = \frac{1}{2}(u + v)t$

$v^2 = u^2 + 2as$

$s = ut + \frac{1}{2}at^2$

momentum $p = mv$

$F = \Delta(mv)/\Delta t$

for constant m, $F = ma$

impulse $F\,\Delta t = \Delta(mv)$

work done $W = Fs \cos \theta$

$E_k = \frac{1}{2}mv^2$

$\Delta E_p = mg\,\Delta h$

power $P = \Delta W/\Delta t = Fv$

moment $= Fd$

efficiency = useful power output/power input

MATERIALS

density $\rho = m/V$

Hooke's law $F = k\,\Delta L$

tensile stress $= F/A$

tensile strain $= \Delta L/L$

Young modulus = tensile stress/tensile strain

energy stored $E = \frac{1}{2}F\,\Delta L$

ELECTRICITY

$I = \Delta Q/\Delta t$

$V = W/Q$

$R = V/I$

resistivity $\rho = RA/l$

resistors in series $R_{tot} = R_1 + R_2 + R_3 + \ldots$

resistors in parallel $R_{tot} = 1/R_1 + 1/R_2 + 1/R_3 + \ldots$

power $P = VI = I^2R = V^2/R$

emf $\varepsilon = E/Q$

$\varepsilon = I(R + r)$ where r is the internal resistance

alternating current:

$I_{rms} = I_0/\sqrt{2}$

$V_{rms} = V_0/\sqrt{2}$

for a transformer $N_s/N_p = V_s/V_p$

transformer efficiency $= I_sV_s/I_pV_p$

CIRCULAR MOTION

$\omega = v/r$

$\omega = 2\pi f$

centripetal acceleration $a = v^2/r = \omega^2 r$

centripetal force $F = mv^2/r = m\,\omega^2 r$

SIMPLE HARMONIC MOTION

$a = -\omega^2 x$

$x = A \cos \omega t$

$v = \pm\omega\sqrt{A^2 - x^2}$

$v_{max} = \omega A$

$a_{max} = \omega^2 A$

$T_{spring} = 2\pi\sqrt{(m/k)}$

$T_{pendulum} = 2\pi\sqrt{(l/g)}$

THERMAL PHYSICS

for a change of temperature $Q = mc\,\Delta\theta$

for a change of state $Q = ml$

ideal gases:

$pV = nRT$ where T is the kelvin temperature

$pV = NkT$

kinetic theory:

$pV = \frac{1}{3}Nm(c_{rms})^2$

$\frac{1}{2}m(c_{rms})^2 = \frac{3}{2}kT = 3RT/2N_A$

GRAVITATIONAL FIELDS

$F = Gm_1m_2/r^2$

$g = F/m$

$\Delta W = m\,\Delta V$

for radial field $g = GM/r^2$

$V = -GM/r$

$G = -\Delta V/\Delta r$

$M_{Sun} = 1.99 \times 10^{30}$ kg

$R_{Sun} = 6.96 \times 10^{8}$ m

$M_{Earth} = 5.97 \times 10^{24}$ kg

$R_{Earth} = 6.37 \times 10^{6}$ m

ELECTRIC FIELDS AND CAPACITORS

$F = Q_1Q_2/4\pi\varepsilon_0r^2$

$F = EQ$

$\Delta W = Q\,\Delta V$

for uniform field $E = V/d$

for radial field $E = Q/4\pi\varepsilon_0r^2$

$V = Q/4\pi\varepsilon_0r$

$E = \Delta V/\Delta r$

$C = Q/V$

$C = A\varepsilon_0\varepsilon_r/d$

energy stored $= \frac{1}{2}QV = \frac{1}{2}CV^2 = \frac{1}{2}Q^2/C$

time constant $= RC$

discharging $Q = Q_0\,e^{-t/RC}$

charging $Q = Q_0\,(1 - e^{-t/RC})$

MAGNETIC FIELDS

force on a current $F = BIl$

force on a moving charge $F = BQv$

$\Phi = BA$

flux linkage by a coil $N\Phi = BAN\cos\theta$

magnitude of induced emf $\varepsilon = N\,\Delta\Phi/\Delta t$

in a rotating coil $\varepsilon = BAN\omega\sin\omega t$

NUCLEAR PHYSICS

for gamma radiation $I = k/x^2$

for radioactive decay $\Delta N/\Delta t = -\lambda N$

$N = N_0\,e^{-\lambda t}$

activity $A = \lambda N$

$T_{1/2} = (\ln 2)/\lambda$

nuclear radius $R = R_0\,A^{1/3}$ where A = nucleon number

$E = mc^2$

1 u = 931.5 MeV

OPTION UNITS

ASTROPHYSICS

1 astronomical unit (AU) $= 1.50 \times 10^{11}$ m

1 light year $= 9.46 \times 10^{15}$ m

1 parsec = 206 265 AU $= 3.08 \times 10^{16}$ m = 3.26 light year

magnification

$$M = \frac{\text{angle subtended by image at eye}}{\text{angle subtended by object at unaided eye}}$$

for refracting telescope in normal adjustment

$$M = \frac{f_o}{f_e}$$

Rayleigh criterion $\theta \approx \lambda/D$

magnitude equation $m - M = 5 \log (d/10)$

Wien's law $\lambda_{max} T = 2.9 \times 10^{-3}$ m K

Stefan's law $P = \sigma A T^4$

escape velocity $v_{esc} = \sqrt{\frac{2GM}{R}}$

Schwarzschild radius $R_S \approx \frac{2GM}{c^2}$

Doppler shift (for $v << c$) $z = \frac{\Delta f}{f} = -\frac{\Delta \lambda}{\lambda} = \frac{v}{c}$

Hubble's law $v = Hd$

Hubble constant $H = 65$ km s^{-1} Mpc^{-1}

MEDICAL PHYSICS

lens power $= 1/f$

lens equation $\frac{1}{f} = \frac{1}{u} + \frac{1}{v}$

magnification $m = v/u$

threshold of hearing $I_0 = 1.0 \times 10^{-12}$ W m^{-2}

intensity level $= 10 \log (I/I_0)$

absorption of (ultra)sound $I_x = I_0 e^{-\mu x}$

acoustic impedance $Z = \rho c$

for reflection of ultrasound at a boundary

$$\frac{I_r}{I_i} = \left(\frac{Z_2 - Z_1}{Z_2 + Z_1}\right)^2$$

half-lives of a radiopharmaceutical:

$$\frac{1}{T_E} = \frac{1}{T_B} + \frac{1}{T_P}$$

ENGINEERING PHYSICS

moment of inertia $I = \sum mr^2$

angular kinetic energy $E_k = \frac{1}{2} I \omega^2$

equations of angular motion:

$\omega_2 = \omega_1 + \alpha t$

$\omega_2^2 = \omega_1^2 + 2\alpha\theta$

$\theta = \omega_1 t + \frac{1}{2}\alpha t^2$

$\theta = \frac{1}{2}(\omega_1 + \omega_2)t$

torque $T = Fr = I\alpha$

angular momentum $= I\omega$

angular impulse $T\Delta t = \Delta(I\omega)$

work done $W = T\theta$

power $P = T\omega$

first law of thermodynamics $Q = \Delta U + W$ where W is work done by system

adiabatic change pV^γ = constant

isothermal change pV = constant

heat engines:

work done per cycle = area of loop

input power = calorific value × fuel flow rate

indicated power = area of loop × number of cycles per second × number of cylinders

output or brake power $= T\omega$

friction power = indicated power − brake power

efficiency = output (brake) power/input power = thermal efficiency × mechanical efficiency

for an idealised heat engine:

thermal efficiency $\varepsilon = \frac{W}{Q_H} = \frac{Q_H - Q_C}{Q_H}$

maximum theoretical thermal efficiency

$$\varepsilon_{max} = \frac{T_H - T_C}{T_H}$$

for a refrigerator:

$COP_{ref} = Q_C/W = Q_C/(Q_H - Q_C)$

for a heat pump:

$COP_{hp} = Q_H/W = Q_H/(Q_H - Q_C)$

TURNING POINTS IN PHYSICS

electrons in fields:

$F = eE = eV/d$

$F = Bev$

$r = mv/Be$

$\frac{1}{2}mv^2 = eV$

Millikan's experiment:

for stationary drop $\frac{QV}{d} = mg$

resistive force on falling drop $F = 6\pi\eta rv$

electromagnetic waves:

Maxwell's formula $c = \frac{1}{\sqrt{\mu_0 \varepsilon_0}}$

wave–particle duality:

de Broglie wavelength of electrons $\lambda = \frac{h}{\sqrt{2meV}}$

special relativity:

$$t = \frac{t_0}{\sqrt{1 - \frac{v^2}{c^2}}}$$

$$l = l_0\sqrt{1 - \frac{v^2}{c^2}}$$

$$m = \frac{m_0}{\sqrt{1 - \frac{v^2}{c^2}}}$$

$$E = \frac{m_0 c^2}{\sqrt{1 - \frac{v^2}{c^2}}}$$

ELECTRONICS

LC circuits:

resonant frequency $f_0 = \frac{1}{2\pi\sqrt{LC}}$

Q factor $Q = \frac{f_0}{f_B}$

operational amplifiers:

open loop $V_{out} = A_{OL}(V_+ - V_-)$

inverting amplifier $V_{out}/V_{in} = -R_f/R_{in}$

non-inverting amplifier $V_{out}/V_{in} = 1 + R_f/R_1$

summing amplifier $V_{out} = -R_f\left(\frac{V_1}{R_1} + \frac{V_2}{R_2} + \frac{V_3}{R_3} + \ldots\right)$

difference amplifier $V_{out} = (V_+ - V_-)\frac{R_f}{R_1}$

Boolean algebra:

$\bar{A}$ = **NOT** A

$A.B = A$ **AND** B

$A + B = A$ **OR** B

Signal modulation:

AM bandwidth $= 2f_H$ where f_H is the highest audio frequency

FM bandwidth $= 2 \times (\Delta f + f_M)$ where is f_M is the peak frequency of the modulated signal

INDEX

D

E

F

G

H

I

J

K

L

M

N

O

P

R

S

ACKNOWLEDGEMENTS

The publishers wish to thank the following for permission to reproduce photographs. Every effort has been made to trace copyright holders and to obtain their permission for the use of copyright materials. The publishers will gladly receive any information enabling them to rectify any error or omission at the first opportunity.

Practical work in physics

p2, top: Peter Ginter/Getty Images; p2, Bottom: indukas/Thinkstock; p3, Phil Boorman/Science Photo Library; p4, top: www.philipharris.co.uk; p4, bottom: Martyn F Chillmaid/Science Photo Library

Chapter 1

p5, Fig 1: NASA; p5, Fig 2: JuliusKielaitis/Shutterstock; p9, Fig A1: CHEN WS/Shutterstock; p11, Fig 12: Pavel L Photo and Video/Shutterstock; p12, Fig 14: Perspectives – Jeff Smith/ Shutterstock; p12, Fig 15: Viktor Keremedchiev/Shutterstock; p14, Fig 17: EvrenKalinbacak/Shutterstock, p15, Fig A1: gregdx/Shutterstock

Chapter 2

p18, background: Eliks/Shutterstock; p18, Fig 1: SpeedKingz/Shutterstock; p27, Fig 10: Dorling Kindersley/UIG/ Science Photo Library; p33, Fig 19: used with kind permission from Norman Hossack, www.hossack-design.com; p34, Fig 20: Library of Congress/Science Photo Library; p35, Fig 25: Elena Elisseeva/Shutterstock

Chapter 3

p39, Fig 1: NSF/Steffen Richter/Harvard University/Science Photo Library; p39, Fig 2: kavram/Shutterstock; p40, Fig 4: Fouad A. Saad/Shutterstock; p44, Fig 5: pryzmat/Shutterstock; p46, Fig 8: IBM Research/Science Photo Library; p50, Fig 12: Royal Astronomical Society/Science Photo Library; p51, Fig P2: Andrew Lambert Photography/ Science Photo Library; p62, Fig 22: www.a3bs.com

Chapter 4

p71, background: koya979/Shutterstock; p71, Fig 1: European Space Agency/P Carril/Science Photo Library; p71, Fig 2: European Space Agency/GOCE/Science Photo Library; p72, Fig 3: American Institute of Physics/Science Photo Library; p74, Fig 6: NASA/APL/SwRI/ Science Photo Library; p77, Fig 10: Bruce Weaver; p82, Fig 17: NASA/JHU/ APL/Science Photo Library; p85, Fig 19: NASA/JPL-Caltech/Science Photo Library; p85, Fig 20: this image shows the sulphur dioxide emissions from Iceland's Bardarbunga volcano as of about midday on 4 September 2014, as detected by the GOME-2 instrument on the MetOp-A and -B satellites. ESA's Volcanic Ash Strategic Initiative Team (VAST) and Support to Aviation Control Service (SACS) are monitoring the situation closely, and have detected sulphur dioxide emissions since early September. SACS and VAST uses multiple satellites – including Europe's MetOp and Meteosat missions – to provide early warning information about volcanic eruptions. When an eruption occurs, an alert is sent to interested users, most notably to Volcanic Ash Advisory Centres and airlines, and public, maps are generated showing the extent and intensity of the volcanic plumes. The Support to Aviation Control Service (SACS; http://sacs.aeronomie.be) is hosted by the Belgian Institute for Space aeronomy (BIRA-IASB)

Chapter 5

p90, background: Johan Swanepoel/ Shutterstock; p90, Fig 1: Aerial Archives/ Alamy Stock Photo; p90, Fig 2: Peter Ginter/Alamy Stock Photo; p91, Fig 3: Mike Dunning/Getty Images; p91, Fig 4: sciencephotos/Alamy Stock Photo; p91, Fig 5: GIPhotostock/Science Photo Library; p91, Fig 6: charistoone-images/ Alamy Stock Photo; p92, Fig 7: Emilio Segre Visual Archives/American Institute of Physics/Science Photo Library; p92, Fig 9: Science Photo Library; p98, Fig 15: IgorGolovniov/Shutterstock

Chapter 6

p111, background: Vasin Lee/ Shutterstock; p111, Fig 1: Fingerhut/ Shutterstock; p111, Fig 2: Maxwell Technologies, Inc.; p112, Fig 3: yurazaga/ Shutterstock; p113, Fig 7: hsagencia/ Shutterstock; p113, Fig 9: GIPhotostock/ Science Photo Library; p115, Fig 11: SeDmi/Shutterstock; p116, Fig 15: Jeffrey B. Banke/Shutterstock; p118, Fig 18: Baloncici/Shutterstock

Chapter 7

p133, Fig 1: NASA; p133, Fig 2: Jamen Percy/Shutterstock; p134, Fig 3: Awe Inspiring Images/Shutterstock; p134, Fig 4: GIPhotostock/Science Photo Library; p141, Fig 14: CERN; p145, Fig 16: Lawrence Berkeley Lab/Science Photo Library

Chapter 8

p150, background: Albert Lozano/ Shutterstock; p150, Fig 1: Wisanu Boonrawd/Shutterstock; p150, Fig 2: image used with kind permission from Evatran and Plugless Power.; p151, Fig 5: Stock Montage/Getty Images; p160, Fig 24: Christina Richards/ Shutterstock; p163, Fig P1: image used with kind permission from Timstar.; p164, Fig P4: Bernd Juergens/ Shutterstock

Chapter 9

p172, background: C Powell, P. Fowler & D. Perkins/Science Photo Library; p172, Fig 1: muratart/Shutterstock; p172, Fig 2: NASA/LANL; p173, Fig 3: Professor Peter Fowler/Science Photo Library; p179, Fig 11: Everett Historical/ Shutterstock; p183, Fig P1: Martyn F. Chillmaid/Science Photo Library; p183, Fig P2: Maurice Savage/Alamy Stock Photo; p185, Fig 17: Guy Cali/Alamy Stock Photo; p187, Fig 20: Joyce Mar/ Shutterstock; p188, Fig A1: Trevor Clifford Photography/Science Photo Library; p193, Fig 26: NASA/Alamy Stock Photo

Chapter 10

p196, background: sakkmesterke/ Shutterstock; p196, Fig 1: Haydn Denman/Alamy Stock Photo; p203, Fig 13: BSIP/Getty Images; p213, Fig 25: Ria Novosti/Science Photo Library; p213, Fig 26: Patrick Landmann/Science Photo Library; p215, Fig 30: Patrick Aventurier/Getty Images; p216, Fig 31: Martin Bond/Science Photo Library